From childhood, each of us develops our own personal set of theories and beliefs about the world in which we live. Given the impossibility of knowing about every event that can ever take place, we use cognitive short cuts to try and predict and make sense of the world around us. One of the fundamental pieces of information we use to predict future events, and make sense of past events, is 'frequency' - how often has such an event happened to us, or how often we have observed a particular event? With such information we will make inferences about the likelihood of its future appearance. We will make judgements, assess risk, or even consumer decisions, on the basis of this information. We also form associations between events that frequently occur together, and even (often incorrectly) assume causality between one event and the other as a result of their simultaneous appearance.

How is it though that we process such information? How does our brain deal with information on frequencies? How does such information influence our behaviour, beliefs, and judgements? Important new findings on this topic have come from research across social and experimental psychology, though until now, never brought together in a single volume.

This is the first book to bring together two disparate literatures on this topic - drawing on research from both cognitive psychology and social psychology. Including contributions from world leaders in the field, this is a timely, and long overdue volume on this topic.

ETC.
FREQUENCY PROCESSING AND COGNITION

ETC. FREQUENCY PROCESSING AND COGNITION

PETER SEDLMEIER

and

TILMANN BETSCH

This book has been printed digitally and produced in a standard specification in order to ensure its continuing availability

OXFORD
UNIVERSITY PRESS

Great Clarendon Street, Oxford OX2 6DP
Oxford University Press is a department of the University of Oxford.
It furthers the University's objective of excellence in research, scholarship,
and education by publishing worldwide in
Oxford New York
Auckland Cape Town Dar es Salaam Hong Kong Karachi
Kuala Lumpur Madrid Melbourne Mexico City Nairobi
New Delhi Shanghai Taipei Toronto
With offices in
Argentina Austria Brazil Chile Czech Republic France Greece
Guatemala Hungary Italy Japan South Korea Poland Portugal
Singapore Switzerland Thailand Turkey Ukraine Vietnam

Oxford is a registered trade mark of Oxford University Press
in the UK and in certain other countries

Published in the United States
by Oxford University Press Inc., New York

© Oxford University Press 2002

The moral rights of the author have been asserted

Database right Oxford University Press (maker)

Reprinted 2010

All rights reserved. No part of this publication may be reproduced,
stored in a retrieval system, or transmitted, in any form or by any means,
without the prior permission in writing of Oxford University Press,
or as expressly permitted by law, or under terms agreed with the appropriate
reprographics rights organization. Enquiries concerning reproduction
outside the scope of the above should be sent to the Rights Department,
Oxford University Press, at the address above

You must not circulate this book in any other binding or cover
And you must impose this same condition on any acquirer

ISBN 978-0-19-850863-2

CONTENTS

List of contributors *page vii*

1 Frequency processing and cognition: introduction and overview *page 1*
  Peter Sedlmeier, Tilmann Betsch, and Frank Renkewitz

Theoretical Models and Perspectives

2 Frequency processing: a twenty-five year perspective *page 21*
  Rose T. Zacks and Lynn Hasher

3 Encoding, representing, and estimating event frequencies: a multiple strategy perspective *page 37*
  Norman R. Brown

4 In the year 2054: innumeracy defeated *page 55*
  Gerd Gigerenzer

5 Frequency judgements and retrieval structures: splitting, zooming, and merging the units of the empirical world *page 67*
  Klaus Fiedler

6 Experiential and contextual heuristics in frequency judgement: ease of recall and response scales *page 89*
  Norbert Schwarz and Michaela Wänke

7 Tversky and Kahneman's availability approach to frequency judgement: a critical analysis *page 109*
  Tilmann Betsch and Devika Pohl

8 A memory models approach to frequency and probability judgement: applications of Minerva 2 and Minerva DM *page 121*
  Michael R. P. Dougherty and Ana M. Franco-Watkins

9 Associative learning and frequency judgements: The PASS model *page 137*
  Peter Sedlmeier

10 Frequency, contingency and the information processing theory of conditioning *page 153*
  C. R. Gallistel

Essential Empirical Results

11 Effects of processing fluency on estimates of probability and frequency *page 175*
  Rolf Reber and Natasha Zupanek

12 Frequency judgements of emotions: the cognitive basis of personality assessment *page 189*
    Ulrich Schimmack

13 Online strategies versus memory-based strategies in frequency estimation *page 205*
    Susanne Haberstroh and Tilmann Betsch

14 Frequency learning and order effects in belief updating *page 221*
    Martin Baumann and Josef F. Krems

15 The psychophysics metaphor in calibration research *page 239*
    Gernot D. Kleiter, Michael E. Doherty, and Gregory L. Brake

Practical Implications

16 Frequency effects in consumer decision making *page 259*
    Joseph W. Alba

17 Free word associations and the frequency of co-occurrence in language use *page 271*
    Manfred Wettler

18 Technology needs psychology: how natural frequencies foster insight in medical and legal experts *page 285*
    Ralph Hertwig and Ulrich Hoffrage

19 Frequency processing and cognition: stock-taking and outlook *page 303*
    Tilmann Betsch and Peter Sedlmeier

Index *page 319*

LIST OF CONTRIBUTORS

Joseph W. Alba, Department of Marketing, 212 Bryan Hall, University of Florida, Gainesville, FL 32611-7155, USA

Martin Baumann, Department of Psychology, Chemnitz University of Technology, 09107 Chemnitz, Germany

Tilmann Betsch, Psychological Institute, University of Heidelberg, Hauptstr. 47-51, 69117 Heidelberg, Germany

Gregory L. Brake, Microsoft Corp., Microsoft Way, Redmond, WA, USA

Norman R. Brown, Department of Psychology, Biological Sciences Building, University of Alberta, Edmonton, Alberta, Canada T6G 2E9

Michael E. Doherty, Department of Psychology, Bowling Green State University, Bowling Green, Ohio, USA

Michael R. P. Dougherty, Department of Psychology, University of Maryland, College Park, MD 20742-4411, USA

Klaus Fiedler, Psychological Institute, University of Heidelberg, Hauptstr. 47-51, 69117 Heidelberg, Germany

Ana M. Franco-Watkins, Department of Psychology, University of Maryland, College Park, MD 20742, USA

C. R. Gallistel, Rutgers Center for Cognitive Science, 152 Frelinghuysen Road, Piscataway, NJ 08854-8020, USA

Gerd Gigerenzer, Center for Adaptive Behavior and Cognition, Max Planck Institute for Human Development, Lentzeallee 94, 14195 Berlin, Germany

Susanne Haberstroh, Psychological Institute, University of Heidelberg, Hauptstr. 47-51, 69117 Heidelberg, Germany

Lynn Hasher, Department of Psychology, University of Toronto, 100 St. George St., Toronto, Ontario, Canada M5S 3G3

Ralph Hertwig, Psychology Department, Columbia University, Schermerhorn Hall, 1190 Amsterdam Avenue, New York City, NY 10027, USA

Ulrich Hoffrage, Center for Adaptive Behavior and Cognition, Max Planck Institute for Human Development, Lentzeallee 94, 14195 Berlin, Germany

Gernot D. Kleiter, Department of Psychology, University of Salzburg, Hellbrunnerstr. 34, A-5020 Salzburg, Austria

Josef F. Krems, Department of Psychology, Chemnitz University of Technology, 09107 Chemnitz, Germany

Devika Pohl, University of Heidelberg, Psychological Institute, Hauptstrasse 47-51, 69117 Heidelberg, Germany

Rolf Reber, University of Berne, Department of Psychology, Muesmattstrasse 45, CH-3000 Bern 9, Switzerland

Frank Renkewitz, Department of Psychology, Chemnitz University of Technology, 09107 Chemnitz, Germany

Ulrich Schimmack, Department of Psychology, UTM, Erindale College, 3359 Mississauga Road North, Mississauga, Ontario, Canada L5L 1C6

Norbert Schwarz, Institute for Social Research, University of Michigan, 426 Thompson St., Ann Arbor, MI 48106-1248, USA

Peter Sedlmeier, Department of Psychology, Chemnitz University of Technology, 09107 Chemnitz, Germany

Michaela Wänke, Department of Psychology, University of Erfurt, Postfuch 900221, 99105 Erfurt, Germany

Manfred Wettler, FB 2 – Psychology, 33095 Paderborn, University of Paderborn, Germany

Rose T. Zacks, Dept. of Psychology, Mich. State Univ., East Lansing, MI 48824-1117, USA

Natasha Zupanek, University of Berne, Department of Psychology, Muesmattstrasse 45, CH-3000 Bern 9, Switzerland

CHAPTER 1

FREQUENCY PROCESSING AND COGNITION: INTRODUCTION AND OVERVIEW

PETER SEDLMEIER, TILMANN BETSCH, AND FRANK RENKEWITZ

Why this book?

Frequencies and how they are processed is currently the subject of much debate in psychological research, as evidenced in recent conference programmes and publications in major journals. It is apparent that when scholars talk and write about frequencies, they are often not concerned with the same issues, they take different perspectives, and use different methodological approaches. When we organized some symposia on the topic, we found that different subdisciplines in psychology hold markedly different views about how frequencies of events are processed and how this influences several aspects of cognition. Coming from different subdisciplines ourselves—social psychology (Tilmann Betsch) and cognitive psychology (Frank Renkewitz, Peter Sedlmeier)—we learned that researchers working in the two fields seldom pay much attention to theoretical models and empirical results from each other. This experience prompted us to think about editing a book that provides an up to date overview of the theoretical and empirical results in the fields that deal with the processing of frequencies in one way or another, that highlights existing controversies, and offers promising directions for further research. This book intends to present the state of the art and to further fruitful exchange among all researchers working in fields dealing with frequency processing and cognition by bringing together a large variety of different approaches and viewpoints. Among the contributors are many of the leading scholars in the respective fields, including researchers from several different subdisciplines of psychology. Thus, we have tried to create a balanced and representative overview that can be the basis for advancing our knowledge about frequency processing and cognition.

Some background

The topic of how repeatedly occurring events are processed and how the resulting frequencies of events influence cognition, although very much in vogue these days, is certainly not a new one. The idea that the frequency with which two events—A and B—co-occur determines how strongly we associate A when we hear B, or the idea that the frequency with which objects are encoded determines how well we can remember them, dates from the beginning of psychology (e.g. James 1890; Ebbinghaus 1913 [1885]). Later, the frequency with which events co-occur was a core part of models of memory (e.g. Hebb 1949) and played the central role in the probability learning paradigm (e.g. Estes 1950). It is an interesting fact that the mere exposure or repetition of items, that is, the frequency with which items are encountered, strongly influences the liking of these items (Bornstein 1989; Zajonc 1968). At present, there are several areas in which the topic of frequency processing and cognition is prominent. Surprisingly, scholars working in different areas come to quite different conclusions.

Consider the case of estimates of frequencies and relative frequencies. In *cognitive psychology*, people are usually found to make quite adequate judgements (for an early review see Hasher & Zacks 1984) and there are several detailed memory models that simulate these judgements well (Hintzman 1988; Dougherty *et al.* 1999; Sedlmeier 1999). In these models, frequency judgements are seen as some function of actually encountered frequencies. In the fields of *social psychology* and *judgement and decision making*, however, the view prevails that frequency judgements are often inadequate (Tversky & Kahneman 1973). The model most often used in this research, the availability heuristic, says that frequency judgements of some event are not based on actually encoded frequencies but on the ease with which instances of that event come to one's mind (Schwarz *et al.* 1991; Wänke *et al.* 1995).

For a long time these two views of adequate and inadequate frequency judgements coexisted without much exchange. Only recently have some studies begun to compare models (Betsch, Siebler *et al.* 1999; Manis *et al.* 1993; Sedlmeier *et al.* 1998) or tried to create models that account for both kinds of results (Dougherty *et al.* 1999). Recently another theoretical approach, stemming from *evolutionary psychology*, has been proposed. It postulates the existence of cognitive algorithms that are specifically tuned to frequency information (Gigerenzer 1996a, see also Kahneman & Tversky 1996, for a contrasting view) and exploit the structure of that information in the environment (Hertwig *et al.* 1999). This approach has strong practical implications because it holds that the use of frequency information as an input for decision tasks yields much better results than the use of probability information (Gigerenzer 1996b). Research on frequency processing and cognition also led to other practical implications. Memory models relying solely on the frequency of occurrence and co-occurrence of words in large text corpora are beginning to be used in automatic text understanding or product marketing (Burgess 1998; Wettler & Rapp 1993). Finally, research on *animal learning* (Gallistel 1990) has very seldom entered other areas in psychology (but see Whalen *et al.* 1999), although this research has produced an abundance of results and sophisticated models of frequency processing which might prove very useful for human research.

In sum, the topic of frequency processing and cognition plays a crucial role in several areas of psychological research and there exists a wealth of models and empirical results. At present, one of the main problems that hinders faster progress in our understanding of how frequencies of events are processed and how this influences cognition is that the exchange of information does not work adequately among the different areas. Some of the differences arose because researchers were interested in different questions, but other differences seem to be due to a missing exchange of ideas, models, and experimental results.

In the remainder of this chapter, we will first give a brief overview about the research on how frequency processing affects cognition and then discuss briefly the content of the remaining chapters.

How does the processing of frequencies affect cognition?

It is very rare that we experience an object or event only once; otherwise humankind would, for instance, not have been able to develop a sign system to communicate about objects and events: language. We usually have to deal with repeatedly occurring objects or events, and these repeated occurrences—frequencies—permeate all aspects of our lives. Thus the topic of frequency processing is relevant whenever we perceive, reason, think, judge, and make decisions, that is, it is relevant for many aspects of cognition. Figure 1.1 gives an overview of the processes and possible situations involved.

The input to the processing of frequencies is, of course, repeatedly encountered events or objects. These events or objects may be directly encountered, that is, one by one, or indirectly, that is, heard or read about in an aggregated form (e.g. 'I saw five young people with green hair'). Some contributions in this book will show that the way frequencies are encoded can make a big difference (Hertwig & Hoffrage, Chapter 18; Sedlmeier, Chapter 9). If events or objects are directly encountered they first have to be categorized; otherwise each perception would be idiosyncratic.

In the next section, we will briefly review the findings on the relationship between frequency processing and categorization and representation in memory. Then we will deal with judgements of frequency, relative frequency and probability; and finally, we will discuss more complex judgements that are based on frequencies.

Categorization and representation

Talking about the frequencies of events or objects is not trivial. At a certain level of analysis all sensory experiences might be seen as unique, that is, all frequencies would be confined to a frequency of 1. To be able to process frequencies, our mind first has to categorize or classify an experience into an event or object. Humans are extremely good at this task (Thorpe & Imbert 1989). In fact, the ability to match one pattern to another might be regarded as '... *the* essential component to most cognitive behaviour'

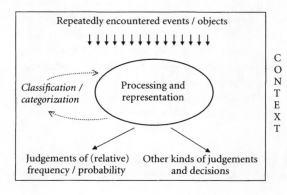

Figure 1.1 A schematic overview of frequency processing and cognition.

(Rumelhart *et al.* 1986, p. 44). Smith and Medin (1981) have summarized the vast research on the topic of categorization and come up with three different views of how humans categorize events. The input used for categorizing an object in all three views is a featural description of that object. They discard the *classical* view which holds that a concept can be defined by singly necessary and jointly sufficient features because of numerous theoretical and empirical problems attached to it. Remaining are the *exemplar* and the *prototype*, or probabilistic views. The exemplar view (e.g. Estes 1986; Hintzman 1986; Nosofsky 1989) holds that no single description of a category exists. The category is represented by a collection of separate descriptions of some or all category members. The prototype view in contrast assumes that prior encounters with an object have led to a composite trace, a summary description, which represents a prototype. The features of a new object entering into the summary description are probabilistically related to the latter. This approach to categorization has been successfully modelled with neural or connectionist networks (e.g. Anderson & Mozer 1981; Estes *et al.* 1989; Gluck & Bower 1988; Shanks 1990, 1991).

In both the exemplar and the prototype view, categorization is intimately linked to memory representation, because categorization in fact relies on representation. Several chapters in this book use such memory models to describe how the mind categorizes events by processing the frequency with which collections of features are encountered (Dougherty & Franco-Watkins, Chapter 8; Sedlmeier, Chapter 9, Wettler, Chapter 17; see also Zacks & Hasher, Chapter 2).

Judgements of frequencies, relative frequencies and probabilities

In the last four decades, numerous studies investigated how accurate judgements of frequencies, relative frequencies and probabilities are and how such judgements might be shaped by several factors. In some of these studies, researchers asked for the frequency or probability of events or objects from everyday life, for example the frequency of letters (Attneave 1953), pairs of letters (Underwood 1971), words (Shapiro 1969), diseases and causes of death (Lichtenstein *et al.* 1978), karate techniques performed in a training session (Bedon & Howard 1992) or visits to a library or to a doctor (Thompson & Mingay 1991). However, in most of the research on frequency and probability judgements stimuli were generated in the laboratory. In these experiments, participants were usually presented a list of items (mostly words). Afterwards, they had to make some kind of frequency or probability judgement. This judgement was either about the relative or absolute size of a certain category in the list (e.g. 'Did the list contain more males or more females?'), or about the frequency of occurrence of specific items in the list (e.g. 'How often did the word "Mary" appear in the list?'). Despite the similarity of those studies, researchers often referred to each other only vaguely, and the results were contradictory. Studies in the tradition of the heuristics and biases programme (Kahneman *et al.* 1982) concluded that frequency and probability estimates are error prone. However, numerous studies in the memory literature came to a completely contrasting conclusion. Typically, the participants there showed considerable sensitivity to frequencies.

Biased judgements

The point of view that frequency and probability estimates are susceptible to biases was based on a few very well known demonstrations of faulty estimates. The famous names study (Tversky & Kahneman 1973) may serve as an example here: in this study, participants were presented with a list of famous names of one sex and slightly more non-famous names of the other sex. In a subsequent frequency judgement the vast majority of participants was of the opinion that the list contained more names of the 'famous sex'. Faulty frequency estimates like this one are usually explained with recourse to the availability heuristic (Tversky & Kahneman 1973), according to which judgements of the frequency of an event are based on the ease with which instances of that event come to one's mind. Availability is commonly measured by the number of instances recalled (e.g. Beyth-Marom & Fischhoff 1977; Bruce et al. 1991; Greene 1989; but see Schwarz et al. 1991, for a different approach). In accordance with the availability hypothesis, the participants in the famous names study not only overestimated the frequency of the famous sex but also recalled more famous names. The availability heuristic and the issue of biases in frequency judgements is addressed in various chapters in this volume (Betsch & Pohl, Chapter 7; Fiedler, Chapter 5; Reber & Zupanek, Chapter 11; Schwarz & Wänke, Chapter 6).

Valid judgements

Despite their impact on public opinion, studies that produced biased frequency estimates were always the exception. Valid estimates were found both for the frequency of events from everyday life (e.g. Attneave 1953; Lichtenstein et al. 1978; Shapiro 1969) and for list frequencies in laboratory settings (e.g. Flexser & Bower 1975; Hasher & Chromiak 1977; Hintzman 1969; Hintzman & Block 1971). Initially, these results primarily referred to frequency of occurrence judgements; later they could also be generalized to category size judgements (Alba et al. 1980; Barsalou & Ross 1986; Brooks 1985; Watkins & LeCompte 1991). Hasher and Zacks (1984) provided an early review of these studies. Moreover, these authors observed that frequency estimates were not influenced by numerous task variables (such as instructions and training) and person variables (such as age and ability). From these results they concluded that the encoding of frequency information is a fairly automatic process; that is, a process that requires little or no attentional capacity and that works without any awareness or intention (see also Hasher & Chromiak 1977; Hasher & Zacks 1979). According to this view, people can rely on some kind of automatically encoded frequency knowledge and do not need to construct estimates by the retrievability of relevant instances when they are giving frequency judgements (see Haberstroh & Betsch, Chapter 13; Zacks & Hasher, Chapter 2).

Relationship between recall and judgements

Several experiments examined the relationship between the number of events recalled and judgements of their frequency. This relationship is important because the most common version of the availability heuristic holds that recall is the basis for estimates. For that, two different methods were used: one method studied whether various manipulations cause parallel effects (e.g. group means) in both recall and estimates (like in the famous names study). The other method checked whether substantial correlations exist between individual recall and frequency estimates. Both methods led to contradictory results. Parallel effects in frequency estimates and recall performances were found consistently for some manipulations. For instance, a prolongation of the presentation time per stimulus results in both better recall performances and more valid frequency estimates (Lewandowsky & Smith 1983; Williams & Durso 1986). Similarly, in a recall test, self-generated items have an advantage over items that are merely copied; moreover, self-generated items receive higher frequency estimates (Greene 1988, 1989). The results for other manipulations, however, are less clear. Higher typicality (Barsalou & Ross 1986; Williams & Durso 1986) and blocked presentation of category exemplars (Alba et al. 1980; Barsalou & Ross 1986; Greene 1989) consistently improved recall but only sometimes led to improved frequency estimates. Furthermore, for some domains such as positional letter frequencies (Sedlmeier et al. 1998; but see Tversky & Kahneman 1973) and property frequencies (Barsalou & Ross 1986; Freund & Hasher 1989) estimates turned out to be almost completely independent of availability measures. Results from studies in which the correlation between recall and frequency judgements was determined were also somewhat inconsistent. While substantial correlations could sometimes be found (Williams & Durso 1986; Freund & Hasher 1989; Manis et al. 1993), the correlations in other cases were low (Lewandowsky & Smith 1983) or close to zero (Bruce et al. 1991). Altogether, these studies render it plausible that under some circumstances the retrievability of instances may have an impact on frequency estimates. Nevertheless, the fundamental outcome of these studies may be the insight that the availability heuristic is certainly not the only cognitive process underlying frequency estimates. This is particularly evident in studies indicating that none of several recall measures can account for the accuracy of the corresponding frequency estimates (Watkins & LeCompte 1991) or that estimates still show substantial correlations with actual frequencies when recall is partialled out (Betsch, Siebler et al. 1999; Bruce et al. 1991).

Factors influencing frequency processing and judgement

Further research on the automatic coding hypothesis focused on Hasher and Zacks' (1984) assertion that frequency judgements are not influenced by most task and person variables. Particularly, the impact of instruction and age was investigated many times. For both variables, this led to inconclusive results. Thus early studies did not find any age differences in the frequency estimates of kindergarten children and third graders (Hasher & Zacks 1979), second graders and college students (Hasher & Chromiak 1977) and college students and older adults (Attig & Hasher 1980; Kausler & Puckett 1980). Later studies, however, found that the accuracy of frequency judgements increased from kindergarten to third grade (Lund et al. 1983; Ghatala & Levin 1973) and decreased from early to late adulthood (Kausler et al. 1981, 1982, 1984). Similarly, there is ample evidence that participants who know in advance that they will be tested on the frequency of items do no better on a frequency test than do participants who are told to expect a free recall test or an unspecified memory test (e.g. Attig & Hasher 1980; Flexser & Bower 1975; Greene 1984; Hasher & Chromiak 1977; Howell 1973; Kausler et al. 1982; Kausler & Puckett 1980; Rose & Rowe 1976; but see Williams & Durso 1986). Nevertheless, the announcement of a frequency test can improve frequency estimates: Differences between experimental (announcement) and control (no announcement) groups typically occurred when the encoding times were relatively short (Fisk & Schneider 1984; Naveh-Benjamin & Jonides 1986) or whenever procedures were applied which made sure that the participants in the control group learned the tested material incidentally (Greene 1984, 1986; Sanders et al. 1987; see also Kausler et al. 1984).

While studies which examined the role of instruction and age came up with mixed results, it could be demonstrated that several other factors have a reliable impact on frequency estimates. For instance, deeper processing of items results in superior judgements of frequencies than does shallower processing (Erickson & Gaffney 1985; Greene 1984, 1986; Jonides & Naveh-Benjamin 1987; Maki & Ostby 1987; Naveh-Benjamin & Jonides 1986; Rose & Rowe 1976; Rowe 1974; see Freund & Hasher 1989 and Hasher & Zacks 1984 for an attempt to explain this effect within the automatic coding framework). Items with spaced repetitions receive higher estimates than items with massed ones (Hintzman 1969; Hintzman et al. 1973; Jacoby 1972; Rowe & Rose 1977; Rose & Rowe 1976). Finally, higher estimates are also given when items are repeated in the same context rather than in different contexts (Brown 1995; Hintzman & Stern 1978; Jacoby 1972; Rowe 1973a, 1973b; Rose 1980; see also Brown, Chapter 3 this volume).

On one hand, this body of research shows clearly that memory of frequencies is less invariant than was originally suggested by the automatic coding hypothesis (Hasher & Zacks 1984). On the other hand, much evidence has been accumulated which supports the view that people are sensitive to frequencies across a wide range of conditions. The effects of the applied manipulations are mostly small. Even under the most disadvantageous circumstances participants often give reasonable estimates (Greene 1984), and wrong rank orderings of actual frequencies—as they were found in the famous names study—remain the very rare exception. However, people are not sensitive to all kinds of frequencies. Thus, after the presentation of a list of different items, it is not only possible to ask how many of these items belong to a superordinate category (e.g. 'How many different kinds of vegetables were on the list?') but also how many of these items share a certain property (e.g. 'How many things that are red were on the list?'). People show little or no sensitivity to these property frequencies unless they are directed to focus on the relevant properties (e.g. by presenting the items blocked by properties) at the time of encoding (Barsalou & Ross 1986; Freund & Hasher 1989). Generally, the degree to which participants focus on those attributes of the stimuli that are appropriate for the later frequency judgements seems to have an impact on the accuracy of the judgements. For example, Betsch, Siebler et al. (1999) could demonstrate that if participants are led to focus on superordinate categories, they give less accurate frequency of occurrence judgements.

Explanatory models

As already mentioned, there are several detailed memory models that hold specific assumptions about the representation of frequency knowledge (e.g. Dougherty et al. 1999; Fiedler 1996; Hintzman 1988; Sedlmeier 1999). Among these models, Minerva 2 (Hintzman 1988) has presumably been applied to results from the research on frequency judgements most often. Minerva 2 simulates the usual finding of valid frequency estimates quite well. In addition, it can explain some of the effects observed in frequency estimates such as the spacing effect or the context effect. However, for a long time Minerva 2 has not been applied to results coming from the heuristics and biases research. Generally, results and theoretical approaches from the heuristics and biases programme and results and approaches from memory literature existed side by side with little connection and exchange. This situation changed recently after a modification of the Minerva 2 model (Dougherty et al. 1999). Another candidate model that explains both valid and invalid frequency judgements but relies on a different memory representation is Sedlmeier's (1999) PASS model. Both models are described in detail in this volume. As an explanation for frequency processing in animals, a model based on information theory is proposed by Gallistel (Chapter 10). Brown's (1997, Chapter 3 this volume) multiple strategy perspective represents an alternative approach, situated at a more abstract level. He assumes that people use a variety of strategies to generate frequency estimates. The choice of a certain strategy is assumed to affect magnitude and accuracy of frequency judgements (see also Haberstroh & Betsch, Chapter 13).

Effects of event frequency in other domains of judgement and decision making

Judgements of frequency are sometimes relevant per se, but often, they are the basis for other kinds of judgements. These include judgements about the confidence one has in an estimate, the strength of one's belief, or in the intensity with which one experiences emotions. Frequencies also determine the strength of habits. Moreover, frequencies or probabilities (calculated from relative frequencies) are often the basis for more complex judgements. Because the frequency with which objects and events co-occur is the basis for associative learning, all memory processes might be said to rely to a certain degree on frequency processing. This is most apparent in memory tasks that require heavy use of associations.

Confidence judgements

Relative frequencies obtained in a (random) sample are often used to make judgements about population proportions. It is quite common that different samples from the same population (e.g. opinion polls done by different companies) result in different relative frequencies. All other things being equal, one should have the highest confidence in the relative frequency obtained from the largest sample (e.g. Freedman *et al.* 1991). Do people's judgements reflect this statistical property? For a considerable time it seemed that the results were inconsistent: in some studies people took sample size into account but in others they did not. In a review of the research, Sedlmeier and Gigerenzer (1997; 2000) concluded that people indeed behave as if they followed an intuitive version of the 'empirical law of large numbers' that is, the observable fact that larger samples tend to lead to more exact estimates than smaller samples. It turned out that studies that had found negative evidence had mostly used tasks that required participants to make judgements about the variance of sampling distributions, that is, about distributions of relative frequencies, not about a given relative frequency. Independent evidence supported this distinction between the two kinds of tasks (Sedlmeier 1998). A possible explanation for how confidence judgements may result as a by-product of associative learning is given by Sedlmeier (1999; Chapter 9 this volume).

Frequencies vs. probabilities

One issue that currently receives much attention in psychological research is whether the format in which frequency information is presented, that is, whether it is given in frequencies, relative frequencies or probabilities, makes a difference. This is especially relevant in more complex tasks, which often deal with probability judgements of conjunctive events or judgements of conditional probabilities. A prominent example is probability revision: in a simple probability revision task the original probability (e.g. the probability that a woman in a certain age group has breast cancer is 1%: p[cancer] = 0.01) is revised in the light of a new result (e.g. a mammography test came out positive [pos]) and two conditional probabilities, the hit rate (e.g. p[pos|cancer] = 0.8) and the false alarm rate (e.g. p[pos|no cancer] = 0.1). Applying Bayes's formula, one obtains a revised probability of 7.5%:

p[cancer|pos] = p[cancer]p[pos|cancer]/(p[cancer]p[pos|cancer] + p[no cancer] p[pos|no cancer] = 0.01*0.8/(0.01*0.8 + 0.99*0.1) = 0.075.

If, in such tasks, information is given in probabilities, the solution rates are very low. If, in contrast, the same information is given in natural frequencies, solution rates soar (Gigerenzer & Hoffrage 1995). Using natural frequencies, one would begin with a sample of say 1000 women of whom 10 (1%) have breast cancer. Of those 10 women, 8 (80%) could expect a (correct) positive test result and 99 (90% of the remaining 990 healthy women) could also expect a (wrong) positive test result. Thus, the conditional probability being sought p[cancer|pos] can be calculated as 8 out of 8 + 99 which, of course, leads to the same result as the one obtained with Bayes's formula. The chapters by Gigerenzer and Hertwig and Hoffrage will deal extensively with some results and theoretical implications of this kind of research.

Frequency of co-occurrence and judgements based on associative memory

With the advent of powerful computer technologies, it is possible to construct associative memories by counting the co-occurrences of words from large machine-readable text corpora. The sole input to such artificial associative memories is the frequency with which words co-occur. Memories were built by using counting procedures of various degrees of sophistication, but despite the simplicity of this 'learning process' results are very promising (e.g. Burgess 1998; Landauer & Dumais 1997; Wettler & Rapp 1993). Such models, which capitalize on global lexical co-occurrence, have explanatory power that captures a broad range of cognitive phenomena such as assertions about semantic or grammatical properties of words and texts to be used in automatic text understanding systems. In Chapter 17 Wettler describes such a model in detail and points out some practical implications.

Role of frequency in affect, decision making and choice

The frequency of certain aspects of experience can have a substantial impact on affect, decision making and choice. In a seminal paper, Zajonc demonstrated that increases in exposure frequency can yield increasingly positive affective reactions to initially neutral stimulus objects (Zajonc 1968). In general, organisms are especially sensitive to the frequency of experiences with hedonic relevance, such as reinforcing events. One of the cornerstone assumptions of historical and contemporary choice theories states that behavioural tendencies reflect the frequency of prior reinforcements (Davis *et al.* 1993; Thorndike 1898). For example, the effectiveness of a constant amount of reward on learning increases with an increase in the number of reinforcement episodes (Wolfe & Kaplon 1941). In a similar vein, attitude research provides evidence that humans are remarkably sensitive for the frequency of value-charged experiences with objects. Even when concrete memories about experiences with an object are lost or can no longer be accessed in memory, intuitive judgements still reflect the frequency (and intensity) of the entire amount of prior encounters (Betsch, Plessner *et al.* 2001). The frequency of behaviour repetition itself influences future behaviour (Bentler & Speckart 1979). For example, a high frequency of behaviour repetition increases the likelihood that the behaviour will be frozen into habit, and will be automatically chosen in subsequent decisions (Aarts & Diksterhuis 2000; Oulette & Wood 1998; Ronis *et al.* 1988; Triandis 1977). Frequently repeated behaviours are likely to be maintained even when the decision maker has acquired new information indicating that the behaviour is inadequate (Betsch, Haberstroh *et al.* 2001; Betsch, Brinkmann *et al.* 1999; Betsch *et al.* 1998). In deliberate decisions, the depth and elaborateness of information search decreases with increasing frequency of prior behaviour repetition (Verplanken *et al.* 1997). Moreover, an information search tends to be biased towards the frequently performed behaviour, such that disconfirming evidence about a routine behaviour is neglected or even avoided (Betsch, Haberstroh *et al.* 2001). The role of frequency of occurrence on affect, decision making and choice is highlighted in several contributions to this book (Alba, Chapter 16; Baumann & Krems, Chapter 14; Haberstroh & Betsch, Chapter 13; Schimmack, Chapter 12).

Contents of the book

This book does not cover all aspects outlined in Fig. 1.1 to the same extent. It is largely confined to (1) how frequencies are processed, (2) how judgements of frequency, relative frequency and probability come about and how they are shaped by contextual influences, and (3) how these judgements implicitly or explicitly influence other kinds of judgement, and decision making. The book consists of three parts. The first discusses theoretical models and perspectives, the second presents summaries of empirical results of some research groups, and the third contains some examples of practical implications. The emphasis of the book is not on brand new research results (although occasionally such results will be reported) but on overviews of topics with which the contributors have spent considerable time.

Part I: Theoretical models and perspectives

Part I contains contributions that focus on theoretical approaches and perspectives rather than on detailed empirical results or applications. It begins with Chapter 2 by Rose Zacks and Lynn Hasher, who reflect on the impact of their seminal work on the literature concerning frequency processing and cognition and conclude that their general theoretical framework still remains valid. In Chapter 3, Norman Brown describes his theoretical perspective, which holds that people represent event frequency in several ways and use a variety of estimation strategies when they are required to judge event frequencies. In addition, he posits that strategy selection is restricted by the frequency-relevant contents of memory and that it can affect magnitude and accuracy of frequency judgements, as well as the time required to produce them.

In Chapter 4, Gerd Gigerenzer introduces an evolutionary psychology approach to frequency processing and cognition and discusses issues of information processing and external representation in the form of a fictitious podium discussion taking place in the year 2053. Some of the points he makes are that frequencies specify the reference class which is left undefined in single-event probabilities, that frequencies clarify the term 'probable' which is polysemious in natural languages, and that natural frequencies (but not relative frequencies) simplify Bayesian computations. In Chapter 5, Klaus Fiedler explains his view that the process of sampling and the kinds of samples one takes have a strong impact on frequency judgements of all sorts. He argues that the accuracy of frequency judgements heavily depends on the retrieval–cue structure.

The next chapters deal with the availability approach as an explanation for different kinds of frequency based judgements. In Chapter 6, Norbert Schwarz and Michaela Wänke review conditions for when individuals rely on the subjective experience of ease of recall when performing frequency estimates. They then address what respondents learn from frequency scales provided by the researcher in arriving at frequency judgements as well as other kinds of judgements. In Chapter 7, Tilmann Betsch and Devika Pohl discuss the availability approach and postulate the development of more detailed theories on frequency processing.

Such detailed theories based on memory models—one based on the exemplar and one the prototype view—are described next. Chapter 8, by Mike Dougherty and Ana Franco-Watkins, gives a tutorial on the MINERVA-DM model that ties many of the heuristics and biases together within the context of a multiple trace memory model for frequency judgements. In Chapter 9, Peter Sedlmeier describes an alternative model based on associative learning mechanisms. Randy Gallistel, in his theoretical account of how conditioning processes work in animals, questions the central role of associative learning and instead introduces an information processing theory in Chapter 10. He argues that what matters in the conditioning process is not the individual events but rather the information they convey.

Part II: Essential empirical results

The second part of the book reflects the different theoretical views outlined above, describes a collection of quite diverse research programmes, and also presents some new findings. The contributions deal with several aspects of frequency and probability judgements per se but also with other kinds of judgements that are based on frequency processing.

It begins with Chapter 11 by Rolf Reber and Natasha Zupanek, who examine the influence of processing fluency on frequency estimates when the actual frequency cannot be enumerated. They summarize the conditions under which processing fluency makes a difference and highlight its specific impact. Chapter 12, written by Ulrich Schimmack, examines possible links between meta-memory judgements and frequency judgements. In particular, he is interested in when people rely on recall of instances and when they bypass the recall process to estimate frequencies. In his experiments, he uses frequencies of emotional experiences. Susanne Haberstroh and Tilmann Betsch report research on how people judge the frequency of their own behaviour in Chapter 13. Specifically, they identify conditions moderating the accuracy of such frequency judgements. They find that frequency judgements are based on automatically encoded information when made under a 'spontaneous guess' instruction or under time pressure. However, frequency judgements are influenced by the availability of information when an accuracy instruction is applied or when participants are paid contingent upon their performance.

In Chapter 14, Martin Baumann and Josef Krems examine the conditions under which the content of new-information items and the order by which they are encountered influence the updating of beliefs. According to the prevailing psychophysics metaphor, subjective probability judgements are evaluated for their validity by comparing them with the relative frequencies in which the probabilistic guesses are true. In Chapter 15, the final chapter of Part II, Gernot Kleiter, Michael Doherty and Gregory Brake give an overview of the corresponding research and argue that the psychophysics metaphor cannot be upheld.

Part III: Practical implications

The last part of the book gives three examples of when frequency processing may have some direct practical implications. Josef Alba gives an overview of how the frequency of occurrence of a variety of product-related stimuli can affect product judgements and purchase decisions in Chapter 16. In Chapter 17, Manfred Wettler shows how one can build an associative memory by extracting co-occurrences of words from a large machine-readable text corpus and how predictions based on such an artificial memory may be used in marketing research. Finally, in Chapter 18, Ralph Hertwig and Ulrich Hoffrage report evidence that the way in which statistical evidence is presented, that is, whether in the form of frequencies or probabilities, has a strong impact on how well it is understood by experts such as physicians and lawyers, which in turn can have far reaching consequences in medical and juridical settings.

Conclusions

This book addresses students and scholars in the fields of judgement and decision making, cognitive psychology, social psychology, animal learning and applied psychology. Moreover, it should be of use for practitioners in judgement and decision making and advertising. To help this diverse audience in drawing conclusions from the heterogeneous topics discussed, Chapter 19, written by the editors, tries to summarize briefly what has been achieved in the field thus far and to point out existing problems. From an analysis of the current state of affairs, it offers some tentative guidelines for further research.

Acknowledgements

This book arose from several symposia on frequency processing. The most influential one took place in Heidelberg on 26–28 November 1999, with leading researchers from different subdisciplines in psychology whose work centres on frequency processing and cognition. The German Science Foundation generously funded this small group meeting via the national research grant, SFB 504 at the University of Mannheim. The meeting could not have taken place had we not received help from our research assistants Claudia Bäumer, Connie Höhle, Bronwyn Bosse and Devika Pohl. We also are very grateful to Bronwyn Bosse for proofreading, and Kate Smith and Martin Baum of OUP for their competent and friendly assistance. Last but not least, we thank the contributors of this book who took on the burden to review each other's contributions. The book has benefited enormously from this review process.

References

Aarts, H. & Dijksterhuis, A. (2000). Habits as knowledge structures: Automaticity in goal-directed behavior. *Journal of Personality and Social Psychology, 78*:53–63.

Alba, J. W., Chromiak, W., Hasher, L. & Attig, M. S. (1980). Automatic encoding of category size information. *Journal of Experimental Psychology: Human Learning and Memory, 6*:370–378.

Anderson, J. A. & Mozer, M. C. (1981). Categorization and selective neurons. In G. E. Hinton & J. A. Anderson (eds) *Parallel models of associative memory* (pp. 251–274). Hillsdale, NJ: Lawrence Erlbaum Associates.

Attig, M. & Hasher, L. (1980). The processing of frequency of occurrence information by adults. *Journal of Gerontology, 35*:66–69.

Attneave, F. (1953). Psychological probability as a function of experienced frequency. *Journal of Experimental Psychology, 46*:81–86.

Barsalou, L. W. & Ross, B. H. (1986). The role of automatic and strategic processing in sensitivity to superordinate and property frequency. *Journal of Experimental Psychology: Learning, Memory and Cognition, 12*:116–134.

Bedon, B. G. & Howard, D. V. (1992). Memory for the frequency of occurrence of karate techniques: A comparison of experts and novices. *Bulletin of the Psychonomic Society, 30*:117–119.

Bentler, P. M. & Speckart, G. (1979). Models of attitude-behavior relations. *Psychological Review, 86*:452–464.

Betsch, T., Brinkmann, J., Fiedler, K. & Breining, K. (1999). When prior knowledge overrules new evidence: Adaptive use of decision strategies and the role of behavioral routines. *Swiss Journal of Psychology, 58*:151–160.

Betsch, T., Fiedler, K. & Brinkmann, B. J. (1998). Behavioral routines in decision making: The effects of novelty in task presentation and time pressure on routine maintenance and deviation. *European Journal of Social Psychology, 28*:861–878.

Betsch, T., Haberstroh, S., Glöckner, A., Haar, T. & Fiedler, K. (2001). The effects of routine strength on adaptation and information search in recurrent decision making. *Organizational Behavior and Human Decision Processes, 84*:25–53.

Betsch, T., Plessner, H., Schwieren, C. & Gütig, R. (2001). I like it but I don't know why: a value-account approach to implicit attitude formation. *Personality and Social Psychology Bulletin, 27*:242–253.

Betsch, T., Siebler, F., Marz, P., Hormuth, S. & Dickenberger, D. (1999). The moderating role of category salience and category focus in judgments of set size and frequency of occurrence. *Personality and Social Psychology Bulletin, 25*:463–481.

Beyth-Marom, R. & Fischhoff, B. (1977). Direct measures of availability and judgments of category frequency. *Bulletin of the Psychonomic Society, 9*:236–238.

Bornstein, R. F. (1989). Exposure and affect: Overview and meta-analysis of research, 1968–1987. *Psychological Bulletin, 106*:265–289.

Brooks, J. E. (1985). Judgments of category frequency. *American Journal of Psychology, 98*:363–372.

Brown, N. R. (1995). Estimation strategies and the judgment of event frequency. *Journal of Experimental Psychology: Learning, Memory and Cognition, 21*:1539–1553.

Brown, N. R. (1997). Context memory and the selection of frequency estimation strategies. *Journal of Experimental Psychology: Learning, Memory, and Cognition, 23*:898–914.

Bruce, D., Hockley, W. E. & Craik, F. I. M. (1991). Availability and category-frequency estimation. *Memory and Cognition, 19*:301–312.

Burgess, C. (1998). From simple associations to the building blocks of language: Modeling meaning in memory with the HAL model. *Behavior Research Methods, Instruments and Computers, 30*:188–198.

Davis, D. G. S., Staddon, J. E. R., Machado, A. & Palmer, R. G. (1993). The process of recurrent choice. *Psychological Review, 100*:320–341.

Dougherty, M. R. P., Gettys, C. F. & Ogden, E. E. (1999). MINERVA-DM: A memory process model for judgments of likelihood. *Psychological Review, 106*:180–209.

Ebbinghaus, H. (1913). *Memory: A contribution to experimental psychology* (H. A. Ruger and C. E. Bussenues, trans.). New York: Teachers College, Columbia University. (Original work published 1885.)

Erickson, J. R. & Gaffney, C. R. (1985). Effects of instructions, orienting task, and memory tests on memory for words and word frequency. *Bulletin of the Psychonomic Society, 23*:377–380.

Estes, W. K. (1950). Toward a statistical theory of learning. *Psychological Review, 57*:94–107.

Estes, W. K. (1986). Array models for category learning. *Cognitive Psychology, 18*:500–549.

Estes, W. K., Campbell, J. A., Hatsopoulos, N. & Hurwitz, J. B. (1989). Base-rate effects in category learning: A comparison of parallel network and memory storage-retrieval model. *Journal of Experimental Psychology: Learning, Memory, and Cognition, 15*:556–571.

Fiedler, K. (1996). Explaining and simulating judgment biases as an aggregation phenomenon in probabilistic, multiple-cue environments. *Psychological Review, 103*:193–214.

Fisk, A. D. & Schneider, W. (1984). Memory as a function of attention, level of processing, and automatization. *Journal of Experimental Psychology: Learning, Memory, and Cognition, 10*:181–197.

Flexser, A. J. & Bower, G. H. (1975). Further evidence regarding instructional effects on frequency judgments. *Bulletin of the Psychonomic Society, 6*:321–324.

Freedman, D., Pisani, R., Purves, R. & Adhikari, A. (1991). *Statistics*, 2nd edn. New York: Norton.

Freund, J. S. & Hasher L. (1989). Judgments of category size: Now you have them, now you don't. *American Journal of Psychology, 102*:333–352.

Gallistel, C. R. (1990). *The organization of learning*. Cambridge, MA: MIT Press.

Ghatala, E. S. & Levin, J. R. (1973). Developmental differences in frequency judgments of words and pictures. *Journal of Experimental Child Psychology, 16*:495–507.

Gigerenzer, G. (1996a). On narrow norms and vague heuristics: A reply to Kahneman and Tversky (1996). *Psychological Review, 103*:592–596.

Gigerenzer, G. (1996b). The psychology of good judgment: Frequency formats and simple algorithms. *Journal of Medical Decision Making, 16*:273–280.

Gigerenzer, G. & Hoffrage, U. (1995). How to improve Bayesian reasoning without instruction: Frequency formats. *Psychological Review, 102*:684–704.

Gluck, M. A. & Bower, G. H. (1988). From conditioning to category learning: An adaptive network model. *Journal of Experimental Psychology: General, 117*:227–247.

Greene, R. L. (1984). Incidental learning of event frequency. *Memory and Cognition, 12*:90–95.

Greene, R. L. (1986). Effects of intentionality and strategy on memory for frequency. *Journal of Experimental Psychology: Learning, Memory, and Cognition, 12*:489–495.

Greene, R. L. (1988). Generation effects in frequency judgment. *Journal of Experimental Psychology: Learning, Memory, and Cognition, 14*:298–304.

Greene, R. L. (1989). On the relationship between categorical frequency estimation and cued recall. *Memory and Cognition, 17*:235–239.

Hasher, L. & Chromiak, W. (1977). The processing of frequency information: An automatic process? *Journal of Verbal Learning and Verbal Behavior,* 16:173–184.

Hasher, L. & Zacks, R. T. (1979). Automatic and effortful processes in memory. *Journal of Experimental Psychology: General,* 108:356–388.

Hasher, L. & Zacks, R. T. (1984). Automatic processing of fundamental information: The case of frequency of occurrence. *American Psychologist,* 39:1372–1388.

Hebb, D. O. (1949). *The organisation of behavior.* New York: Wiley.

Hertwig, H., Hoffrage, U. & Martignon, L. (1999). Quick estimation: Letting the environment do some of the work. In G. Gigerenzer, P. M. Todd, and the ABC Research Group *Simple heuristics that make us smart* (pp. 209–234). New York: Oxford University Press.

Hintzman, D. L. (1969). Apparent frequency as a function of frequency and the spacing of repetitions. *Journal of Experimental Psychology,* 80:139–145.

Hintzman, D. L. (1986). 'Schema abstraction' in a multiple trace memory model. *Psychological Review,* 93:411–428.

Hintzman, D. L. (1988). Judgments of frequency and recognition memory in a multiple-trace memory model. *Psychological Review,* 95:528–551.

Hintzman, D. L. & Block, R. A. (1971). Repetition and memory: Evidence for a multiple-trace hypothesis. *Journal of Experimental Psychology,* 88:297–306.

Hintzman, D. L., Block, R. A. & Summers, J. (1973). Modality tags and memory for repetitions: Locus of the spacing effect. *Journal of Verbal Learning and Verbal Behavior,* 12:229–238.

Hintzman, D. L. & Stern, L. D. (1978). Contextual variability and memory for frequency. *Journal of Experimental Psychology: Human Learning and Memory,* 4:539–549.

Howell, W. C. (1973). Storage of events and event frequencies: A comparison of two paradigms in memory. *Journal of Experimental Psychology,* 98:260–263.

Jacoby, L. L. (1972). Context effects on frequency judgments of words and sentences. *Journal of Experimental Psychology,* 94:255–260.

James, W. (1890). *The principles of psychology.* New York: Dover.

Jonides, J. & Naveh-Benjamin, M. (1987). Estimating frequency of occurrence. *Journal of Experimental Psychology: Learning, Memory, and Cognition,* 13:230–240.

Kahneman, D., Slovic, P. & Tversky, A. (eds) (1982). *Judgment under uncertainty: Heuristics and biases.* Cambridge: Cambridge University Press.

Kahneman, D. & Tversky, A. (1996). On the reality of cognitive illusions. *Psychological Review,* 103:582–591.

Kausler, D. H., Hakami, M. K. & Wright, R. E. (1982). Adult age differences in frequency judgments of categorial representations. *Journal of Gerontology,* 37:365–371.

Kausler, D. H., Lichty, W. & Hakami, M. K. (1984). Frequency judgments for distractor items in a short-term memory task: Instructional variation and adult age differences. *Journal of Verbal Learning and Verbal Behavior,* 23:660–668.

Kausler, D. H. & Puckett, J. (1980). Frequency judgments and correlated cognitive abilities in young and elderly adults. *Journal of Gerontology,* 35:376–382.

Kausler, D. H., Wright, R. E. and Hakami, M. K. (1981). Variation in task complexity and adult age differences in frequency-of-occurrence judgments. *Bulletin of the Psychonomic Society,* 18:195–197.

Landauer, T. K. & Dumais, S. T. (1997). A solution to Plato's problem: The latent semantic analysis theory of acquisition, induction, and representation of knowledge. *Psychological Review,* 104:211–240.

Lewandowsky, S. & Smith, P. W. (1983). The effect of increasing the memorability of category instances on estimates of category size. *Memory and Cognition, 11*:347–350.

Lichtenstein, S., Slovik, P., Fischhoff, B., Layman, M. & Combs, B. (1978). Judged frequency of lethal events. *Journal of Experimental Psychology: Human Learning and Memory, 4*:551–581.

Lund, A. M., Hall, J. W., Wilson, K. P. & Humphreys, M. S. (1983). Frequency judgment accuracy as a function of age and school achievment (learning disabled versus non-learning-disabled) patterns. *Journal of Experimental Child Psychology, 35*:236–247.

Maki, R. H. & Ostby, R. S. (1987). Effects of levels of processing and rehearsal on frequency judgments. *Journal of Experimental Psychology: Learning, Memory, and Cognition, 13*:151–163.

Manis, M., Shedler, J., Jonides, J. & Nelson, T. E. (1993). Availability heuristic in judgments of set-size and frequency of occurrence. *Journal of Personality and Social Psychology, 65*:448–457.

Naveh-Benjamin, M. & Jonides, J. (1986). On the automaticity of frequency coding: Effects of competing task load, encoding strategy, and intention. *Journal of Experimental Psychology: Learning, Memory, and Cognition, 12*:378–386.

Nosofsky, R. M. (1989). Further tests of an exemplar-similarity approach to relating identification and categorization. *Perception and Psychophysics, 45*:279–290.

Oulette, J. A. & Wood, W. (1998). Habit and intention in everyday life: The multiple processes by which past behavior predicts future behavior. *Psychological Bulletin, 124*:54–74.

Ronis, D. L., Yates, J. F. & Kirscht, J. P. (1988). Attitudes, decisions and habits as determinants of repeated behavior. In A.R. Pratkanis, S.J. Breckler and A.G. Greenwald (eds) *Attitude structure and function* (pp. 213–239). Hillsdale, NJ: Erlbaum.

Rose, R. J. (1980). Encoding variability, levels of processing, and the effects of spacing of repetitions upon judgments of frequency. *Memory and Cognition, 8*:84–93.

Rose, R. J. & Rowe, E. J. (1976). Effects of orienting task and spacing of repetitions on frequency judgments. *Journal of Experimental Psychology: Human Learning and Memory, 2*:142–152.

Rowe, E. J. (1973a). Context effects in judgment of frequency. *Bulletin of the Psychonomic Society, 2*:231–232.

Rowe, E. J. (1973b). Frequency judgments and recognition of homonyms. *Journal of Verbal Learning and Verbal Behavior, 12*:440–447.

Rowe, E. J. (1974). Depth of processing in a frequency judgment task. *Journal of Verbal Learning and Verbal Behavior, 13*:638–643.

Rowe, E. J. & Rose, R. J. (1977). Effects of orienting task, spacing of repetitions, and list context on judgments of frequency. *Memory and Cognition, 5*:505–512.

Rumelhart, D. E., Smolenky, P., McClelland, J. L. & Hinton, G. E. (1986). Schemata and sequential thought processes in PDP models. In J. L. McClelland, D. E. Rumelhart (eds) *Parallel distributed processing, Vol. 2: Psychological and biological models* (pp. 7–57). Cambridge, MA: MIT Press.

Sanders, R. E., Gonzalez, E. G., Murphy, M. D., Liddle, C. L. & Vitina, J. R. (1987). Frequency of occurrence and the criteria for automatic processing. *Journal of Experimental Psychology: Learning, Memory, and Cognition, 13*:241–250.

Schwarz, N., Bless, H., Strack, F., Klumpp, G., Rittenauer-Schatka, H. & Simons, A. (1991). Ease of retrieval as information: Another look at the availability heuristic. *Journal of Personality and Social Psychology, 61*:195–202.

Sedlmeier, P. (1998). The distribution matters: Two types of sample-size tasks. *Journal of Behavioral Decision Making, 11*:281–301.

Sedlmeier, P. (1999). *Improving statistical reasoning: Theoretical models and practical implications.* Mahwah: Lawrence Erlbaum Associates.

Sedlmeier, P. & Gigerenzer, G. (1997). Intuitions about sample size: The empirical law of large numbers. *Journal of Behavioral Decision Making, 10*:33–51.

Sedlmeier, P. & Gigerenzer, G. (2000). Was Bernoulli wrong? On intuitions about sample size. *Journal of Behavioral Decision Making, 13*:133–139.

Sedlmeier, P., Hertwig, R. & Gigerenzer, G. (1998). Are judgments of the positional frequencies of letters systematically biased due to availability? *Journal of Experimental Psychology: Learning, Memory, and Cognition, 24*:754–770.

Shanks, D. R. (1990). Connectionism and the learning of probabilistic concepts. *Quarterly Journal of Experimental Psychology, 42A*:209–237.

Shanks, D. R. (1991). Categorization by a connectionist network. *Journal of Experimental Psychology: Learning, Memory, and Cognition, 17*:433–443.

Shapiro, B. J. (1969). The subjective estimation of relative word frequency. *Journal of Verbal Learning and Verbal Behavior, 8*:248–251.

Smith, E. E. & Medin, D. L. (1981). *Categories and concepts*. Cambridge, MA: Harvard University Press.

Thompson, C. P. & Mingay, D. (1991). Estimating the frequency of everyday events. *Applied Cognitive Psychology, 5*:497–510.

Thorndike, E. L. (1898). Animal intelligence: An experimental study of the associative processes in animals. *Psychological Review, Monograph Supplement 2*.

Thorpe, S. J. & Imbert, M. (1989). Biological contraints on connectionist modelling. In R. Pfeifer, Z. Schreter, F. Fogelman-Soulie and L. Steels (eds). *Connectionism in perspective* (pp. 63–92). Amsterdam: Elsevier.

Triandis, H.C. (1977). *Interpersonal behavior*. Monterey, CA: Brooks/Cole Publishing Company.

Tversky, A. & Kahneman, D. (1973). Availability: A heuristic for judging frequency and probability. *Cognitive Psychology, 4*:207–232.

Underwood, B. J. (1971). Recognition Memory. In H. H. Kendler and J. T. Spence (eds), *Essays in neobehaviorism: A memorial volume to K. W. Spence* (pp. 313–335). New York: Appleton-Century-Crofts.

Verplanken, B., Aarts, H. & van Knippenberg, A. (1997). Habit, information acquisition, and the process of making travel mode choice. *European Journal of Social Psychology, 27*:539–560.

Wänke, M., Schwarz, N. & Bless, H. (1995). The availability heuristic revisited: Experienced ease of retrieval in mundane frequency judgments. *Acta Psychologica, 89*:83–90.

Watkins, M. J. & LeCompte, D. C. (1991). Inadequacy of recall as a basis for frequency knowledge. *Journal of Experimental Psychology: Learning, Memory, and Cognition, 17*:1161–1176.

Wettler, M. & Rapp, R. (1993). Associative text analysis of advertisements. *Marketing and Research Today, 21*:241–246.

Whalen, J., Gallistel, C. R. & Gelman, R. (1999). Nonverbal counting in humans: The psychophysics of number representation. *Psychological Science, 10*:130–137.

Williams, K. W. & Durso, F. T. (1986). Judging category frequency: Automaticity or availability? *Journal of Experimental Psychology: Learning, Memory, and Cognition, 12*:387–396.

Wolfe, J. B. & Kaplon, M. D. (1941). Effect of amount of reward and consummative activity on learning in chickens. *Journal of Comparative Psychology, 31*:353–361.

Zajonc, R. B. (1968). Attitudinal effects of mere exposure. *Journal of Personality and Social Psychology, 9* (2, Part 2): 1–28.

Part I

THEORETICAL MODELS AND PERSPECTIVES

CHAPTER 2

FREQUENCY PROCESSING: A TWENTY-FIVE YEAR PERSPECTIVE

ROSE T. ZACKS AND LYNN HASHER

Abstract

We first proposed that frequency of occurrence information is 'automatically' encoded in the context of a general theoretical framework relating attention and memory encoding (Hasher & Zacks 1979). This chapter begins with a description of the origins of that framework, focusing on earlier evidence indicating that people of all ages and under a very broad range of circumstances reliably and unintentionally encode information about the relative frequencies of events. Notwithstanding challenges to the automatic encoding view, we believe this empirical generalization remains valid today. Additionally, we describe recent examples of findings from research on language processing and statistical reasoning that add to earlier evidence of the critical contribution of frequency knowledge to cognitive and social functioning. Finally, we note that, in a number of respects, the broader intellectual climate in psychology today is more consistent with our approach to memory encoding than was the intellectual climate of the 1970s.

In the early 1970s, there was already substantial evidence that people are both acutely sensitive to frequency of occurrence information and that they also use that information to solve a wide range of cognitive and behavioural problems (cf. Hasher & Zacks 1979, 1984, for reviews). These observations, together with some data of our own, initially led us to speculate that frequency of occurrence information is encoded 'automatically'. In this paper, we consider some of this early evidence along with more recent findings; we briefly review the theoretical framework we developed (the Automatic and Effortful framework); and consider the value of frequency information for a wide array of behaviours. In a final section, we discuss the intellectual climate at the time we developed this framework and the relevance of its theoretical claims to contemporary issues. In concluding we note that the available evidence continues to support the observations we started with 25 years ago: frequency of occurrence is a fundamental aspect of the information that people code about their experiences in the world and that information in turn plays a fundamentally important role in a wide range of behaviours.

Encoding of frequency of occurrence information

Early data on processing of frequency of occurrence included a number of salient demonstrations of people's pervasive sensitivity to how often the same event was repeated both in their everyday life and in the laboratory.[1] For example, evidence was available that people could actually reliably (and rather surprisingly) rank order the frequencies of individual words, of single letters, and even of pairs of letters in English (see Hasher & Zacks 1979, p. 369). These findings demonstrated that information in memory tracked the frequencies of naturally occurring events, even for such apparently meaningless units as letter pairs. In the laboratory, several studies had shown that participants could not only accurately estimate the relative frequency of presentation of words in a list, they could also accurately keep track of how often the same words occurred in two successive lists (Hintzman & Block 1971) and of how often a sentence was repeated in a gist versus a verbatim form (Gude & Zechmeister 1975). Furthermore, frequency judgements were known to be independent of both temporal recency of presentation and duration of presentation (cf. Hintzman 1976). Additionally, other studies (e.g. Hasher & Chromiak 1977) consistently found that the accuracy of frequency judgements was no greater when participants were explicitly informed prior to list presentation that they would be tested on frequency than when they were given only general memory instructions. There was also evidence that frequency judgement accuracy was not improved by practice or by feedback on performance (Hasher & Chromiak 1977).

Data reported in our 1979 paper added to this general picture of pervasive encoding of frequency information. In one of two experiments that compared groups of different ages, the participants were children from kindergarten and grades 1, 2, and 3. Following a general instruction that their memory would be tested, the children were shown pictures of common objects, each of which occurred 1, 2, 3, or 4 times in a randomly ordered sequence. They then saw each picture again and were asked to say how often it had appeared in the preceding list. The means of these judgements as a function of actual presentation frequency for each age group are reprinted in Fig. 2.1. For all four age groups, judged frequency increased regularly with increasing presentation frequency. Indeed, the frequency judgement functions were very similar across ages. This absence of age differences across the kindergarten through third grade age span contrasts sharply with the typical finding of large improvements on most other explicit memory tasks over this age range. Furthermore, Hasher and Chromiak (1977) had

---

[1] We note that the research we describe here stems from an era that predates the rise of interest in implicit learning and memory. Consequently, despite our claim of automatic encoding of frequency information and our informal observations that people were generally unaware of the degree to which they had encoded frequency of occurrence data, our research on frequency, like that of others, used *direct* (explicit) measures of frequency knowledge. That is, we used tasks that asked participants to directly judge absolute or relative frequency (e.g. to rank order the frequencies of a set of items). To the extent that implicit measures of some types of knowledge are more sensitive than explicit measures (e.g. Nissen and Bullemer 1987), the older findings as well as ours can be seen, in retrospect, even more striking than they appeared at the time.

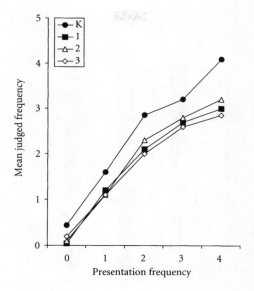

Figure 2.1 Estimated frequency of occurrence as a function of actual frequency of occurrence for children from kindergarten through third grade. Reprinted from Hasher and Zacks (1979). David Goldstein was a collaborator on this experiment.

found no age differences testing participants ranging in age between second graders and college students, suggesting that the ability to encode frequency information is fully functional quite early in life, as more recent work has demonstrated even more compellingly and at an even younger age (see the description below of the work of Saffran and colleagues). Finally, we also found no evidence of an age difference in the patterns of frequency judgements between college students and healthy older adults.

At this time, a particularly compelling finding for us emerged from additional analyses done on several sets of frequency judgements. We computed correlations between judged and actual frequency of occurrence for individual participants and found that the great majority of these correlations were strongly positive. For example, in one study (Hasher & Zacks 1979, Experiment 2) all but one (of 80) of the correlations was significant. Indeed, in that paper, the mean of the *individual* correlations was 0.77. Thus, the sensitive encoding of frequency of occurrence characterizes the performance of *each individual participant* and is not an artifact of averaging across participants, some of whom do process frequency and some of whom do not.

That ability differences also do not impact on frequency processing is suggested by findings from our final example of past research on processing of frequency of occurrence. One of the experiments reported by Zacks, Hasher, and Sanft (1982) included groups of college students who differed by approximately 140 points in verbal Scholastic Aptitude Test (SAT) scores.[2] Across a variety of instructional conditions, there was no hint that the higher and lower verbal SAT groups differed in memory for frequency of occurrence.

---

[2] The SAT is a widely administered test of verbal and mathematical reasoning skills that is used by many US colleges and universities to make admission decisions. This size difference is a difference of approximately 1.4 standard deviation units.

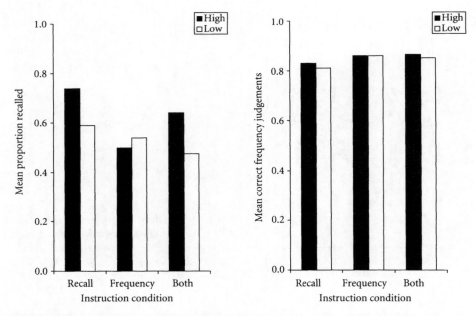

Figure 2.2 Effects of verbal ability and instruction on free recall (left panel) and relative frequency judgements (right panel). Groups of participants tested on both memory tasks were instructed prior to study that they would receive a recall test, a frequency test, or both. The filled bars are for the higher ability groups and the unfilled bars for the lower ability groups. Figure 2.2 is based on results reported in Zacks, Hasher, and Sanft (1982).

These examples indicated that people of a wide range of ages and abilities and under a wide range of conditions reliably encode a record of the frequency with which particular events occur. Importantly, the pattern of findings that we and others had obtained on memory for frequency (and to some degree for memory for spatial location and temporal order; see Hasher & Zacks 1979) is strikingly different than findings that have long been known to occur on other, more standard explicit memory tests including paired-associate, serial, and free recall, and recognition tests. On such tasks, age, verbal ability, instructions, practice, and feedback *do* make a difference (Hasher & Zacks 1979, 1984; Zacks et al. 1982). For example, the Zacks et al. (1982) experiment included groups receiving a free recall test rather than a frequency test. On that memory test, there were robust effects of ability and instructions—higher ability and test-appropriate instructions were associated with better free recall performance (see Fig. 2.2).

Automatic and Effortful encoding framework

Our 1979 and 1984 papers set out a theoretical framework that encompassed these strikingly divergent patterns of memory performance for frequency of occurrence information on the one hand and for item memory on the other. The primary distinction in the framework is between two types of encoding mechanisms, automatic and effortful. *Automatic* encoding mechanisms make minimal demands on attentional resources. They handle fundamental attributes of experience, including frequency, spatial location, and temporal order, and ensure that, for attended events,[3] these attributes are 'continually registered in memory, whatever the age, the ability, the education, or the motivation of an individual' (Hasher & Zacks 1984, p. 1372). (In a similar vein, Underwood [1969] had earlier argued that spatial, temporal, and frequency attributes are encoded as part of the memory of an event regardless of the type of encoding activity.) By contrast, *effortful* encoding mechanisms require considerable attentional capacity. They include such optional encoding operations as conscious rehearsal, organization, imagery, and semantic elaboration that are critical to good performance on most standard episodic memory tests. Because available attentional resources can vary across individuals and situations (Kahneman 1973), the use and effectiveness of effortful encoding operations and the resulting memory performance can vary with age, ability, educational experiences, and motivation.

A major focus of our work on this framework was on the encoding of frequency information (cf. Hasher & Zacks 1984). At the time, automatic processing was a lively topic in the cognitive literature, but with little consensus on exactly how to define automaticity (Shiffrin 1988). Along with most other theorists (but not all; see Allport 1989), we believed that *the* fundamental characteristic of automatic processing was the minimal demands such processing placed on our limited capacity attentional system. This notion, along with other considerations (e.g. the presumption that there are systematic differences in attentional capacity as function of age and other individual difference factors), allowed us to generate an operational definition of automatic encoding. This definition took the form of a set of criteria that together served as converging operations for concluding that a particular encoding operation was automatic. The specific criteria were that automatic encoding mechanisms (a) operate without intention; (b) do not benefit from intention to encode the particular attribute; (c) do not benefit from training to encode the attribute or from feedback; (d) show minimal individual differences; (e) show minimal age differences; and (f) show minimal impact of state (e.g. arousal level) and situational (e.g. divided attention demands) factors that otherwise impact on available attentional capacity. By contrast, effortful encoding operations were defined by the opposite pattern on each of these criteria.

In part because our main goal was to highlight the unique properties of encoding of frequency, spatial location, and temporal information, our original formulation of the Automatic and Effortful framework did not make specific claims about how these types of information were represented in memory. However, in our 1984 paper on frequency of occurrence, we did briefly review arguments favouring a multiple-trace view over alternatives such as strength-based or frequency counter views (Hasher & Zacks 1984, pp. 129–130). More specifically, it seemed to us and to others (Hintzman 1976; Greene

---

[3] Hasher and Zacks (1979) assumed that attending to an event was a precondition for the occurrence of automatic as well as effortful encoding operations. (This point has sometimes been ignored.) Automatic encoding operations require no further attentional resources, whereas effortful encoding operations do require additional resources.

1992) easier to reconcile the evidence that we continually and unintentionally store a fine-grained record of the repetition of individual events with a memory system in which each attended occurrence of an event results in the establishment of an independent memory trace rather than with a system in which each occurrence produces an increment in the strength of a single memory trace or a counter. Nonetheless, we recognized that such arguments are not definitive and that alternatives to the multiple-trace view could not be entirely ruled out. In fact, Alba, Chromiak, Hasher, and Attig (1980) suggested that encoding of superordinate category frequency involved incrementing a frequency tag or counter for the category label as each new category instance was presented. Contributions to the current volume demonstrate that this issue remains unresolved. The positions of different authors on how frequency information is mentally represented include both strength-type (Whalen, Gallistel & Gelman 1999) and multiple-trace-type (Fiedler, this volume) views. In addition, a number of the current authors (e.g. Brown, this volume; Haberstroh & Betsch, this volume) propose combined views in which frequency information is assumed to be represented in multiple ways in memory. Given that years of research have yielded at least partial support for several different proposals about how frequency is coded in memory, this latter approach may have some face validity. But adoption of a multiple representation approach would not be the end of the matter. As authors like Brown (this volume) and Haberstroh and Betsch (this volume) are well aware, other questions would arise, including ones about how situational factors influence reliance on different frequency representations.

Automatic encoding criteria: need for revision?

We have already described several examples of the earlier (before 1984) research that supported our contention that encoding of frequency information conformed to each of our automatic encoding criteria. In our view, more recent research continues to demonstrate people's striking and largely unintentional sensitivity to frequency of occurrence information. Nonetheless, findings have been reported that, by at least some accounts, call into question the validity of one or more of our automatic criteria with respect to frequency encoding. For example, the literature contains findings of significant individual and developmental differences in memory for frequency, including cases in which older adults were found to be slightly less proficient at encoding frequency of occurrence than younger adults (e.g. Kausler, Lichty & Hakami 1984). There are also reports of significant effects of encoding variables such as level or processing and, occasionally, benefits of prewarning participants of a memory test (for a summary of such findings, see Greene 1992, pp. 143–145).

Additional questions have been raised about the *accuracy* of frequency judgements, particularly as measured in tasks that require participants to assign numerical frequency estimates to items varying in actual presentation frequency. Although, in general, these estimates accurately track the *relative* frequencies of the items (i.e. judged frequency increases regularly with increasing actual frequency), the *absolute* accuracy of the provided estimates is subject to various distortions. In particular, low frequencies tend to be overestimated and high frequencies tend to be underestimated (see Fig. 2.1); and the estimates are affected by such factors as item retrievability (the 'availability heuristic') and specific characteristics of the judgement task (e.g. whether or not a range of frequencies is indicated and if it is, the specific low and high values; cf. Schwarz & Wänke, this volume; Zechmeister & Nyberg 1982).

That factors influencing accuracy and bias in frequency judgements continue to be of interest to cognitive and social cognition researchers is clear from a number of contributions to this volume. One example is the work described by Schwarz and Wänke relating to the availability heuristic. Their findings indicate that the subjective 'ease' of retrieval of target instances can influence frequency estimates independently of amount recalled, but their findings also show that relevant situational and instructional cues can lead to discounting of the ease factor. Another example is Fielder's research (this volume) showing a strong effect of retrieval cue structure on the accuracy of frequency judgements.

With respect to the hypothesis of automatic encoding of frequency information, one critical point illustrated by the work just described is the fact that multiple factors—not just stored information—influence frequency judgements. That is, to understand how people arrive at frequency estimates under even relatively sterile laboratory conditions, let alone in the complexity of real life, requires consideration of different judgement heuristics, retrieval factors, contextual variables, and so on. Admittedly then, a framework such as ours that concerned only *encoding* mechanisms provides, at best, only a partial account of how people arrive at the frequency estimates they produce on frequency judgement tasks. On the other hand, the multifaceted nature of frequency estimation processes permits the suggestion that inaccuracies and biases in frequency judgements, developmental differences, and so on are not necessarily contrary to the automatic *encoding* view. At least some of these effects could be a consequence of postencoding factors. Indeed, as suggested above, it appears that relative accuracy measures (e.g. rank order correlations between judged and actual frequency) are less likely to show effects not predicted by the automaticity view than are absolute accuracy measures (e.g. mean deviation of judged from true frequency). Arguably, this pattern is due to the lower sensitivity of the relative measures to potential sources of postencoding bias such as anchoring effects and scaling distortion in number assignment. In general, we remain convinced that the bulk of the literature indicates a robust sensitivity to the relative frequency of events across a wide range of circumstances and subject variables. Perhaps even more convincing on this point is evidence that demonstrates sensitivity to frequency of occurrence information through indirect or implicit measures, as is the case in the studies described in the next section.

The uses of frequency information

In addition to establishing the principle of automatic encoding of frequency (and of spatial location and temporal order) information, a major goal of our 1979 and 1984 papers was to consider the purposes to which automatically encoded information might be put. That is, having suggested that frequency, spatial location, and temporal order information were automatically encoded, we wanted to explore their contributions to cognition. At a general level, it seemed reasonable to consider the possibility that frequency, space, and time were fundamental organizing dimensions of behaviour.

Rather than attempting a broad discussion of the more specific uses of all types of automatically encoded information, the 1984 paper (Hasher & Zacks 1984) focused on the case of frequency. In particular, that paper summarized research demonstrating that frequency knowledge plays a critical role in event memory, in the representation and acquisition of new knowledge and skills, in decision making, and in cognitive and social development. We noted, for example, the considerable evidence indicating that category representations are based, in part at least, on stored information about feature frequency. We also noted evidence suggesting that children's sensitivity to the frequency with which members of their own versus the opposite sex display particular attitudes and behaviours contributes to their acquisition of gender-appropriate social roles. As well, we found that one source of beliefs is the frequency with which assertions are made (Hasher, Goldstein & Toppino 1977). Such findings supported our conclusion that frequency information was critical to at least cognitive functioning but probably was useful across a broad range of behavioural functions. More recent findings, including ones described throughout this volume, place our conclusion about the importance of frequency information to behaviour on even firmer grounds now than in 1984. Consider some dramatic examples from the recent literature.

Our first example comes from research on early language acquisition. Before a child learns the meanings of words, or the syntactic roles words play in sentences, the child must be able to break the speech record up into words. But how does a child pick out individual words from a speech stream that can be characterized as 'mostly continuous, without consistent pauses or other acoustic cues marking word boundaries' (Saffran, Newport, Aslin, Tunick & Barrueco 1997, p. 102)? An impressive series of studies by Saffran and colleagues (Aslin, Saffran & Newport 1998; Saffran, Aslin & Newport 1996; Saffran, Newport & Aslin 1996; Saffran *et al.* 1997) indicates that sensitivity to the frequency with which different sounds follow each other in speech helps babies (and children and adults) to segment the speech stream into words.

To arrive at this conclusion, Saffran and colleagues use a paradigm in which participants as young as eight months of age are exposed to brief samples of an artificial language consisting of several three-syllable nonsense 'words' (e.g. *bupada* and *tubitu*). Random repetitions of the words are strung together in a continuous auditory stream containing no acoustic or prosodic cues to the boundaries between words. In this situation, the primary cue to word boundaries is the greater frequency of two-syllable sequences that occur within a word (e.g. *bupa* and *pada*) as compared to two-syllable sequences that occur between two words (e.g. *datu* and *tubu*). Use of this frequency cue as a basis for word segmentation has been demonstrated in infants using a selective attention paradigm (Aslin *et al.* 1998; Saffran, Aslin & Newport 1996) and in 6–7 year old children and adults (Saffran, Newport & Aslin 1996; Saffran *et al.* 1997) using forced-choice recognition procedures. Other important findings are that children and adults are equally adept at this kind of learning (Saffran, Newport & Aslin 1996; Saffran *et al.* 1997) and that intentional and incidental learning of the words are equally effective (Saffran *et al.* 1997). The apparent age invariance in these results, the lack of a benefit from intentional encoding, and most importantly, the demonstration of the exquisite sensitivity to frequency information (including under the impoverished conditions of this paradigm) are all aspects of the Saffran *et al.* findings that echo the sorts of observations that originally got us interested in the processing of frequency—and, that confirm the suggestion that frequency serves as a fundamental attribute for organizing the physical world.

It can be noted that in contrast to the highly artificial speech samples used by Saffran and colleagues, normal speech provides additional frequency-based cues that people use to help them pick out word boundaries (Kelly & Martin 1994). For example, it has been shown that word segmentation is influenced by the knowledge that certain syllables (e.g. *the, to*) are more frequently followed by the start of a new word than others (e.g. *a, in*) and by the knowledge that words more frequently begin with a syllable containing a fully realized vowel than with a syllable containing a reduced vowel.

Segmentation of words is not the only area of language to which frequency information makes a contribution. For example, Hasher and Zacks (1984) reviewed results suggesting that sensitivity to 'single-letter positional frequency' (the frequency with which particular letters occur in specific positions within words) plays an implicit role in written wording decoding (see also Sedlmeier, Hertwig & Gigerenzer 1998.) Indeed, it is this knowledge that may contribute to the well-known word superiority effect, by which a letter within a word is more easily identified than is a single letter presented alone.

More recently, important claims have been made about the role of frequency information in syntactic parsing, that is, in the determination of the syntactic structure of a sentence. So-called 'constraint-based views' of syntactic parsing (e.g. Trueswell 1996) claim that parsing is guided by multiple types of information, including knowledge of the frequencies with which particular content words, especially verbs, occur in different syntactic roles. These influences can be demonstrated with respect to the resolution of temporary syntactic ambiguities. Consider, for example, sentences containing verbs (e.g. *searched, accepted*) that are the same whether they are functioning as a past tense verb (e.g. *The thief searched the room*) or as a past participle (e.g. *The thief searched by the police was indicted*). As a result of this overlap in function, some verbs create sentences that are temporarily ambiguous until after a critical word is presented (e.g. 'by' in the second example sentence above). Trueswell (1996) showed that the amount of difficulty people have in arriving at a correct interpretation of such sentences depends in part on the frequency with which a particular verb occurs as a main verb versus as a participle. In particular, sentences containing the less preferred (and overall, less frequent) participle form produce relatively little difficulty for comprehenders when the verb involved has a high participle frequency (e.g. *accepted*) as compared to cases in which the verb involved has a low participle frequency (e.g. *searched*).

Turning now the importance of frequency information in a nonlanguage domain, we consider work on statistical reasoning. In sharp contrast to results deriving from the 'heuristics and biases' approach initiated by Kahneman and Tversky (1973), several studies published in the1990s have demonstrated that people possess good statistical intuitions and that they display these intuitions when the relevant information is presented to them as frequencies rather than as probabilities. In particular, 'base rate neglect' is a prominent feature of reasoning on Bayesian reasoning problems when a probabilistic format is used to present the relevant data. That is, people are generally insensitive to the prior probability or base rate of a particular hypothesis (e.g. that a woman has breast cancer) when they estimate the probability of that hypothesis given that a particular fact is known (e.g. she has a positive mammogram test). When base rates are low, as they are for most specific diseases, gross overestimates are made of the probability that a positive test result means a positive diagnosis. By contrast, Cosmides and Tooby (1996) and Gigerenzer and Hoffrage (1995) (see also Hertwig & Hoffrage, this volume) have found that the majority of participants (up to 92% in Cosmides & Tooby's data) do take base rate appropriately into account when a frequency format is used for the presentation of the critical information (for more details, see Chapter 18). Cosmides and Tooby argue that such results support the 'frequentist hypothesis' which states that 'some of our inductive reasoning mechanisms do embody a calculus of probability, but they are designed to take frequency information as input and produce frequencies as output' (1996, p. 3). Likewise, Gigererenzer and Hoffrage (1995, p. 697) have argued that 'the (human) mind is attuned to frequencies'. And we would of course agree.

Descriptions of the uses of frequency information in many additional domains are found throughout this book. To take just one example, in chapter 16 Alba presents a picture of pervasive effects of frequency information in consumer decision making. A particularly striking instance of these effects comes from studies suggesting that the relative frequency of lower prices for particular products is a determining factor in people's judgements that one store is overall cheaper than its competitors. More specifically, when asked to pick the store that is cheaper on a market basket of items, there is a strong tendency to pick the store that is cheaper on *more* of the items. This is true even when the other store has a large price advantage on the items on which it is cheaper, with the result that the total cost of the market basket is actually the same across the two stores.

In summary, like the claim that frequency of occurrence is continually encoded into memory, the claim that frequency information serves a wide range of cognitive and social functions remains on firm ground. If anything, evidence such as that just reviewed places our claim about the uses of frequency information on broader and firmer ground than it was 25 years ago. In fact, very similar claims were recently highlighted in a paper by Kelly and Martin (1994). These authors argue that animals and humans share a 'fine-grained sensitivity to probabilistic (i.e., relative frequency) patterns in their environment' (Kelly & Martin 1994, p. 105, parenthetical comment added), and that this sensitivity facilitates numerous behaviours in both animals (e.g. conditioning, foraging) and humans (e.g. depth perception, language processing).

The ideas outlined in Hasher and Zacks (1979, 1984) have proved their staying power. Of course, our ideas developed in a particular intellectual climate in (mostly) experimental psychology and in the next section of the paper we explore the relation between the Automatic and Effortful framework and the intellectual climate in psychology 25 years ago and now.

Historical context

At the time our first paper was published, research in human memory was firmly in the grasp of the 'cognitive revolution' (e.g. Lachman, Lachman & Butterfield, 1979), a presumed advance over the associationist/functionalist approach to memory (known then as 'verbal learning'[4]) that had dominated the field of human experimental psychology in North America for 40–50 years. Despite the revolution, some critical intellectual characteristics (including pre-theoretical assumptions) of the earlier tradition remained intact in the new cognitive approach to the study of memory. Several of these characteristics are worth noting because our own ideas either explicitly or implicitly rejected them. In particular, our viewpoint disagreed with the continued pursuit of 'general laws' of learning, memory, and cognition, with the continued study of memory as independent from other cognitive processes, particularly attention, and with the focus on deliberate forms of learning. We develop these points in the following paragraphs.

A primary goal of the associationist tradition, whether humans or other animals were the focus of study, was the pursuit of general laws of learning and memory. In our view, an overly stringent application of this goal led to a number of shortcomings, including the failure to note that behaviour has an ecological and evolutionary component to it. Because the goal of classic experimental psychology was (and of course to some degree, still is) to establish general laws of learning, differences among species, their development, and their ecological niche were largely considered nuisance factors (or at least ones to be looked at after the general laws had been specified) rather than issues to be studied in their own right. The result was an absence of diversity in the behaviours and subjects studied: In human learning and memory laboratories, young college students (often sophomores, mostly attending elite research institutions) were the focus of research. In animal laboratories, it was rats and pigeons. In both types of laboratories, the range of learning paradigms was similarly restricted.

The limitations of so narrow a focus became apparent in some animal laboratories in the late 1960s and early 1970s with the recognition of critical boundary conditions for basic learning in nonhuman animals. For example, one standard principle of classical conditioning was that acquisition of an association between the conditioned and unconditioned stimuli (CS and US, respectively) is strongly dependent on the CS-US interval. Much evidence had shown that if more than a few seconds separates the CS and the US, conditioning proceeds very slowly, if at all (Kimble 1961). However, research in the 1960s (e.g. Garcia, Ervin & Koelling 1966) indicated that this general rule does not apply in the case of a type of learning that has survival value for the organism, specifically, learning to avoid food that has in the past made one sick, the 'bait shyness' effect. In research that brought such learning into the lab, rats (and other organisms) were found to readily acquire an aversion to a taste (e.g. the taste of saccharine-sweetened water) that was paired with sickness (e.g. from exposure to X-ray radiation) even though the delay between the CS (sweet taste) and US (sickness) was an hour or more. Furthermore, such taste aversion learning was found to require only a single learning trial, and aversions were not formed to nontaste stimuli (e.g. a flashing bright light or a clicking noise) even after many pairings with sickness (Garcia *et al.* 1966).

This and other work was making it apparent at that time that members of a species are 'prepared' by evolutionary forces to learn some associations easily but not others (e.g. Seligman 1970). The laws of learning established for unprepared associations do not necessarily apply to prepared ones. Moreover, because prepared learning mechanisms presumably evolved to solve the particular problems that a species has in meeting basic needs (food, water, safety, procreation), some of these will be species specific, dependent on the particular ecological niche that the species occupies.

By the early 1970s, evolutionary considerations and notions of adaptive function had not much filtered into the study of human learning and memory (or other cognitive functions for that matter). Our own views were heavily influenced by the animal research findings. We believed that adaptive function was an important consideration for human as well as nonhuman research. Similar to Seligman's (1970) notion of preparedness, we specifically proposed that humans have an innate ability to automatically encode frequency, spatial location, and temporal order because these are fundamental attributes of experience that define the flow of events in our environment and that enable us to solve critical problems (e.g. see Hasher & Zacks 1979; p. 360). Our views represent an early example of what has since come to be known as evolutionary psychology (cf. Barkow, Cosmides & Tooby 1992). Our views also antedate subsequent work with nonhuman animals that also suggests the fundamental importance of, and sensitivity to frequency (Shettleworth 1998, pp. 363–377).

In addition to the lack of attention to adaptive function, the pursuit of general laws of learning restricted research to a limited number of subject types—in human research, primarily the college sophomore. Prompted by research on memory development in animals (e.g. Campbell & Spear 1972), by the then more limited research on young children and older adults, and by some of our own data comparing individuals of different ages and abilities (see above), we came to believe that much could be learned by using developmental and individual difference approaches to test our theoretical notions. In this view, we were particularly influenced by Underwood's (1975) notion that consideration of individual differences can play a beneficial role in theory development.

A second property of North American psychology in the 1970s that we disagreed with was its tendency to treat learning and memory as topics that were easily separable from other cognitive domains such as language, problem solving, and attention. Our own view was (and remains) that memory is central to the study of any number of allied cognitive fields, in that both its implicit and explicit components limit behaviour. However, it was (and remains) our view that memory and attention are actually integral aspects of each other and neither is likely to be successfully studied without explicit recognition of the other (see Hasher, Zacks & May 1999, for our more recent views on this topic). To a large degree, the development of our views on this point coincided with the publication of Kahneman's book, *Attention and effort*, in 1973. In that book, Kahneman forcefully argued that cognitive processes have varying attentional resource demands and that although attentional capacity is limited, the limit is not a rigid one but varies significantly both between and within individuals (e.g. as a result of mood, arousal, etc.). These ideas were central to the Automatic and Effortful framework.

Finally, there was at the time what we saw as an exaggerated emphasis on the importance of intentional learning (though see Postman's [1964] work on incidental memory) and associated memory strategies (such as rehearsal, imagery mnemonics, etc.) for guaranteeing that experienced information left a residue in memory. Several books and papers outside of the human memory tradition had made the point in one way or another that much of behaviour is actually either acquired implicitly or used implicitly, or both. For example, Goffman (1963) suggested that social and cultural rules were acquired in this way. Skinner (1971) argued that much of human behaviour was under the control of stimuli that people were unaware of. And there were a few startling findings in the human literature, notably the early suggestion of implicit retrieval by amnesics and their controls of lists of words they could not recall explicitly (Warrington &

Weiskrantz 1968), that seemed quite telling with respect to the overemphasis on deliberate processes for learning and memory. With respect to frequency information, the evidence that people knew the relative frequencies of single letters, pairs of letters, syllables and other units of experience all suggested to us that much information could be acquired without the deliberate intention of the learner. Indeed, we were impressed that participants in our studies inaccurately denied their ability to judge frequency reliably, especially in the case of incidentally encoded information. Our reviews of the ways in which frequency knowledge was used (e.g. to establish the truth of assertions, to make decisions, and so on) was in keeping the implicit memory findings of Warrington and Weiskrantz (1968) and the incidental memory findings of Postman (1964), and of course is in keeping with the current wide interest in implicit acquisition and utilization of knowledge.

In many ways, cognitive psychology in the early 2000s fits more tightly with the ideas raised by the Automatic and Effortful framework. In particular, much research is directed at the study of implicit knowledge and at its implicit utilization. There is direct consideration in the literature of adaptive functions of cognition, and there are burgeoning literatures on the development of memory and on its pattern of decline in normal and abnormal aging. There are also many who are now concerned with the integral relationship between attention and memory.

Summary and reflections

From the vantage point of hindsight, it is clear that we might have done a few things differently. First, our choice of the term 'automatic' created a number of problems that stemmed from confusions between our use of the term (as described above) and its quite different use in several widely cited papers (Posner & Snyder 1975; Shiffrin & Schneider 1977). For example, Shiffrin and Schneider used the term 'automatic' processing to describe the result of extensive training under consistent mapping conditions, whereas we used the same term to refer to a largely innate mechanism that results in the inevitable encoding of certain attributes of attended events. We would have avoided some confusion and the excess theoretical baggage that tended to get attached to our view if we had used a term like 'obligatory' encoding.

Secondly, we may have diverted attention from our central points (sensitivity to and importance of frequency information) by trying to specify the criteria for automatic encoding. Our goal was definitional precision, but this effort fostered an emphasis in subsequent research on whether or not *all* the criteria applied to a particular encoding process. This focus on the criteria probably detracted from the basic idea of incidental, obligatory encoding of critical aspects of experience. Additionally, we might well have thought more about variables that influence levels of accuracy in frequency judgements and what these variables meant for our notion of automaticity. Likewise, we acknowledge our minimal contribution to debates on the underlying representation of frequency information. (Both issues have justifiably received considerable attention, but judging from other contributions to this volume, answers remain incomplete.)

These considerations notwithstanding, we believe that the Automatic and Effortful view was basically correct in its proposals about the encoding of frequency of occurrence information (and to a lesser degree, temporal order and spatial location). That is, we still believe the following:

1. The encoding of frequency information is an inevitable consequence of attending to events and, in that sense, is obligatory;
2. Frequency knowledge plays a critical role in many implicit cognitive functions;
3. Research limited to one sample type (in North America, the college sophomore) is likely to be less informative than is research that adds age as a variable and that takes into consideration nonnormal samples and factors such as mood and arousal changes; and
4. Memory and attention are interdependent cognitive functions that cannot be adequately understood in isolation from each other.

We close with an observation that we made in the course of the reading we did to prepare this chapter. We believe there is (e.g. Kelly & Martin 1994) and has been for some time (e.g. Hintzman 1976; Underwood 1969), considerable consensus about our acute sensitivity to frequency information and about the critical roles that frequency information plays in cognition. Surprisingly, this knowledge is unlikely to be conveyed to students of cognitive psychology—an informal recent survey of cognitive psychology textbooks suggests that few of them pay more than passing attention to the processing of frequency information. This seems curious to us based on our reading of the literature as well as other chapters in this volume.

Acknowledgements

The preparation of this chapter was supported by Grant Number AGO 4306 from the National Institute on Aging. We express our continued gratitude to David Goldstein for his assistance with our early work with children.

References

Alba, J. W., Chromiak, W., Hasher, L. & Attig, M. S. (1980). Automatic encoding of category size information. *Journal of Experimental Psychology: Learning, Memory, and Cognition,* 4:370–378.

Allport, A. (1989). Visual attention. In M. I. Posner (ed.) *Foundations of cognitive science* (pp. 631–682). Cambridge, MA: MIT Press.

Aslin, R. N., Saffran, J. R. & Newport, E. L. (1998). Computation of conditional probability statistics by 8-month-old infants. *Psychological Science,* 9:321–324.

Barkow, J. H., Cosmides, L. & Tooby, J. (1992). *The adapted mind: Evolutionary psychology and the generation of culture.* New York: Oxford University Press.

Campbell, B. A. & Spear, N. E. (1972). Ontogeny of memory. *Psychological Review,* 79:215–236.

Cosmides, L. & Tooby, J. (1996). Are humans good intuitive statisticians after all? Rethinking some conclusions from the literature on judgement under uncertainty. *Cognition,* 58:1–73.

Garcia, J., Ervin, F. R. & Koelling, R. A. (1966). Learning with prolonged delay of reinforcement. *Psychonomic Science,* 5:121–122.

Gigerenzer, G. & Hoffrage, U. (1995). How to improve Bayesian reasoning without instruction: Frequency formats. *Psychological Review,* 102:684–704.

Goffman, E. (1963). *Behaviour in public places: Notes on the social organization of gatherings.* New York: Free Press.

Greene, R. L. (1992). *Human memory: Paradigms and paradoxes.* Hillsdale, NJ: Lawrence Erlbaum Associates.

Gude, C. & Zechmeister, E. B. (1975). Frequency judgements for the 'gist' of sentences. *American Journal of Psychology, 88*:385–396.

Hasher, L. & Chromiak, W. (1977). The processing of frequency information: An automatic mechanism? *Journal of Verbal Learning and Verbal Behaviour, 16*:173–184.

Hasher, L., Goldstein, D. & Toppino, T. (1977). Frequency and the conference of referential validity. *Journal of Verbal Learning and Verbal Behaviour, 16*:107–112.

Hasher, L. & Zacks, R. T. (1979). Automatic and effortful processes in memory. *Journal of Experimental Psychology: General, 108*:356–388.

Hasher, L. & Zacks, R. T. (1984). Automatic processing of fundamental information: The case of frequency of occurrence. *American Psychologist, 39*:1372–1388.

Hasher, L., Zacks, R. T. & May, C. P. (1999). Inhibitory control, circadian arousal, and age. In A. Koriat and D. Gopher (eds.) *Attention and performance XVII. Cognitive regulation of performance: Interaction of theory and application* (pp. 653–675). Cambridge, MA: MIT Press.

Hintzman, D. L. (1976). Repetition and memory. In G. H. Bower (ed.) *The psychology of learning and motivation,* vol. 10 (pp. 47–91). New York: Academic Press.

Hintzman, D. L. & Block, R. A. (1971). Repetition and memory: Evidence for a multiple-trace hypothesis. *Journal of Experimental Psychology, 88*:297–306.

Kahneman, D. (1973). *Attention and effort.* Englewood Cliffs, NJ: Prentice-Hall.

Kahneman, D. & Tversky, A. (1973). On the psychology of prediction. *Psychological Review, 80*:237–251.

Kausler, D. H., Lichty, W. & Hakami, M. K. (1984). Frequency judgements for distractor items in a short-term memory task: Instructional variation and adult age differences. *Journal of Verbal Learning and Verbal Behaviour, 23*:660–668.

Kelly, M. H. & Martin, S. (1994). Domain-general abilities applied to domain-specific tasks: Sensitivity to probabilities in perception, cognition, and language. *Lingua, 92*:105–140.

Kimble, G. A. (1961). *Hilgard and Marquis's conditioning and learning.* New York: Appleton-Century-Crofts.

Lachman, R., Lachman, J. L. & Butterfield, E. C. (1979). *Cognitive psychology and information processing: An introduction.* Hillsdale, NJ: Lawrence Erlbaum Associates.

Nissen, M. J. & Bullemer, P. (1987). Attentional requirements of learning: Evidence from performance measures. *Cognitive Psychology, 19*:1–32.

Posner, M. I. & Snyder, C. R. R. (1975). Attention and cognitive control. In R. L. Solso (ed.) *Information processing and cognition: The Loyola symposium* (pp. 55–85). Hillsdale, NJ: Lawrence Erlbaum Associates.

Postman, L. (1964). Short-term memory and incidental learning. In A. Melton (ed.) *Categories of human learning* (pp. 146–201). New York: Academic Press.

Saffran, J. R., Aslin, R. N. & Newport, E. L. (1996). Statistical learning by 8-month-old infants. *Science, 274*:1926–1928.

Saffran, J. R., Newport, E. L. & Aslin, R. N. (1996). Word segmentation: The role of distributional cues. *Journal of Memory and Language, 35*:606–621.

Saffran, J. R. Newport, E. L., Aslin, R. N., Tunick, R. A. & Barrueco, S. (1997). Incidental language learning: Listening (and learning) out of the corner of your ear. *Psychological Science, 8*:101–105.

Sedlmeier, P., Hertwig, R. & Gigerenzer, G. (1998). Are judgements of positional frequencies of letters systematically biased due to availability? *Journal of Experimental Psychology: Learning, Memory, and Cognition, 24*:754–770.

Seligman, M. E. P. (1970). On the generality of the laws of learning. *Psychological Review, 77*:406–418.

Shettleworth, S. J. (1998). *Cognition, evolution, and behaviour.* New York: Oxford University Press.

Shiffrin, R. M. (1988). Attention. In R. C. Atkinson, R. J. Herrnstein, G. Linszey, and R. D. Luce (eds.) *Steven's handbook of experimental psychology: vol. 2. Learning and cognition* (pp. 739–811). New York: John Wiley and Sons.

Shiffrin, R. M. & Schneider, W. (1977). Controlled and automatic human information processing: II. Perceptual learning, automatic encoding, and a general theory. *Psychological Review, 84*:127–190.

Skinner, B. F. (1971). *Beyond freedom and dignity.* New York: Alfred A. Knopf.

Trueswell, J. C. (1996). The role of lexical frequency in syntactic ambiguity resolution. *Journal of Memory and Language, 35*:566–585.

Underwood, B. J. (1969). Attributes of memory. *Psychological Review, 76*:559–573.

Underwood, B. J. (1975). Individual differences as a crucible in theory construction. *American Psychologist, 30*:128–134.

Warrington, E. K. & Weiskrantz, L. (1968). A new method of testing long-term retention with special reference to amnesic patients. *Nature, 217*:972–974.

Whalen, J., Gallistel, C.R. & Gelman, R. (1999). Nonverbal counting in humans: The psychophysics of number representation. *Psychological Science, 10*:130–137.

Zacks, R. T., Hasher, L. & Sanft, H. (1982). Automatic encoding of event frequency: Further findings. *Journal of Experimental Psychology: Learning, Memory, and Cognition, 8*:106–116.

Zechmeister, E. B. & Nyberg, S. E. (1982). *Human memory: An introduction to research and theory.* Monterey, CA: Brooks/Cole Publishing Co.

CHAPTER 3

ENCODING, REPRESENTING, AND ESTIMATING EVENT FREQUENCIES: A MULTIPLE STRATEGY PERSPECTIVE

NORMAN R. BROWN

Abstract

People use several formats to represent frequency information and they use several strategies to generate frequency judgements. These observations are at the core of an approach to event frequency called the Multiple Strategy Perspective *(MSP). This chapter provides an overview of the MSP and a review of pertinent research methods and findings. The overview catalogues the common frequency formats, estimation strategies, and performance patterns. It also identifies a set of systemic relations which first links encoding factors and event properties to frequency representations, then links these frequency representations to strategy selection, and finally links strategy selection to estimation performance (i.e. speed, accuracy, and bias). Strategy selection and implication for basic and applied research are considered in the concluding sections of the chapter.*

Introduction

How many sexual partners have you had in your life time?
During the past month, how often have you shopped at a department store?
How many times did the word METAL appear on the list that you just studied?

Questions like the first two are asked on surveys by sociologists, economists, and market researchers who hope to obtain an accurate information about the frequency of potentially important, but undocumented, behaviours; questions like the third are put to experimental participants by psychologists interested in understanding how repeated experiences affect memory and performance. Of course, in content, these questions could hardly be more different. Nonetheless, the processes used to answer them and the knowledge that these processes operate on have much in common. Sometimes people retrieve and count relevant instances; sometimes they recall stored tallies, rates or impressions; and sometimes they gauge frequency by assessing some aspect of memory performance or the memory trace. In other words, regardless of whether the to-be-estimated items are arbitrary words presented on a list or meaningful personal events, people represent frequency information in a number of different ways, and they use a variety of estimation strategies to generate frequency judgements. Moreover, event properties and encoding factors are systemically related to the way that frequency information is represented. Furthermore, the representation of frequency information and the strategies used to judge event frequencies are related in a systematic manner, as are strategy selection and estimation performance.

An approach called the *Multiple Strategy Perspective* (*MSP*) has been developed to summarize the regularities linking encoding, representation, process, and performance and to catalogue common frequency formats and estimation strategies. This chapter provides an overview of this approach and a selective review of pertinent research methods and findings. The main premises of the MSP are as follows:

(A) Information about event frequency can be represented in several different ways;

(B) People use multiple strategies to estimate event frequency;

(C) Encoding factors and event properties influence the way frequency is represented;

(D) The frequency-relevant contents of memory restrict strategy selection;

(E) Motivational factors and response context also affect strategy selection;

(F) Strategy selection affects estimation speed and accuracy;

(G) In turn, patterns of performance reflect strategy choice.

The sequence of stages implied by premises C, D, and F is sketched in Fig. 3.1. In this figure, the arrows connecting the stages are dashed to express the idea that there is variability, and at times redundancy, in the way that people encode frequency relevant information and that there also is certain amount of flexibility in the way that people respond when they are confronted with multiple sources of frequency-relevant information. In other words, the MSP is not strictly deterministic. Rather, it is assumed that a given set of encoding conditions may foster some types of frequency-relevant representations and not others, and that as a result some strategies will be applicable, but not others. The implication here is that a single set of frequency judgements may include estimates produced by several different estimation processes.

These premises, with appropriate substitution of terms, can be or have already been applied to a wide variety of estimation and problem-solving tasks (e.g. Siegler 1987). This is in keeping with the notion that human cognition is both flexible and opportunistic,

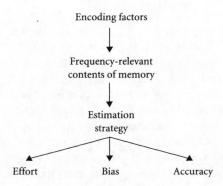

Figure 3.1 Overview of the core elements of the Multiple Strategy Perspective (MSP).

but that it is also constrained by the nature of accessible task-relevant information. What links this general view to the present topic is an empirically grounded catalogue of estimation strategies and a growing understanding of the ways that encoding, representation, process, and performance are related in this domain. Figs. 3.2 and 3.3 summarize this information. Fig. 3.2 presents a process-based taxonomy of the commonly observed strategies, and Fig. 3.3 lists several of these strategies along with the encoding factors, representational formats, and performance patterns that accompany their use.

The material presented in these summary illustrations is discussed at length in the next sections. Before turning to this material, some background is necessary. There are two related points; both concern the nature of the research that led to the development of the MSP, and the differences between this line of research and conventional laboratory-based work on event frequency. The first point is that MSP studies have often dealt with real-world events rather than word lists. These events were emphasized because many of the researchers interested in this topic were specifically concerned with understanding how survey respondents answer *behavioural frequencies questions*. These questions require the

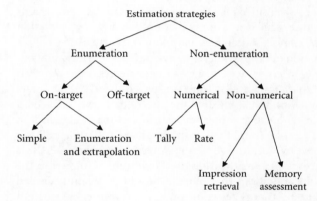

Figure 3.2 A taxonomy of common frequency estimation strategies.

Encoding	Content	Strategy	Performance
'Memorable' events	'On-target' instances	On-target enumeration	■ RT ↑ freq ■ underestimation
Regularity	Rate	Rate retrieval	■ Fast, flat RT ■ heaping
Intention and time	Tally	Tally retrieval	■ Fast, flat RT ■ accurate(?)
Frequent presentation	Vague quantifiers	Impression retrieval	■ Fast, flat RT ■ overestimation
Similar instances	Coherent traces	Memory assessment	■ Fast, flat RT ■ overestimation
Encoding/test mismatch	'Off-target' instances	Off-target enumeration	■ SLOW, flat RT ■ regressive estimates

Figure 3.3 Summary of the relation between encoding factors, frequency-relevant memory contents, and performance characteristics associated with several common frequency estimation strategies.

respondent to indicate how often he or she has engaged in a particular activity during a given reference period. For example, a consumer survey might ask respondents how often they have purchased laundry detergent in the past month, or a health survey might ask about the number of doctor's visits in the past year. The immediate goals here were to determine whether people could provide accurate unbiased answers to such questions and, assuming performance was less than perfect, to identify event properties and task conditions that affect estimation accuracy.

The second feature that differentiates event frequency studies that have contributed to the development of the MSP from those that have not is a fairly obvious one. In addition to measures that reflect the magnitude and accuracy of the obtained frequency judgements, all MSP studies have included at least one process-sensitive measure. Several different methods have been used. These include on-line measures such as response times (RTs) and concurrent verbal reports, and off-line measures such as immediate retrospective reports, effort ratings, and strategy menus. Although each method has its limitations, it is common for MSP studies to employ more than one measure and for these measures to converge on a conclusion.

With few exceptions (Hintzman & Gold 1983; Hintzman *et al.* 1981; Hockley 1984; Marx 1985; Voss *et al.* 1975), process-sensitive measures have not appeared in the mainstream frequency memory literature. This is not to say that process and representation have not been an issue. Indeed, memory researchers have long speculated that people may represent frequency information in several different ways and that they may use several different estimation processes to generate frequency judgements (Bruce *et al.* 1991; Hasher & Zacks 1984; Hintzman 1976; Howell 1973; Johnson *et al.* 1979; Watkins & LeCompte 1991). However, in retrospect, it seems that the almost exclusive reliance on accuracy measures, coupled with the relative simplicity of their test materials, made it difficult for these researchers to profit from their insights. In contrast, the catholic approach to methodology adopted by *Cognitive Aspects of Survey Methodology* (CASM) researchers and the complex nature of the real-world events that were of interest to the survey community established the conditions that made it possible to identify several estimation strategies and to associate strategy selection with event proprieties. Much of the research reviewed below demonstrates that the findings reported in the behavioural frequency literature could be corroborated and extended using experimental methods, and that the two approaches could be coordinated in a way that combined rigour of the laboratory with the insights gleaned from observing real-world performance.

A taxonomy of estimation strategies

Enumeration-based estimates

The main division in taxonomy presented in Fig. 3.2 is between strategies that involve enumeration and those that do not. An estimate is based on enumeration when individual items or events are retrieved and counted, and the count arrived at by this retrieval process serves as the basis for a frequency judgement (Barsalou & Ross 1986; Begg *et al.* 1986; Blair & Burton 1987; Brown 1995, 1997; Bruce *et al.* 1991; Burton & Blair 1991; Conrad & Brown 1994; Conrad *et al.* 1998; Conrad *et al.* 2001; Greene 1989;

Williams & Durso 1986). Enumeration is considered to be *on-target* when (almost) all of the retrieved items are members of the to-be-estimated event category; when a high percentage of retrieved items do not belong to the target category, enumeration is considered to be *off-target* (Conrad et al. 2001). In principle, the distinction between on- and off-target enumeration could be a fuzzy one. In practice, this is not a problem, at least in the laboratory; depending on the task, enumeration protocols and corresponding cued-recall tests indicate that intrusion rates are either very high, which means that enumeration is off-target, or very low which means it is on-target; there simply is no middle ground (see below).

There are two types of on-target enumeration: *simple enumeration* and *enumeration and extrapolation*. An enumerated response is assigned to the former category when the number of items retrieved during the estimation process and the magnitude of the frequency judgement that follows are equal, and to the latter category when estimated frequency is greater than the number of retrieved instances. Other things being equal, simple enumeration becomes less common and enumeration and extrapolation more common as presentation frequency increases (Brown 1995, 2001). Interestingly, participants do not extrapolate when they are off-target.

On-target enumeration

On-target enumeration has been observed in both laboratory and real-world settings. It is characterized by two distinctive features. First, the time required to decide on a specific numerical response increases with event frequency; second, people tended to underestimate event frequency, and this tendency is more pronounced for frequently presented items than for less frequently presented items. Fig. 3.4 provides examples of these tendencies. The data plotted in this figure are drawn from Brown (1995). In this study, participants were

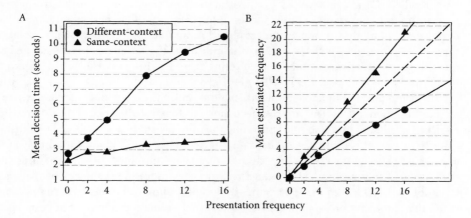

Figure 3.4 Mean response times (Panel A) and mean estimated frequencies (Panel B) and for the different-context group (solid circle) and the same-context group (empty circle). In Panel B, the solid lines represent the best linear fit for the means, and the dashed line represents the actual frequencies. These data are drawn from Brown (1995), Experiment 2.

presented with word pairs consisting of a target word (a category label, e.g. CITY) and a context word (a category exemplar, e.g. London). One group studied a *different-context* list and the other a *same-context* list. On the different-context list, target words were paired with unique context words on each presentation (e.g. CITY—Boston, CITY—Cleveland, CITY—London, etc.), and on the same-context list, a target word were always paired with same context word (e.g. CITY—London, CITY—London, CITY—London, etc.), and on both lists, presentation frequencies for the target words ranged from 2 to 16. Following the study list, target words (and non-presented category labels) were presented one at a time, and participants either thought aloud as they generated their estimates (Experiment 1) or were timed as they silently judged the list frequency of each (Experiment 2). The verbal protocols collected in Experiment 1 indicated that different-context participants frequently enumerated (57%) and the same-context participants never enumerated. The decision time data collected in Experiment 2 indicated that RT increased sharply with presentation frequency, but only in the different-context condition. In both experiments, participants who studied the different-context list (i.e. participants who relied on enumeration) tended to underestimate event frequencies and those who studied the same-context lists overestimated them.

This pattern of performance has been replicated many times (Brown 1995, 1997, 2001; Conrad & Brown 1994; Conrad et al. 2001) and is easy to understand. RT increases with presentation frequency because enumeration is a serial process. Assuming that the time to retrieve an instance is constant (or increases with each additional instance, Bousfield & Sedgewick 1944; Indow & Togano 1970) and that participants retrieve more instances when responding to high frequency items than to low frequency items, it follows that it should take more time to respond to the former than to the latter. Underestimation occurs because relevant instances may be permanently forgotten, because output interference causes some instances to become temporally inaccessible, and because retrieval efforts are sometimes terminated before all relevant instances have been recalled. Concurrent verbal protocols and immediate and delayed retrospective reports provide converging support for this interpretation. When participants are tested under conditions that produce underestimation and steep RT functions, inevitably, these process measures include information indicating that estimates were generated by retrieving instances one at a time and counting them up (Brown 1995, 2001; Conrad & Brown 1994; Conrad et al. 1998, 2001; Marx 1985).

An obvious precondition for enumeration is the availability in memory of relevant, readily accessible event traces. This implies that conditions that foster the encoding, storage, and retrieval of these traces should promote the use of enumeration-based strategies. Consistent with this claim several behavioural frequency studies have found a positive association between the rated distinctiveness of repeated instances of an event class and the likelihood that respondents enumerate when they judge the frequency of that class (Conrad et al. 1998; Menon 1993). Similarly, as noted above, laboratory studies demonstrate that enumeration is common when repeated events are unique, but completely absent when the repeated events were identical (Brown 1995). The usual explanation of this finding is that highly similar event instances are readily assimilated into a pre-existing scheme, and that as a consequence, access to the individual event traces is lost, whereas dissimilar instances maintain their individuality in memory (Menon 1993).

Subsequent research has demonstrated that distinctiveness is a necessary, but not a sufficient condition for enumeration. In a series of experiments, Brown (1997) presented participants with lists comprised of target-context pairs. The target words were repeated across the study list; however each context word appeared only once making each pair unique, and hence distinctive. After studying the lists, participants estimated the presentation frequency of the target words and then recalled context words.

Study time (6 sec per pair versus 2 sec per pair) and target-context relatedness (category labels paired with either category exemplars or unrelated words) were manipulated between experiments. Study-phase instructions were manipulated within experiment; participants in one group were instructed to commit the context words to memory; participants in a second group were instructed to focus on the frequency of the target words; and participants in a third group were instructed to study the word pairs for a memory test, but were not informed of the nature of the test.

This set of manipulations produced large differences in memory for the context words and large differences in the steepness of the functions that related RT to presentation frequency. More importantly, there was a strong systematic relation between context memory and response time; when context memory was best, estimation time increased sharply with presentation frequency, and the steepness of this estimation time-presentation frequency function decreased with context memory. More concretely, when participants were informed of the upcoming cued-recall test and were given 6 sec per pair to study the related context lists, RT increased from 3 sec to 11 sec as presentation frequency increased from 2 to 16; in a subsequent cued recall test, participants in this group recalled 46% of the context words. In contrast, RT increased from 3 sec to 4 sec across the same range when participants studied the unrelated-pair list under comparable encoding conditions. Cued recall for this group was 15%.

These results indicate that enumeration was common when instance memory was good and that the use of this strategy declined as relevant memory traces became increasing difficult to recall. More generally, this study demonstrated that factors which increase the memorability of event instances increase the use of enumeration and these factors encompass both encoding conditions (i.e. study time and study phase instructions) and event properties (i.e. similarity of event instances and memorability of differentiating contexts). Several other factors, unrelated to event memory, are also known to play a role in determining whether people enumerate. The influence of these factors on strategy selection is taken up in the last section of this chapter.

Off-target enumeration

Both off-target enumeration and on-target enumeration engage a serial retrieval process and produce frequency judgements that are based on the outcome of this process. Despite these similarities the two strategies are used under very different conditions and produce very different patterns of performances. A recent study by Conrad illustrates these points (Conrad *et al.* 2001). In this study, participants in one group, the *instance-plus-property* group, studied 109 word pairs consisting of a property label and a related noun (e.g. chocolate-BROWN, garbage-SMELLY, etc.). These pairs were constructed from published property norms (Battig & Montague 1969; McEvoy & Nelson 1982; Underwood & Richardson 1956). As in the different-context conditions described above, property labels were presented multiple times across the list, but each noun was presented only once. Participants in a second group, the *instance-only* group, studied the same set of nouns, but in the absence of the property labels. Following the study phase, all participants were presented with the property labels and were timed as they estimated how many of the studied nouns possessed the target property.

Participants in the two groups responded very differently during the frequency judgement task (see Fig. 3.5). In the instance-plus-property condition, RT increased with presentation frequency, frequency judgements were quite accurate and high frequency properties tended to be underestimated. For reasons outlined above, these results indicate that participants in this condition made heavy use of standard on-target enumeration-based estimation strategies. In contrast, in the instance-only condition, RTs were very slow and the speed of responding was unrelated to the presentation frequency. Moreover, these effortful responses were highly inaccurate; judged frequencies for property-bearing instances were uncorrelated with the normative property frequencies and participants tended to overestimate the frequency uncommon property nouns and underestimate the frequency of the common property nouns. Other researchers have looked at the property frequency task and have also found that that the instance-only condition yields very poor frequency discrimination (Barsalou & Ross 1986; Brooks 1985; Freund & Hasher 1989).

A protocol study (Conrad *et al.* 2001, Experiment 1) shed light on the origins of this 'worst of both worlds' performance observed in the instance-only condition. In this experiment, participants who had studied the instance-only list thought aloud as they estimated the frequency of the target properties. An analysis of their verbal reports indicated that these participants retrieved and counted list items on 80% of the trials and that the use of enumeration was unrelated to the presentation frequency. However, it was more common for the protocols to include more list items that were normatively *incorrect* than ones that were normatively correct; on average, 2 out of 3 enumerated

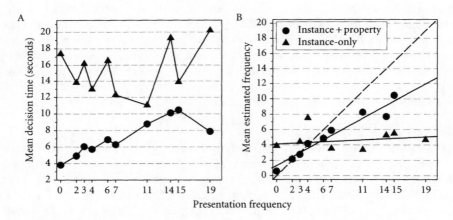

Figure 3.5 Mean response times (Panel A) and mean estimated frequencies (Panel B) and for the instance-plus-property group (solid circle) and the instance-only group (empty circles). In Panel B, the solid lines represent the best linear fit for the means, and the dashed line represents the actual frequencies. These data are drawn from Conrad *et al.* (2001).

items fell into the former category. This is why the application of enumeration under these circumstances is termed 'off-target'. Apparently, in this situation people recall instances in a more or less haphazard manner and use a liberal criterion to determine whether a retrieved item should be accepted as a member of the target category.

The use of on-target enumeration in the instance-plus-property condition and off-target enumeration in the instance-only condition results also rules out the possibility that it is the nature of the frequency judgement task (i.e. taxonomic frequency vs. property frequency) that determines whether enumeration is on- or off-target. Instead, it appears that people rely on off-target enumeration when there is a mismatch between the way that retrievable items are categorized at encoding and the way memory is subsequently probed. This mismatch can be avoided by indicating the appropriate category at the time of study or by probing memory in a way that captures the organization that people naturally impose on diverse instances. This suggests that it might be possible to use a frequency judgement task, supplemented with the appropriate processes measures, to determine whether or not a proposed organizational scheme is natural, i.e. employed without prompting at encoding to classify instances and at retrieval to access them. Natural schemes should support on-target enumeration, whereas people will have to rely on off-target enumeration when they are asked to respond to 'unnatural' schemes.

One final point. This line of research indicates that people do not naturally encode events by their properties, and suggests that under normal circumstances the assessment of property frequency is a futile endeavour. In frequentist models of judgement and decision making property frequencies are equivalent to base rates (Dougherty *et al.* 1999; Gigerenzer & Hoffrage 1995). Given that people are very poor at assessing property frequencies, it is not surprising that they usually neglect base rates when generating conditional likelihood estimates. On this view, base rate is neglected at least in part because it cannot be assessed with any certainty.

Nonenumeration strategies

In Fig. 3.2, the nonenumeration strategies are divided into two groups. The numerical strategies, *retrieved rate* and *retrieved tally*, are grouped together because, in both cases, prestored *numerical* information is retrieved and used as the basis for a frequency judgement. Access to prestored nonnumerical frequency information also plays a key role in the application of *impression retrieval* strategy. However, this strategy is grouped with memory assessment because both share a common response process, one that converts a frequency judgement from a relative or qualitative format to a numerical one and because it is difficult if not impossible to distinguish them using process measures (Conrad & Brown 1996).

Numerical strategies
Obviously, a prerequisite for the application of the retrieval strategies is access to facts that specify frequency information, and the absence of such facts precludes their use. Beyond this commonality, these strategies display substantial differences. The simplest of the retrieval strategies is retrieved tally. People who have stored and can recall a tally are able to respond to a frequency question by stating its value. Assuming that potentially relevant

events are classified correctly and that a relevant counter has been updated with the occurrence of each new instance of the target category, these responses should be quite accurate.

Brown and Sinclair (1999) found that tally-based responses are fairly common when people are asked to estimate how many sexual partners they have had in their life time. Furthermore, Brown (1997) has used RT data to argue that people who were informed of an upcoming frequency test prior to exposure to a study list sometimes counted presentation frequencies and used these counts as the basis for subsequent frequency judgements (see below). Interestingly, the use of retrieved tallies has not been reported in other studies that have looked at behavioural or laboratory frequency. Taken together, these findings suggest that counts can be used to represent event frequencies, but that this is the expectation rather than the rule. Apparently, people sometimes count events, but only when the events are particularly important, interesting, or noteworthy. In addition, research participants may attempt to track event frequencies, but only when they are instructed to do so. Otherwise, it can be assumed that people do not keep count of most activities and hence are typically unable to use a retrieved-tally strategy when asked to estimate event frequencies.

Many behavioural frequency studies have documented the use of rate-based strategies (Burton & Blair 1991; Conrad *et al.* 1998; Means & Loftus 1991; Menon 1993). Typically, this strategy is used when a respondent is asked about a common activity that occurs on a regular basis. It makes sense that the regularity of an activity is associated with knowledge of its rate of occurrence. When the period that defines the rate and the period specified by the question are the same, the retrieved rate can be stated as an estimate. However, when there is a mismatch between the two periods, respondents must multiply (or divide) the rate by an appropriate value to arrive at a response.

It is this latter process that produces the spiked or 'clumped' pattern in the distribution of the estimated frequencies that is characteristic of rate-based responding. A study by Conrad *et al.* (1998) provides an example of this tendency. In this study, respondents in a telephone survey were asked to estimate the number of times they had participated in several common activities (shopped for groceries, purchased gasoline, etc.) during the past month. In addition, they were required to provide a retrospective strategy report following each estimate. These reports made it possible to construct the *response distributions* presented in Fig. 3.6. These distributions display the percentage of frequency judgements produced by a given strategy that were assigned a given value (also see Brown, Buchanan & Cabeza 2000; Hintzman & Curran 1995). As Fig. 3.6 (Panel A) indicates, the most frequent response made by people using a rate-based strategy was 4 (once a week). The only other common rate-based estimates were 2 (once every two weeks), 8 (twice a week), and 30 (every day). This pattern was in sharp contrast to ones produced from enumerated estimates and general impression estimates (i.e. estimates produced by a memory assessment strategy or qualitative retrieval); the former (Fig. 3.6, Panel B) displayed an exponential decrease across the low end of the range, and the latter (Fig. 3.6, Panel C) displayed a strong clumping of responses at 'prototypical' values (i.e. 10, 15, 20).

Nonnumerical strategies

As mentioned above, two processes are also required to produce a frequency judgement when a participant uses a qualitative retrieval strategy. The first is a memory retrieval

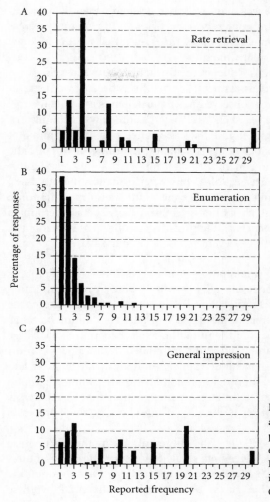

Figure 3.6 Distribution of frequency estimates as a function of estimation strategy. Panel A plots rate-based estimates, Panel B plots enumeration-based estimated strategy, and Panel C plots estimates that included general impression terms. These data are drawn from Conrad *et al.* (1998).

process that succeeds in accessing a fact or impression from memory that expresses frequency relevant information in a nonnumerical manner (e.g. 'The word CITY showed up *many times* on the preceding list' or 'I have *rarely* been cross-country skiing'). The second process involves selecting a numerical value to represent the quantity expressed by the retrieved fact.

One reason for proposing that people sometimes encode and later retrieve this type of nonnumerical information is the finding that vague quantifiers (terms like *a lot, many, very few*, etc.) are often included in verbal reports and that the use of these terms increases dramatically with presentation frequency (Brown 1995, Experiment 1). This is what one would expect if the repetition of an item or event increases peoples' awareness that it has been repeated and the likelihood that a fact conveying this awareness is stored in memory. In other words, in the same way that a seasoned academic eventually comes to *know* that he or she has attended many conferences, participants in a memory experiment, particularly one that repeats some items many times, eventually come to *know* that they have seen some of the words many times.

In the absence of verbal reports, it is difficult to distinguish between this direct retrieval strategy and the use of memory assessment. In part this is because both strategies yield fast flat RT functions of the sort observed when participants estimate the frequency of similar or identical instances (e.g. Fig. 3.4, Panel A, same-context condition). In addition, it appears that both strategies engage the same conversion process and that this process produces a characteristic pattern of performance, regardless of whether the input is a vague quantity retrieved from memory or an intuition about relative frequency based on an evaluation of some aspect of memory performance. In both cases, this process must map information about relative frequency on to a numerical response range. When people rely on this process, they often select round numbers as responses (Fig. 3.6, Panel C; see also Brown & Sinclair 1999; Conrad *et al.* 1998; Hintzman & Curran 1995) and they tend to overestimate event frequencies, particularly large ones (Brown 1995; Conrad & Brown 1994).

Overestimation is most pronounced when the upper bound of the response range is unspecified, but it can be eliminated or reversed by providing participants with information that restricts the range or by presenting them with a conservative set of response options. In contrast, enumeration-based and rate-based estimates are immune to range manipulations (Brown 1995, Experiment 3; Menon *et al.* 1995, Experiment 1). These findings suggest that overestimation and conversion are related because people must define a response range without the benefit of explicit numerical information and because the range can be grossly overestimated, but not grossly underestimated. This is not a problem when people base their estimates on information that is inherently numerical (i.e. when they enumerate or retrieve a tally or rate) because these processes do not engage a range-sensitive conversion process.

It should be acknowledged that the MSP does not, at present, specify the exact nature of the memory assessment process used to generate a qualitative evaluation of event frequency. However, it is assumed that each encoding of a repeated event has a monotonic effect on the contents of memory, that access or retrieval is influenced by these frequency-related changes in memory, and that people are sensitive to processing differences brought about by these frequency-related changes. This conceptualization leaves open the possibility that frequency is indexed by retrieval fluency (i.e. availability; Tversky & Kahneman 1973; Betsch & Pohl in this volume), trace strength (Hintzman 1969; Morton 1968), the similarity between a memory probe and the contents of episodic memory (Hintzman 1988; Jones & Heit 1993; Nosofsky 1988), associative strength (Sedlmeier in this volume) or a nonsymbolic mental magnitude (Gallistel this volume; Whalen, Gallistel & Gelman 1999). The difficulty in selecting between these possibilities parallels the more general difficulties encountered in distinguishing impression retrieval and memory assessment. This is another situation where process measures are likely to be uninformative. In particular, regardless of the nature of measure assessment process, the RT function should be fast and flat, and the conversion process should produce similar pattern of estimates regardless of origins of the frequency information.

Strategy selection

Strategy selection, one of the important issues raised by the MSP that has not yet been thoroughly investigated, is the last topic addressed in this chapter. As pointed out above, strategy selection is *restricted* by frequency relevant contents of memory; enumeration is impossible in the absence of retrievable event instances; direct retrieval strategies are impossible in the absence of prestored frequency information; and memory assessment strategies are ineffective when the probe fails to access a coherent representation. What is less obvious is how people select strategies when they have access to multiple sources of potentially relevant information.

One possibility is that people always select the least effortful of the available strategies. If this is position is correct, then it should be difficult to find evidence that people use enumeration, the most effortful of strategies, when they have access to other sources of frequency relevant information. However, the protocols reported for the 1995 article (Brown 1995, Experiment 1) provide two findings that are inconsistent with this prediction. First, enumeration and extrapolation was the most common strategy used in the different context condition, accounting for the 29% of all responses. Of course, in order to extrapolate, one needs to believe that the number of items retrieved is less than the number of items presented. It is possible that the knowledge used to justify extrapolation could also serve as the basis for a nonenumerated response. If so, the frequent use of this strategy is inconsistent with the notion that enumeration is used only as a last resort.

The second finding that argues against the least-effort hypothesis is the occasional appearance of general impression statements (i.e., vague quantifiers) in enumerated responses. For example, in the protocol study, at the highest presentation frequency, 13% of the enumerated estimates included one of these statements. If one assumes that general impression statements are made when a qualitative fact is retrieved from memory or when the product of a memory assessment process is verbalized, then the existence of these dual-strategy responses indicates that people sometimes enumerate even when they have accessed information that could have served as the basis for a less effortful estimate, one based on direct retrieval or memory assessment.

If the least-effort position fails to explain strategy selection, perhaps it can be explained by an enumeration bias. It could be that accessing retrievable traces triggers enumeration and that once enumeration has begun it continues at least until the first chunk or two have been recovered and unpacked. If enumeration is compulsive when retrievable traces are available, then only those factors that affect the availability of these traces should affect the use of the enumeration-based strategies. Counter to this prediction, the application of enumeration-based strategies is influenced by a number of factors that do not necessarily influence the accessibility of enumerable traces. These include: (a) actual frequency—beyond a certain level, enumeration decreases as actual frequency increases. (Note the drop in RT at presentation frequency 19 in Fig. 3.5, Panel A; also see, Bruce & van Pelt 1989; Burton & Blair 1991; Means & Loftus 1991); (b) response deadline—enumeration is less common when response time is restricted than when it is not (Burton & Blair 1991); (c) time frame—longer reference periods produce less enumeration than shorter reference periods (Blair & Burton 1987; Burton & Blair 1991);

(d) the presence of retrievable tallies—holding level of recall constant, the presence of a retrievable tally decreases peoples reliance on enumeration (Brown 1997).

There are good reasons then for rejecting both the least-effort hypothesis and the enumeration-bias hypothesis, though these arguments still leave the strategy selection issue unresolved. The findings outlined above, however, are consistent with the notion that people have a degree of control over strategy selection and that they attempt to balance perceived effort and perceived accuracy when conditions allow for strategy choice (Payne et al. 1993; Russo & Dosher 1983). The assumptions here are that people realize that enumeration is demanding and time consuming; that they prefer concrete information (in the form of instance counts or directly coded frequency facts) to the vague information (in the form of intuitions supplied by memory-assessment processes); and that they are willing to work quite hard to accumulate the former provided they have enough time to do so. As a result, people who have access to enumerable traces may forgo enumeration when the response time is restricted or when they believe that a large number of instances would have to be retrieved to provide a fair indication to the to-be-estimated quantity. Because this belief would have to be based on another source of frequency relevant information, i.e., a pre-stored or computed impressions of frequency, people who decide not to enumerate are still in the position to execute an alternative estimation strategy, albeit that one requires the conversion of a qualitative impression to a numerical value. The preference for retrieved tallies over enumeration reported in Brown (1997) indicates that when two sources of information are equally credible (enumerated counts and directly coded frequency facts), people prefer the more convenient one. In summary, these data suggest that perceived accuracy and convenience are evaluated when an estimation strategy is selected; that accuracy is given more weight when one source of information is clearly more credible; and that convenience is weighted heavily when competing sources are considered equally credible or when the more credible strategy is deemed to be too demanding.

Conclusion

People use different formats to represent event frequency and they use several different strategies to generate frequency judgements. These observations, which are at the core of the MSP, present both challenges and opportunities to researchers in the area. One challenge is to establish an exhaustive catalogue of relevant frequency formats, estimation processes, along with a specification of their prerequisites, conditions of use, and consequences. A second is to develop tasks that isolate estimation processes and methods for relating judgement to process on a case-by-case basis. A third challenge is to develop or modify models of memory and judgement that incorporate or accommodate the central tenets of the MSP.

I believe that the research reviewed above indicates that progress has been made in meeting the first two challenges, but only the first two. Nonetheless, it should be possible to profitably extend the MSP in several directions. One way to extend this approach is to use it to understand (or re-evaluate) problematic findings in the frequency literature (e.g. the diverse impact of encoding contexts and processes on judgements of frequency; cf. Begg et al. 1986; Hintzman & Stern 1978; Jonides & Naveh-Benjamin 1987; Rose 1980; Rowe 1973). The MSP should also be useful to memory psychologists interested in investigating a *single* representation or strategy. The contribution here would be to alert such researchers to the possibility that a single set of frequency judgements may contain estimates that are produced by different strategies and based on different frequency representations. At the same time, if researchers are aware of this pitfall and are willing to take advantage of the available processes-sensitive measures, it should be possible to study a given strategy or format in isolation. Finally, this general approach may provide guidance in designing behavioural frequency questions and interpreting biased or inconsistent responses to such questions (Brown & Sinclair 1999).

In summary, there is now a good deal of evidence indicating that people represent and estimate event frequency in variety of ways and that encoding, representation and performance are linked in a systemic manner. It is at least possible that the MSP will be extended to address a variety of basic and applied issues in memory. If nothing else, this line of research demonstrates the benefits of coordinating real-world and laboratory research efforts and the crucial role that complex decision processes play in memory performance.

Acknowledgement

This project has been supported by the author's NSERC operating grant. I would like to thank Fred Conrad for innumerable discussions of the topics covered in this chapter. These discussions have contributed greatly to the development of the MSP. I would also like to thank Peter Lee for his comments on this chapter.

References

Barsalou, L. W. & Ross, B. H. (1986). The role of automatic and strategic processing in the sensitivity of superordinate and property frequency. *Journal of Experimental Psychology: Learning, Memory, and Cognition, 12*:116–134.

Battig, W. P. & Montague, W. E. (1969). Category norms for verbal items in 56 categories: A replication and extension of the Connecticut category norms. *Journal of Experimental Psychology Monograph, 80* (3, Part 2): 1–46.

Begg, I., Maxwell, D., Mitterer, J. O. & Harris, G. (1986). Estimates of frequency: Attribute or attribution? *Journal of Experimental Psychology: Learning, Memory, and Cognition, 12*:496–508.

Blair, E. & Burton, S. (1987). Cognitive processes used by survey respondents to answer behavioral frequency questions. *Journal of Consumer Research, 14*:280–288.

Bousfield, W. A. & Sedgewick, C. H. (1944). An analysis of sequences of restricted associative responses. *Journal of General Psychology, 30*:149–165.

Brooks, J. E. (1985). Judgements of category frequency. *American Journal of Psychology, 98*: 363–72.

Brown, N. R. (1995). Estimation strategies and the judgement of event frequency. *Journal of Experimental Psychology: Learning, Memory, and Cognition, 21*:1539–1553.

Brown, N. R. (1997). Context memory and the selection of frequency estimation strategies. *Journal of Experimental Psychology: Learning, Memory, and Cognition, 23*:898–914.

Brown, N. R. (2001). Exemplar generation and the use of enumeration-based frequency estimation strategies. Manuscript in preparation.

Brown, N. R., Buchanan, L. & Cabeza, R. (2000). Estimating the frequency of nonevents: The role of recollection failure in false recognition. *Psychonomic Bulletin and Review, 7*:684–691.

Brown, N. R. & Sinclair, R. C. (1999). Estimating number of lifetime sexual partners: Men and women do it differently. *Journal of Sex Research, 36*:292–297.

Bruce, D., Hockley, W. E. & Craik, F. I. M. (1991). Availability and category-frequency estimation. *Memory of Cognition, 19*:301–312.

Bruce, D. & van Pelt, M. (1989). Memories of a bicycle tour. *Applied Cognitive Psychology, 3*:137–156.

Burton, S. & Blair, E. (1991). Task conditions, response formulation processes and response accuracy for behavioral frequency questions in surveys. *Public Opinion Quarterly, 55*:50–79.

Conrad, F. G. & Brown, N. R. (1994). Strategies for estimating category frequency: Effects of abstractness and distinctiveness. *American Statistical Association, Proceedings of the Section on Survey Methods Research* (pp. 1345–1350). Alexandria, VA: American Statistical Association.

Conrad, F. G. & Brown, N. R. (1996). Estimating frequency: A multiple strategy perspective. In D. Herrmann, M. Johnson, C. McEvoy, C. Hertzog and P. Hertel (eds) *Basic and Applied Memory: Research on Practical Aspects of Memory*, vol. 2 (pp. 167–178). Hillsdale, NJ: Erlbaum.

Conrad, F. G., Brown, N. R. & Dashen, M. (2001). Estimating the frequency of events from unnatural categories. Manuscript submitted for publication.

Conrad, F. G., Brown, N. R. & Cashman, E. (1998). Strategies for answering behavioral frequency questions. *Memory, 6*:339–366.

Dougherty, R. P., Gettys, C. F. & Ogden, E. E. (1999). MINERVA-DC: A memory process model of judgements of likelihood. *Psychological Review, 106*:180–209.

Freund, J. S. & Hasher, L. (1989). Judgements of category size: Now you have them, now you don't. *American Journal of Psychology, 102*:333–353.

Greene, R. L. (1989). On the relationship between categorical frequency estimation and cued recall. *Memory and Cognition, 17*:235–239.

Hasher, L. & Zacks, R. T. (1984). Automatic processing of fundamental information: The case of frequency of occurrence. *American Psychologist, 39*:1372–1388.

Gigerenzer, G. & Hoffrage, U. (1995). How to improve Bayesian reasoning without instruction: Frequency formats. *Psychological Review, 102*:506–528.

Hasher, L. & Zacks, R. T. (1979). Automatic and effortful processes in memory. *Journal of Experimental Psychology: General, 108*:356–388.

Hintzman, D. L. (1969). Apparent frequency as a function of frequency and the spacing of information. *Journal of Experimental Psychology, 80*:139–145.

Hintzman, D. L. (1976). Repetition and memory. In G. H. Bower (ed.) *The Psychology of Learning and Motivation*, vol. 10 (pp. 47–91). New York: Academic Press.

Hintzman, D. L. (1988). Judgements of frequency and recognition memory in a multi-trace memory model. *Psychological Review, 95*:528–551.

Hintzman, D. L. & Curran, T. (1995).When encoding fails: Instructions, feedback, and registration without learning. *Memory and Cognition, 23*:213–226.

Hintzman, D. L. & Gold, E. (1983). A congruity effect in the discrimination of presentation frequency: Some data and a model. *Bulletin of the Psychonomic Society, 21*:11–14.

Hintzman, D. L., Grandy, C. A. & Gold, E. (1981). Memory for frequency: A comparison of two multiple-trace theories. *Journal of Experimental Psychology: Human Learning and Memory, 7*:231–240.

Hintzman, D. L. & Stern, L. D. (1978). Contextual variability and memory for frequency. *Journal of Experimental Psychology: Human Learning and Memory, 4*:439–549.

Hockley, W. E. (1984). Retrieval of item frequency information in a continuous memory task. *Memory and Cognition, 12*:229–242.

Howell, W. C. (1973). Representation of frequency in memory. *Psychological Bulletin, 80*:44–53.

Indow, T. & Togano, K. (1970). On retrieving sequence from long-term memory. *Psychological Review, 77*:317–331.

Johnson, M. K., Raye, C. L., Wang, A. Y. & Taylor, T. H. (1979). Fact and fantasy: The role of accuracy and variability in confusing imaginations with perceptual experiences. *Journal of Experimental Psychology: Human Learning and Memory, 5*:220–240.

Jones, C. M. & Heit, E. (1993). An evaluation of the total similarity principle: Effects of similarity on frequency judgements. *Journal of Experimental Psychology: Learning, Memory, and Cognition, 19*:799–812.

Jonides, J. & Naveh-Benjamin, M. (1987). Estimating frequency of occurrence. *Journal of Experimental Psychology: Learning, Memory, and Cognition, 13*:230–240.

McEvoy, C. L. & Nelson, D. L. (1982). Category names and instance norms for 106 categories of various sizes. *American Journal of Psychology, 95*:581–634.

Marx, M. H. (1985). Retrospective reports on frequency judgements. *Bulletin of the Psychonomic Society, 23*:309–310.

Means, B. & Loftus, E. F. (1991). When personal history repeats itself: Decomposing memories for recurring events. *Applied Cognitive Psychology, 5*:297–318.

Menon, G. (1993). The effects of accessibility of information in memory on judgements of behavioral frequencies. *Journal of Consumer Research, 20*:431–440.

Menon, G., Raghubir, P. & Schwarz, N. (1995). Behavioral frequency judgements: An accessibility-diagnosticity framework. *Journal of Consumer Research, 22*:212–228.

Morton, J. (1968). Repeated items and decay in memory. *Psychonomic Science, 10*:219–220.

Nosofsky, R. M. (1988). Similarity, frequency, and category representations. *Journal of Experimental Psychology: Learning, Memory, and Cognition, 14*:54–65.

Payne, J. W., Bettman, J. R. & Johnson, E. J. (1993). *The adaptive decision maker*. New York: Cambridge University Press.

Rose, R. J. (1980). Encoding variability, levels of processing, and the effects of spacing of repetitions upon judgements of frequency. *Memory and Cognition, 8*:84–93.

Rowe, E. J. (1973). Frequency judgements and recognition of homonyms. *Journal of Verbal Learning and Verbal Behavior, 12*:440–447.

Russo, J. E. & Dosher, B. A. (1983). Strategies for multiattribute binary choice. *Journal of Experimental Psychology: Learning, Memory, and Cognition, 9*:676–696.

Siegler, R. S. (1987). The perils of averaging data over strategies: An example children's addition. *Journal of Experimental Psychology: General, 116*:250–264

Tversky, A. & Kahneman, D. (1973). Availability: A heuristic for judging frequency and probability. *Cognitive Psychology, 4*:207–232.

Underwood, B. J. & Richardson, J. (1956). Some verbal materials for the study of concept formation. *Psychological Bulletin, 53*:84–95.

Voss, J. F., Vereb, C. & Bisanz, G. (1975). Stimulus frequency judgements and latency of stimulus frequency judgements as a function of constant and variable response conditions. *Journal of Experimental Psychology: Human Learning and Memory, 3*:337–350.

Watkins, M. J. & LeCompte, D. C. (1991). Inadequacy of recall as a basis for frequency knowledge. *Journal of Experimental Psychology: Learning, Memory, and Cognition, 12*:387–396.

Whalen, J., Gallistel, C. R. & Gelman, R. (1999). Nonverbal counting in humans: The psychophysics of number representation. *Psychological Science, 10*:130–137.

Williams, K. W. & Durso, F. T. (1986). Judging category frequency: Automaticity or availability? *Journal of Experimental Psychology: Learning, Memory, and Cognition, 12*:387–396.

CHAPTER 4

IN THE YEAR 2054: INNUMERACY DEFEATED

GERD GIGERENZER

It is the year 2054. Our great-granddaughters and great-grandsons are celebrating a triple anniversary: the 400th anniversary of the mathematical theory of probability, the 200th of George Boole's *The Laws of Thought*, and the 100th of the publication of Leonard Savage's *Foundations of Statistics*. This year's celebration happens to coincide with the final victory over an intellectual disability that has plagued humankind for centuries: innumeracy, or the inability to think with numbers, specifically numbers that represent uncertainties and risks. Where are we? In Paris. It was in France that probability theory was born, back in the seventeenth century. The Great Hall at the Sorbonne is packed with flowers and guests, and is presided over jointly by the president of France and the president of the World Health Organization. A large curved podium provides the set for the four most distinguished scholars in the social sciences. At least, this is what the programme says. Other scholars in the audience think they should have been asked to speak, but Fortuna was not with them. The topic of this afternoon's panel discussion is 'How the war against innumeracy was won'. The chair is Professor Emile Ecu, an economist at the Sorbonne.

CHAIR: Madame le President, Monsieur le President, dear panel, guests, and audience. We have exactly 30 minutes to reconstruct what is arguably the greatest success of the social sciences in the twenty-first century, the defeat of innumeracy. The twentieth century had eradicated illiteracy, that is, the inability to read and write, at least in France. The challenge to our century was innumeracy. The costs of innumeracy have been a tremendous financial burden to modern economies, as had been those of illiteracy before. This year, the war against innumeracy has been declared won by the World Health Organization. Let us ask our distinguished panelists how this success came about?

SOCIOLOGIST (Paris): It all began with a programmatic statement by the father of modern science fiction, Herbert George Wells, best known, perhaps, as the author of *The Time Machine*. At the beginning of the twentieth century, Wells predicted that 'statistical thinking will one day be as necessary for efficient citizenship as the ability to read and write'. His message spread through all of the influential works that eventually led to the eradication of innumeracy. Here, we have a wonderful case in which literature eventually incited a revolution.

STATISTICIAN (Beijing): Wells never said that. That quotation was made up, most likely by Darrel Huff, who used it as an epigraph in his bestseller, *How to lie with statistics*. Making up quotations was consistent with the title of his book. Always check your sources!

CHAIR: Are you saying that the crusade against innumeracy was started by a fake?

SOCIOLOGIST: No, this is not . . .

STATISTICIAN: Yes, the Wells quote was a fake, and it did not initiate the crusade against innumeracy. It wasn't modern literature. It all started exactly 400 years ago with one of the greatest intellectual revolutions, the probabilistic revolution. At this time a notion of rationality was developed that eventually replaced the old ideal of certainty with a new, modest conception of rational belief that acknowledged uncertainty. The initiators of this revolution were the mathematicians Blaise Pascal and Pierre Fermat who, between July and October of 1654, exchanged letters about problems posed by a notorious gambler and man about town, the Chevalier de Méré. This revolution was not just an intellectual one—the calculus of probability—it was also a moral one, the . . .

CHAIR: What's moral about gambling?

STATISTICIAN: Not gambling, probability. Take Pascal's wager. Before Pascal, people believed in God because they were absolutely certain that He existed. For Pascal, God existed only as a probability, and the decision to believe or not to believe in God should be the outcome of a rational calculation—not of blind faith or stubborn atheism. According to Pascal's reasoning, you could make two errors. If God does not exist but you nevertheless believe, you might forgo some worldly pleasures. That's bad news. But if He does exist and you do not believe, then you face eternal damnation. That's not just bad news; that's a never-ending disaster. Thus, even if there is only a small probability that God exists, the pay-off is infinite—infinite bliss for those who believe and infinite misery for the others who, mistakenly, do not. Here is the beauty of Pascal's wager: The meaning of being moral had changed from blind faith and illusory certainty to rational self-interest and cost-benefit calculations.

CHAIR: I always knew that God was an economist. She must have loved this Pascal. I believe it was he who said, 'The heart has its reasons which reason knows nothing of'.

PSYCHOLOGIST (Boston): Let's get back down to earth. Whatever the moral implications of the calculus of probability, it was soon discovered that ordinary minds didn't understand probabilities, at least most of the time. This phenomenon was called innumeracy.

STATISTICIAN: No, no, no; not so fast. On the contrary, Pierre Laplace and, even earlier, the Enlightenment mathematicians said that probability theory is just common sense reduced to a calculus, and that educated persons—*les hommes éclairé*—have this common sense. And so did George Boole in 1854, when he set out to derive the laws of probability and logic from the laws of thought. It was known that people were occasionally confused by probabilities; Laplace himself described the 'gambler's fallacy' and other errors. But these mistakes were thought to result from the intervention of emotion and wishful thinking into rational processes. The two major rules of probability were definitely believed to be descriptions of actual human reasoning: the law of large numbers by Jacob Bernoulli, and the rule of inverse probabilities by Thomas Bayes.

HISTORIAN OF SCIENCE (Cambridge): Be that as it may, Bayes' rule is not from Bayes. He seems to have copied it from Nicholas Saunderson, who held the most prestigious academic chair in England, the Lucasian chair of Mathematics at Cambridge, which

Newton had held before. My dear friend, you should know Stigler's Law of Eponymy that says that no scientific discovery is named after its original discoverer.[1]

CHAIR: Hmm . . . I wonder who discovered Stigler's law . . .

HISTORIAN OF SCIENCE: One impolite reading of this law is that every scientific discovery is named after the last individual too ungenerous to give due credit to his predecessors.

STATISTICIAN: Why would that point be relevant here?

HISTORIAN OF SCIENCE: A minute ago, you yourself cared about the origins of a quotation attributed to Wells and I became concerned about the origins of a rule attributed to Bayes. Who cares? Pascal's triangle is not from Pascal, Gauss' law is also not from Gauss, and the Pythagorean theorem is not . . .

PSYCHOLOGIST: Can we move on? Innumeracy was described by the mathematician John Allen Paulos, who, in 1988, wrote a bestseller of the same title. For instance, he related the story of a weather forecaster on American television who reported that there was a 50% chance of rain on Saturday and a 50% chance of rain on Sunday, and then concluded that there was a 100% chance of rain that weekend.

SOCIOLOGIST: Paulos did not discover innumeracy. That phenomenon had already been described in the 1970s and 1980s by psychologists, notably Daniel Kahneman and Amos Tversky. In a series of experiments, they showed that people are confused by probabilities. They did not talk about innumeracy, but about 'cognitive illusions'. But it's the same thing: the base rate fallacy, the conjunction fallacy, the . . .

HISTORIAN OF SCIENCE: The conjunction fallacy had actually been described by Bärbel Inhelder and Jean Piaget in 1958 in their book on the early growth of logic in the child; they just used a different term, namely set inclusion rather than conjunction.[2] It's the same phenomenon. The base rate fallacy was actually discovered by the French mathematician . . . [3]

CHAIR: I thought that historians had already given up priority questions in the last century. What's the point of priority if there is no patent, copyright, or other source of income at stake?

PSYCHOLOGIST: OK, by the 1980s, there was ample evidence for a phenomenon eventually labelled innumeracy, the inability to think about uncertainties and risks.

CHAIR: Why was innumeracy diagnosed so late? I have heard that our Statistician and our Sociologist have different opinions on this question.

SOCIOLOGIST: May I go first? In the 1960s, the Western world was confronted with a flood of seemingly irrational behaviour, from the assassinations of John F. Kennedy and Martin Luther King, Jr. in the United States to the violent student revolutions of 1968 in countries all over the world. These events shattered the ideal of reasonable discourse

---

[1] Stigler, S. M. (1983). Who discovered Bayes' Theorem? *American Statistician*, 37:4, 290–296.

[2] Inhelder, B. & Piaget, J. (1958). *Growth of logical thinking: From childhood to adolescence*. New York: Basic Books.

[3] The historian of science seems to refer to Rouanet, H. (1961). Études de décisions expérimentales et calcul de probabilités. [Studies of experimental decision making and the probability calculus]. In *Colloques Internationaux du Centre National de la Recherche Scientifique* (pp. 33–43). Paris, France: Éditions du Centre National de la Recherche Scientifique.

and brought human irrationality to the foreground. Something similar had happened during the French Revolution, the bloody aftermath of which destroyed the idea that common sense would follow the calculus of reason, as probability theory was then called . . .

CHAIR: By the way, if I may interrupt, the events following the French Revolution were one reason why the subjective interpretation of probability—the idea that the laws of probability are about reasonable degrees of belief—was discredited by 1830 and the frequency interpretation began to reign. Degrees of belief came to be thought of as too disorderly to be a proper subject matter for the theory of probability, as opposed to orderly frequencies of things like mortality, suicide, prostitution . . .

SOCIOLOGIST: Fine, but my thesis is that some degree of innumeracy had always been a facet of human minds; it was simply amplified by the political events of the 1960s, as it had been before by those of the French Revolution. Psychologists of the 1970s just took advantage of the political climate and claimed that human disasters of any kind, including racial prejudice and 'hot' social behaviour, could be explained by 'cold' cognitive illusions. Consistent with my hypothesis, errors in statistical reasoning were no big deal in psychology before the political events of the 1960s.

STATISTICIAN: I think that looking for political causes for innumeracy is a bit too far-fetched. Neither the assassinations in the United States nor the political turmoil of the cultural revolution in China, if I may add another event of the 1960s, produced innumeracy. It did not come from outside influences, it came from within the decision theorists themselves. Listen carefully: whether you like it or not, innumeracy was, to some degree, created by decision theorists like Leonard Savage . . .

(Unrest in the audience.)

STATISTICIAN: . . . with their ultra-liberal, one might even say, expansionist policy of extending the laws of probability to everything between heaven and earth. These neo-Bayesians were not satisfied that probabilities mean observable frequencies; no, they claimed that one can and should attach a probability to everything. This created massive confusion in ordinary minds. Please recall what kind of statements confused people. These were statements involving probabilities, specifically single-event probabilities and conditional probabilities! Savage popularized single-event probabilities and Ward Edwards and others brought the message to the social sciences. For our Historian of Science, I add that Savage built on von Neumann and Morgenstern's work in the 1940s, so he does not have to lecture us on that.

(Unrest in the audience finds relief in laughter.)

STATISTICIAN: Since about 1830, probability has been interpreted as a relative frequency in a reference class, or sometimes as a physical propensity, and this gave the laws of probability a well-defined, although modest, realm of application. Unlike many of his followers, Savage was aware of the oddity of his proposal.

CHAIR: You mean extending the laws of probability to messy mental products such as degrees of belief?

STATISTICIAN: Exactly. In his 1954 book, Savage began his chapter on personal probability by saying that he considers it more probable that a Republican president will be elected in 1996 than that it will snow in Chicago sometime in the month of May 1994. And then he added that many people, after careful consideration, are convinced that such subjective probabilities mean precisely nothing, or at best, nothing precisely.

PSYCHOLOGIST: Right! We actually had a Democratic, not a Republican president in 1996 and there was no snow in Chicago in May 1994. So Savage was wrong . . .
SOCIOLOGIST: No, he wasn't. A probability statement about a single event can never be wrong, except when the probabilities are 0 or 1.
HISTORIAN OF SCIENCE: Look, our Statistician's thesis is that the extension of the laws of probability to degrees of belief, including beliefs about singular events, confused people and provided fertile ground for demonstrating reasoning fallacies. In other words, without this extension, there would not have been that magnitude of innumeracy. What she means is that most reasoning fallacies were demonstrated with probabilities rather than with frequencies.
SOCIOLOGIST: Don't be blind to the political dimension! How was it that before 1968 almost all psychologists agreed that man is a good intuitive statistician—pardon the sexist language of those days—and only a few years later, from the 1970s on, the same people embraced the opposite message? That change was not supported by fact, but rather grew out of a new political climate in which irrationality got the applause, and if one looked long enough . . .
CHAIR: Here is the disagreement: our Sociologist argues that the interest in statistical innumeracy was merely amplified by the political events of the 1960s. Our Statistician offers the conjecture that innumeracy itself had already been partly created in 1954 by Savage's extension of the laws of probability beyond frequencies, specifically to singular events. Ordinary people, she assumes, are frequentists.
HISTORIAN OF SCIENCE: Neither of these two interpretations—political turmoil or over-extension of the meaning of mathematical probability—was ever discussed in the 1970s. The explanation usually presented was that people simply suffer from cognitive illusions just as they suffer from visual illusions. Our Statistician's hypothesis should not be misread in the sense that innumeracy was not real; that would be a misunderstanding. The newspapers and the medical textbooks began to use single-event probabilities and conditional probabilities, and citizens and students alike were confused, often without even noticing it. That was all real. There were even court trials over the meaning of single-event probabilities. In one case, the prosecution had offered a defendant a plea bargain, but his lawyer told him that he had a 95% chance of acquittal based on an insanity plea. Based on this probability, the defendant rejected the plea bargain, stood trial, and was sentenced to 20 years in prison for first-degree murder. So he sued his attorney for having given him an unrealistic probability. Courts had to deal with the question, 'Can a single-event probability be wrong?' If it cannot, then what does it mean?
STATISTICIAN: The same confusion emerged in everyday life when institutions started to communicate all kinds of uncertainties in probabilities. In 1965, the US National Weather Service began to express forecasts in probabilities . . .
SOCIOLOGIST: That quantophrenia never occurred in France!
STATISTICIAN: . . . such as that there is a 30% chance of rain tomorrow. Most Americans thought they knew what that meant. However, studies showed that some people understood this statement to mean that it will rain in 30% of the area, others that it will rain 30% of the time tomorrow, and others that it will rain on 30% of the days like tomorrow. A single-event probability leaves, by definition, the reference class open: area, hours, days, or something else. But people, then and now, think in terms of concrete cases and fill a class in.

PSYCHOLOGIST: The insight that people tend to construct reference classes became the basis of the first theory, in the early 1990s, that linked the cognitive processes underlying judgements of confidence with those underlying judgements of frequency; this theory, in turn, revealed how to make a celebrated case of innumeracy, the overconfidence bias, appear and disappear. This was first shown in laboratory research with general-knowledge questions, in which participants were asked, 'What is the probability that your answer is correct?' and then, after a number of questions, 'How many of your answers are correct?' Judgments of probability were systematically higher than judgements of frequencies. Subsequently, the same result was shown for highly consequential decisions such as whether or not a prisoner should be given conditional freedom, that is, probation, bail, or weekend leave. In one influential study, a group of probation officers was asked 'What is the probability that Mr. Smith will commit a violent act if he is discharged?' whereas another was asked 'Think of 100 men like Mr. Smith. How many of them will commit a violent act if they are discharged?' When the average estimate of the probability of harm was 0.30, the frequency estimate was only 20 of 100.[4]

CHAIR: How so?

PSYCHOLOGIST: For the same reason probability judgements were higher in estimations of general knowledge and weather forecasting. Asking for a single-event probability leaves the reference class open. The probation officers themselves need to fill one in: does the probability refer to the situation that Mr. Smith is on weekend release 100 times, or that 100 people like Mr. Smith are on weekend release once, or something else? The answer need not be the same. The frequency question specifies a reference class, the probability question leaves it open and thus can lead to systematic differences in the answers.

CHAIR: If I recall correctly, the idea was initially not well understood that reference classes, as proposed by probabilistic mental models theory, could be at the core of the overconfidence bias, as well as the systematic difference between judgements of single events and frequencies. Many understood that frequency judgements are always right, period; they missed the real issue, namely changing reference classes, which allowed the prediction of when frequency judgements were right or wrong. Some even thought that the difference occurs because frequencies are more frightening or elicit more emotional responses.

STATISTICIAN: In the 1990s, *three* reasons were found to explain why people were confused by probabilities. Unspecified reference classes in single-event statements was the one you mentioned; polysemy and computational complexity were the other two. The three together explained much of the confusion, not all of it, and also provided the key to overcoming it.

---

[4] On the laboratory research see Gigerenzer, G., Hoffrage, U. & Kleinbölting, H. (1991). Probabilistic mental models: A Brunswikian theory of confidence. *Psychological Review*, 98:506–528. On the research with experts in law and psychiatry see Slovic, P., Monahan, J. & MacGregor, D. G. (2000). Violence risk assessment and risk communication: The effects of using actual cases, providing instruction, and employing probability versus frequency formats. *Law and Human Behavior*, 24:271–296.

CHAIR: That's what I meant; often all three explanations were jumbled together as the 'frequency effect'.

PSYCHOLOGIST: I don't think that polysemy is as interesting a reason as the other two. That part was just a game with words with multiple meanings to produce reasoning that looked like innumeracy.

CHAIR: What do you mean?

PSYCHOLOGIST: The game went like this. Experimental participants read a description of a woman called Linda that made her look like a feminist, and then they were asked whether it is more probable that she is a bank teller, or a bank teller and active in the feminist movement; most chose the latter. That was called a conjunction fallacy and was interpreted as demonstrating that the human mind is not built to work with the laws of probability. But when people were asked a frequency question, 'Think of 100 women like Linda. How many are bank tellers? How many are bank tellers and active in the feminist movement?', that alleged fallacy disappeared. Only at that point did it become clear that the problem was with the polysemy of the term probability, that is, with its several legitimate non-mathematical meanings, including 'plausible', 'typical', and 'whether there is evidence'. A frequency question eliminated this polysemy and elicited a different answer. German psychologists were the first to show that the so-called conjunction fallacy disappeared with frequency questions . . . [5]

HISTORIAN OF SCIENCE: No, it was Bärbel Inhelder and Jean Piaget who had shown long before that children can understand conjunctions; they just used the term set inclusions. They put a box containing wooden beads, most of them were brown but two were white, in front of a child and asked: 'Are there more wooden beads or more brown beads in this box?' By the age of eight, a majority of the children responded that there were more wooden beads in the box. Note that Piaget and Inhelder had asked children about frequencies, not probabilities.

CHAIR: At that time, didn't anybody ask why Stanford University students should suffer from the conjunction fallacy when Genevese children, by the age of eight, didn't?

PSYCHOLOGIST: Few researchers asked any questions. At that time, most of them were busy running experiments; that was during the empiricist phase, before psychology finally became a full-blown theoretical science in 2010. There were even scholars who claimed that human disasters of many kinds, including US foreign policy in the late twentieth century, could be explained by the conjunction fallacy;[6] there were scholars who . . .

SOCIOLOGIST: Yeah, yeah, yeah, and there were statesmen who claimed that AIDS was caused by poverty, not by HIV. But I agree that, unlike the other two reasons for public confusion with probability, polysemy had no real practical consequences.

---

[5] The psychologist refers to Fiedler, K. (1988). The dependence of the conjunction fallacy on subtle linguistic factors. *Psychological Research, 50*:123–129; and Hertwig, R. & Gigerenzer, G. (1999). The 'conjunction fallacy' revisited: How intelligent inferences look like reasoning errors. *Journal of Behavioral Decision Making, 12*:275–305. The historian of science refers to Inhelder, B. & Piaget, J. (1958). *Growth of logical thinking: From childhood to adolescence.* New York: Basic Books.

[6] Kanwisher, N. (1989). Cognitive heuristics and American security policy. *Journal of Conflict Resolution, 33*:652–675.

HISTORIAN OF SCIENCE: Uncovering the effect of external representations on statistical thinking started the ball rolling against innumeracy. That is, it was realized that the problem was not just a lack of training in the laws of statistics; innumeracy could be tackled in a much easier way by realizing that some representations of uncertainty help people understand uncertainties, whereas others do not. Unfortunately, the unhelpful representations were the ones generally used in teaching, medicine, by the media, in the court of law . . .

STATISTICIAN: Consistent with my thesis, if I might add, among the representations which tend to confuse people were those promoted by Savage, such as the use of single-event probabilities . . .

HISTORIAN OF SCIENCE: Yes, but also others, such as conditional probabilities and relative risks. In contrast, representations that were found to foster insight included those that specify reference classes, make use of natural frequencies, and communicate absolute risks. Teaching people to use proper representations turned out to be a fast and effective method against innumeracy. Natural frequencies, for example, are much more effective than conditional probabilities; as mentioned earlier, they reduce the computational complexity in making inferences. This was discovered in the mid-1990s by the same two German psychologists who had earlier worked out the crucial role of reference classes in their probabilistic mental models theory, namely . . .

PSYCHOLOGIST: No, no; before that, in 1982, two American scholars, Christensen-Szalanski and Beach, had used natural sampling, of which natural frequencies are the result.[7] Coming under attack by those who did not want to see cognitive illusions disappear, they did not develop the idea further and did not have any influence; but they were first.

HISTORIAN OF SCIENCE: Whoever was first, the insight was that natural frequencies facilitate Bayesian reasoning; that is, they facilitate the estimation of a posterior probability (or a frequency) from observation.[8] This insight turned the question of innumeracy into an ecological, perhaps even an evolutionary, one.

CHAIR: Bayes and evolution?

PSYCHOLOGIST: The argument was that animals and humans have spent most of their evolution in an environment of natural sampling—before the development of probability theory or statistical surveys. Natural frequencies are the result of natural sampling. For instance, take a physician in an illiterate society who is confronted with a new disease. The physician has observed 20 cases of a symptom with the disease and 40 cases of the symptom without the disease. When a new case with the symptom comes in, she can easily compute the Bayesian posterior probability of this patient having the disease: $20/(20 + 40)$, which is one-third. That's how Bayesian inference was done before 1654, that is, before mathematical probabilities were introduced.

---

[7] Christensen-Szalanski, J. J. J. & Beach, L. R. (1982). Experience and the base-rate fallacy. *Organizational Behavior and Human Performance, 29*:270–278.

[8] Gigerenzer, G. & Hoffrage, U. (1995). How to improve Bayesian reasoning without instruction: Frequency formats. *Psychological Review, 102*:684–704; Gigerenzer, G. & Hoffrage, U. (1999). Overcoming difficulties in Bayesian reasoning: A reply to Lewis & Keren and Mellers & McGraw. *Psychological Review, 106*:425–430.

CHAIR: Oh, now I understand the paradox that animals were reported to be good Bayesians but humans were not. It's the representation, not the species. Animals encode natural frequencies, and we poor humans got conditional probabilities . . .

PSYCHOLOGIST: Right. Natural frequencies were transformed into conditional probabilities; for instance, dividing the 20 cases of the symptom with the disease by the total number of disease cases, say 25, results in a probability p(symptom|disease) of .80. The physician in the illiterate society can ignore the base rate of 25—it is not necessary for computing the posterior probability, as we just saw. Not so the medical student who has to multiply the conditional probabilities by the respective base rates in order to get the base rate information back—which amounts to the rather complicated form in which Bayes' rule had been taught, and which most physicians never understood. Thus, the insight was that Bayesian computations depend on the representation of the information in the environment—that's the ecological part. The second insight was that the cure is to use a representation that humans had encountered during most of their evolutionary history, natural frequencies, that is, frequencies that have not yet been normalized with respect to the base rates—that's the historical, or evolutionary, part.

CHAIR: Was this the point where the World Health Organization took over? In educating physicians to understand uncertainties?

HISTORIAN OF SCIENCE: No, this laboratory research had its impact first in the Anglo-Saxon courts, possibly because of their adversarial procedure. Defence lawyers realized that confusions due to probabilities were not in their or their clients' interests, but typically *were* in the prosecution's interests. For many decades, experts had testified in the form of single-event probabilities, for instance, 'The probability that this DNA match occurred by chance is 1 in 100,000'. That made it likely that jurors thought the defendant belonged behind bars. But when experts testified in frequencies instead, the case against the defendant appeared much weaker: 'Out of every 100,000 people, one will show a match'. Mathematically, that's the same, but, psychologically, it made jurors think about how many suspects there might be? In a city with one million adults, there should be 10 who match.[9]

CHAIR: I see. But what took defence teams so long to realize this?

HISTORIAN OF SCIENCE: There were a few attempts to introduce natural frequencies into court proceedings, even before 2000.[10] Based on the laboratory research on natural frequencies, the O. J. Simpson defence team asked Judge Ito not to allow the prosecution's DNA expert, Professor Bruce Weir, to testify in terms of conditional probabilities and likelihood ratios, which are ratios of conditional probabilities. The defence requested frequencies instead. Judge Ito and the prosecution agreed, but the prosecution expert used likelihood ratios anyway! This was in the 'good' old days when statisticians like Weir didn't care about the psychology of jurors and judges.

---

[9] Koehler, J. J. (1996). On conveying the probative value of DNA evidence: Frequencies, likelihood ratios, and error rates. *University of Colorado Law Review*, 67:859–886; Hoffrage, U., Lindsey, S., Hertwig, R. & Gigerenzer, G. (2000). Communicating statistical information. *Science, 290*:2261–2262.

[10] See Gigerenzer, G. (2002). *Calculated risks: How to know when numbers deceive you.* New York: Simon & Schuster.

SOCIOLOGIST: And this is when the French Feminist Association comes in.

STATISTICIAN: Because Simpson had beaten his wife?

SOCIOLOGIST: Ha, ha; very funny. No, it happened in 2010 at this very University. The Feminist Association had a larger goal . . .

HISTORIAN OF SCIENCE: Sorry, but the International Transparent Testimony Act was a few years earlier. After some local rulings by judges in England and the USA disallowing statements involving conditional probabilities or single-event probabilities in testimony, the International Federation of Law ruled that testimony in the courts had to be communicated in terms of natural frequencies rather than probabilities or likelihood ratios. That was in 2006, two decades after DNA fingerprinting was introduced into American criminal investigations. Probability statements about singular events are no longer admissible. The Act helped to bring insight into the court proceedings and get confusion out.

PSYCHOLOGIST: I am not a legal scholar, but isn't there, typically, more evidence than just a DNA match?

HISTORIAN OF SCIENCE: Yes, there is; but jurors and judges need to understand DNA evidence whether or not there is additional evidence, such as eye witness accounts. And since we have had complete DNA data banks in all European and North American countries for twenty years, there is a tendency for police officers to sit at their computers and search in data banks rather than at the scene of a crime.

SOCIOLOGIST: I am not persuaded that that law started the war against innumeracy. The Feminist Association was concerned for years with the harm done to women by misinformation about breast cancer screening. This harm included unnecessary anxiety and unnecessary surgery. For instance, in the year 2005, some 150,000 French women who did not have breast cancer were nevertheless operated on as a consequence of false positive mammogram diagnoses. These women were not informed that some 9 out of 10 positive screening mammograms are actually false positives, but believed that a positive test most likely meant that they had breast cancer. Even after it turned out that the positive result was a false positive, their anxiety remained. The Feminist Association discovered several other ways of presenting statistics with which the medical associations misled women. For instance the use of relative risks as opposed to absolute risks. Your mothers and grandmothers were told that mammography screening reduced mortality in women over 50 years of age by 25%. This is a relative risk. But 25% of what? Again, the reference class was not made transparent. Many women understood it to mean that, out of 1,000 women, 250 were saved from breast cancer. In fact, only 1 out of 1,000 was saved. That's the absolute risk: 0.1%.

CHAIR: How can 25% be the same as 0.1%?

PSYCHOLOGIST: Take 1,000 women who do not participate in screening. Four of them will die from breast cancer within 10 years. Now take 1,000 women who do participate. Three of them will die from breast cancer within 10 years. The difference between 4 and 3 amounts to a relative risk reduction of 25%. In absolute frequencies, however, that's 1 out of 1,000. Despite repeated calls from researchers that the women be given the information in absolute risks, they almost never were, they practically always received it in terms of relative risks. Just as the lack of information about false positives had unnecessarily increased women's anxiety, the information in terms of relative risk reduction falsely increased women's faith in the benefits of screening. The Feminist Association was also concerned that the actual harm mammography screening can do was poorly communicated to women. They finally sued medical associations and industries around the world for intentionally producing confusion and innumeracy that violated the right of women to informed consent.

SOCIOLOGIST: This smart move on the part of the Feminist Association forced the World Health Organization into action. The national health associations and industries were sentenced by the courts to pay 10 billion Euros in damages.

STATISTICIAN: But it took several years before the WHO finally took action. There was a battle over whether the WHO should admit to, or even treat, innumeracy in physicians. Those in favour had all the evidence on their side: In study after study, 80 to 90% of physicians did not understand how to estimate the probability of a disease being present from, for example, standard screening tests.[11] Those in favour also had the therapeutic tools to cure innumeracy. The opponents did not want the medical profession to be associated with innumeracy in the public mind; they feared it would undermine its authority and the public's trust in it. The compromise was that innumeracy was declared a general mental aberration and it was entered into the DSM-VIII, thus it did not look like a doctor's disease alone, which it wasn't.

HISTORIAN OF SCIENCE: Yes, the WHO action was the turning point—that and the changes in the medical curricula in universities. The WHO did enter innumeracy into their catalogues of mental diseases and disabilities, but the definite change came with its Anti Mental Pollution Act, which made clear communication of risks the rule and eliminated forms of communication that had previously 'polluted' minds. Medical students learned how to express risks as absolute risks rather than relative risk reductions, and in natural frequencies rather than conditional probabilities; they learned how to specify reference classes, and so on. At the same time, high school curricula focused on training statistical thinking, and students learned how to play with representations. My own kids loved it when they could confuse people with percentages; this was much more fun than the applications of algebra and geometry.

CHAIR: Don't overlook the economic consequences of the Act. The Anti Mental Pollution Act and the fact that the WHO added innumeracy to their catalogue made millions of dollars available for research and implementation programmes. My estimate is that about 10 billion dollars alone were poured into education in professional schools and high schools.

HISTORIAN OF SCIENCE: But how do we know that all countries are innumeracy-free now? Can we trust these results?

STATISTICIAN: The Sixth International Mathematics and Science Study (SIMSS) focused exclusively on statistical thinking and defined innumeracy operationally in terms of performance on the test. The test items measured the ability to use representations to

---

[11] Hoffrage, U. & Gigerenzer, G. (1998). Using natural frequencies to improve diagnostic inferences. *Academic Medicine, 73*:538–540; Gigerenzer, G. (1996). The psychology of good judgment: Frequency formats and simple algorithms. *Journal of Medical Decision Making, 16*:273–280.

which the human mind is adapted, such as figures and natural frequencies. It also included test items that measured understanding uncertainties in the real world rather than hypothetical situations like urns and balls. SIMSS enabled each country's performance to be measured by the same standard, and, when 99% of twelfth graders passed the test, the country was declared free of innumeracy.

SOCIOLOGIST: What's with the rest of the population? The older people, those with chronic math anxiety, those who read poetry while others discuss baseball statistics?

STATISTICIAN: Worldwide, we reached the SIMSS performance level for the first time in 2054, this year. The professional schools in law, medicine, and business reached that level a decade ago.

SOCIOLOGIST: But still, what's with the older fellows, like you and me?

CHAIR: Time is up. And I still don't understand how 25% can be the same as 0.1%.

Note

By Professor Emile Ecu (Chair)

The editors of this book reminded me that psychology is a discipline that has a short-term rather than a long-term memory. That is, work older than 20 years—before 2034—is likely already out of memory. This is a most unfortunate state of affairs, which we do not have in economics. To compensate for this, I have added footnotes and references to clarify what the panelists refer to.

To begin with, the best analysis of the events 400 years ago that led to the mathematical theory of probability, and of the rise and decline of the classical interpretation of probability (in which frequencies and subjective degrees of belief were merged rather than distinguished), are the books by Lorraine Daston, *Classical probability in the enlightenment*, Princeton, NJ: Princeton University Press (1988) and Ian Hacking, *The emergence of probability*, Cambridge, England: Cambridge University Press (1975). Two hundred years ago, George Boole's *An investigation of the Laws of Thought on which are founded the mathematical theories of logic and probabilities* (New York: Dover 1854/1958) was published, a seminal work in which the laws of logic and probability are derived from the laws of psychology. One hundred years ago, Leonard J. Savage published *The foundations of statistics* (New York: Dover 1954), a seminal book that promoted a subjective view of probability and the revival of personal probabilities during the second half of the twentieth century.

CHAPTER 5

FREQUENCY JUDGEMENTS AND RETRIEVAL STRUCTURES: SPLITTING, ZOOMING, AND MERGING THE UNITS OF THE EMPIRICAL WORLD

KLAUS FIEDLER

Abstract

In this chapter, an attempt is made to explain the discrepancy between the often-cited finding of accurate and robust frequency memory and the existence of serious biases in frequentistic judgements. It is argued that accuracy often depends on the retrieval–cue structure. Accuracy is typically obtained when frequency judgements for different stimuli are elicited by comparable and easy-to-handle retrieval cues. However, strong biases arise when judgement prompts solicit different, misleading retrieval structures. Judges lack the metacognitive devices that would be necessary to correct for the influence of incomparable retrieval cues. The influence of retrieval structures on frequency judgements is illustrated with reference to various paradigms, such as category-split effects, the numerosity heuristic, conjunctions and logical connectives, the base rate fallacy, and inductive judgements of correlated attributes.

The assessment of frequencies in the empirical world is biologically significant and ubiquitous. Our acoustic system is equipped with sensors to assess the frequency of sound pressure waves; our emotional system is sensitive to the frequency of heartbeats and music beats per minute; and learning and conditioning are a function of the frequency of reinforcements. Even primitive organisms are able to learn the rate at which behaviours and situations are associated with pleasant (e.g. food) or unpleasant consequences (e.g. pain).

These elementary examples already illustrate a point that is easily overlooked: what is informative cannot be absolute, unconditional event frequencies per se. Rather, frequency information is only meaningful in the context of specific algorithms applied for frequency assessment. Thus, to perceive pitch, the absolute number of pressure changes must be normalized to some time unit and acoustic waves must co-occur in space. That is, an algorithm is required that assesses the number of pressure changes conditional on temporal and spatial conditions. Likewise, the frequency of heartbeats is only meaningful if it is conditionalized temporally (to a time unit) and spatially (to the same individual's chest). In learning, too, the (relative) frequency of reinforcements is conditionalized on specific stimulus contexts, as manifested in the term 'conditioning'. As a matter of rule, then, the very assessment of frequency is conditional on specific assessment algorithms. This central insight may remind the reader of what Einstein has taught us about the relativity of measuring time, space, and velocity—all concepts derived from frequentistic primitives.

Assets and deficits in cognitive frequency assessment

Turning to contemporary psychology, memory and assessment of frequency is often characterized as well functioning, effortless, quasi-automatic (Hasher & Zacks 1984), independent of explicit attention and intention (Reber 1989), and a natural product of evolution (Gigerenzer 1991; Gigerenzer & Hoffrage 1995). In contrast to these often-cited assets of frequency memory, cognitive functions of frequency information exhibit serious deficits. Among the prominent paradigms of frequency-related illusions are biases in subjective probability judgements (Kahneman et al. 1982), contingency illusions (Jenkins & Ward 1965), over- and underconfidence (Juslin & Olson 1997), unrealistic optimism (Klar & Giladi 1997; Weinstein 1980), and hypothesis confirmation biases (Snyder 1984; Zuckerman et al. 1995). All these highly bias-prone cognitive functions use basically frequentistic input information.

Locus of assets and deficits

How can the lability of cognitive frequency functions on the one hand and the robust accuracy of the original frequency assessment on the other be reconciled? How can this apparent contradiction be resolved? I want to argue that the crucial factor underlying accurate versus inaccurate frequency reports is neither memory capacity nor the amount and completeness of information encoded. Rather, the crucial factor is whether the cognitive algorithm used to retrieve relevant information for frequency judgements is hard or easy to handle. In those experiments that are often cited as evidence for accurate, reliable, quasi-automatic frequency assessment, the cognitive algorithms are much easier to handle and less misleading than in those experiments that evoke shortcomings and frequency illusions.

The structure of retrieval cues

But how do the algorithms differ? One central aspect that can explain a good deal of variance is the structure of retrieval cues used for frequency judgements. In experiments that testify to the sensitivity and accuracy of frequency memory, natural and comparable retrieval prompts are used for all judgement targets. Typically, the categories themselves whose frequency is to be estimated are presented as prompts (e.g. *triangles* is given as a prompt for judging the number of triangles in the stimulus list), and comparable prompts are given for all other targets (*squares*, *circles*, *diamonds*). As the stimulus list used for most episodic memory studies is unstructured and random, no other competing retrieval cues are available that could interfere with the given judgement prompt. Using comparable self-prompts for all targets creates equal chances and increases the likelihood that judgements reflect the relative frequency in episodic memory. In such experiments, judges only have to attend to the mental echo they get when prompting their memory with the target labels. If the quantitative echo is strong, they can respond with high frequency judgements, if the echo is low, they respond with low frequency judgement. No further mental transformation is required. Judges do not have to compare mental reactions to structurally different prompts and algorithms (*triangles* in a short stimulus series with *large red squares* in another, long series). They do not have to perform mental calculations (as in subjective correlation assessment, which is a function of at least four frequencies in a contingency table). Because the episodic memory task involves an arbitrarily selected stimulus series, prior knowledge about the distribution and differential importance of events is ruled out. And because researchers are typically interested only in relative but not in absolute frequency estimates, they do not even have to care about the quantitative scales they use to express their frequency judgements. In such a task situation, a simple algorithm that assesses the associative strength of reactions elicited by judgement prompts is very likely to predict the original (relative) frequencies of encoded stimuli quite accurately.

In contrast, in those experiments that testify to severe biases and inaccurate judgements, the judgement prompts and retrieval algorithms are heterogeneous and complicated by intrusions from prior knowledge and task difficulties. Consider the typical situation in an experiment showing severely biased frequency estimates due to the availability heuristic (Tversky & Kahneman 1973). When judging the prevalence of various causes of death, for instance, people overestimate the relative risk of certain causes, such as lightning, murder, or various catastrophes (Combs & Slovic 1979) that are overrepresented in the media. The common interpretation of this phenomenon in terms of the enhanced availability of memories for events that are highlighted in the media can be paraphrased in terms of retrieval structures. When prompted to judge, say, the prevalence of murders, the effective retrieval structure is not confined to the experimenter's prompt itself, *murder*. Rather, a number of additional retrieval cues are generated spontaneously, based on organized world knowledge and memorized media experience. For example, a whole taxonomy of different types of murder may come to the judge's mind, or a whole fan of murder stories or scripts, affording a rich structure of retrieval cues generated by world knowledge. When judging the risk of another cause of death, say, *heart disease*, a much less differentiated retrieval cue structure may be generated, due to less elaborate knowledge, leading to relative underestimation.

Similar points could be made for other instances of strong biases and ill-calibrated frequency judgements. The overestimation of one's own (as compared with other people's) contributions to mundane activities, usually called an egocentric bias (Ross & Sicoly 1979), can be due to more differentiated retrieval structures for self-related memory. Likewise, a number of other biases, such as the conjunction fallacy (Tversky & Kahneman 1983), stereotypical biases in social psychology (Hamilton & Sherman 1994), or unrealistic optimism (Price 2001; Weinstein 1980) can be understood as elicited by incomparable retrieval structures, causing disproportionate judgements for different target.

Metacognitive blindness

One implicit assumption underlies this interpretation. It is not only assumed here that frequency judgements can be misled by non-comparable retrieval structures. It is also presupposed that judges are blind, metacognitively, to the vicissitudes of incomparable retrieval cues. They ought to understand that the outputs of rich retrieval structures will be different and must be corrected to make them comparable to the outputs from impoverished retrieval structures. But judges do not appear to use such correction procedures. Rather, they appear to be blind to the impact of retrieval structures, and this metacognitive neglect is an important ingredient in the genesis of frequency judgement biases.

Category-split effects

Let us illustrate this point with the following category-split phenomenon (Fiedler & Armbruster 1994). Imagine an experiment in which a hundred slides are presented, each one showing for a few seconds a different type of car. Afterwards, the participants in one experimental condition are asked to estimate the number of *Japanese* cars that they have seen. In another condition, the same stimulus series is shown but participants are asked to estimate the number of *Honda, Mazda, Nissan, Toyota*, and *Mitsubishi* cars, and then to sum up the overall number. In which condition will judges arrive at higher estimates? It seems intuitively clear that a memory scan with five different prompts will be more productive than a scan with only one prompt. This might be noticed by judges themselves, especially when the split manipulation is within judges. However, as already noted, they ignore this obvious influence at the metacognitive level and do not correct their frequency judgements accordingly.

Explanations of category split

To understand and explain category-split effects, it is first of all important to emphasize the active, constructive nature of frequency judgements. However efficiently information is encoded in memory, the memory code cannot be the same as that required for all kinds of judgement tasks that may happen to be presented later on. An individual may have encoded the frequency of Japanese cars and then be asked to estimate the number of Honda, Mazda, Toyota cars etc. Or she may have assessed the rate of Honda, Mazda, Toyota etc. and then be asked to judge even finer subcategories like 'over ten years old, well-conserved Honda Civic'. Thus, rather than reading off the frequentistic quantity required for each and every task context, most frequency judgements must be somehow constructed *ad hoc* when required, starting from a specific retrieval structure. This (re)constructive process leaves considerable latitude for various specific judgement strategies. When motivation is high and there is enough time, an exhaustive scanning of memory (for all traces related to Honda, Mazda, Toyota etc.) might provide the basis for an *enumeration* strategy (Brown 1995), which relies on the number of encoded exemplars that can be found. Alternatively, the process could rely on some measure of *echo strength* elicited by the prompt (e.g. a feeling of familiarity or associative strength elicited by the Honda prompt). If memory is not scanned exhaustively, the process might only try out how well the retrieval cue works (e.g. how easy an instance of Honda comes to mind). In an *availability heuristic* (Schwarz et al. 1991; Tversky & Kahneman 1973), the experienced ease with which instances can be retrieved would be taken as a proxy to frequency estimation.

Many predictions of a retrieval–cue approach remain constant across such different strategies. Whether memory search is exhaustive or not, whether the output is enumeration, memory echo strength, a sense of familiarity, or the experienced ease of recall, a rich and differentiated retrieval structure should lead to higher frequency judgements than impoverished retrieval cues. Moreover, an intriguing new possibility, which suggests itself within this approach, is that direct judgements of the retrieval structure itself are used as a proxy for frequency judgements. Virtually all previous models have presupposed that some kind of retrieval must actually take place to make a frequency judgement. Even the availability heuristic is based on the ease with which a small sample of items, or at least one item, is actually retrieved from memory. At the very least, such a retrieval process must be mentally simulated. However, a strategy based on retrieval–cues might be radically detached from memorized stimulus contents. Without engaging in any retrieval, an algorithm might directly evaluate the retrieval structure itself. Thus, the number, diversity, or differentiation of retrieval cues might be taken as a heuristic of how much can be found in memory without ever engaging in an actual retrieval attempt. Even such a retrieval-independent algorithm would still be sensitive to category-split effects. For example, when estimating the prevalence of causes of death, judges may overestimate *murder* and underestimate *heart disease* simply because they feel a richer retrieval structure (i.e. more links to semantic and autobiographical knowledge) for the former than the latter.

As the Japanese cars example illustrates, the demonstration of category-split effects as a manifestation of the influence of retrieval structures on frequency assessment is simple and intuitively appealing. In the remainder of this chapter, I will first review empirical evidence from this paradigm to illustrate and elaborate the importance of retrieval cue structures. Then I will offer alternative explanations for several judgement biases that are commonly explained differently, in terms of the present theoretical approach. Finally, the summary will reinstate the basic point that ill-controlled algorithms, rather than ineffective stimulus encoding, are the Achilles heel of frequency judgements.

Experimental evidence on category-split effects

Despite their plausibility, category-split effects may not be so easy to explain. That more retrieval cues help to find more items in memory may appear to be obvious. But does this asseveration of obviousness provide a scientific explanation? Is it really obvious that using more search prompts yields more output? Wouldn't it be equally plausible to expect interference between multiple retrieval cues? And what about the boundary conditions? For instance, will category-split effects be stronger for small or large categories? Will the illusion be stronger when the overall retrieval performance is high or low? An informed scientific explanation should not content itself with an 'I knew it all along' feeling, but should rely on clearly spelled out laws, assumptions, and boundary conditions.

The following explanation offers a simple and straightforward account of the pertinent experimental evidence; it is based on the universal principle of regression that underlies all frequency assessment (Fiedler 1991; Sedlmeier 1999). As a matter of rule, subjective reproductions of objectively presented frequency distributions tend to be regressive. That is, actually existing frequency differences shrink in subjective reproductions; large frequencies tend to be underestimated and small frequencies tend to be overestimated. This regressive tendency is more pronounced for extremely large and small frequencies than for moderate frequencies. In general, the degree of regression depends not only on extremity, but also on reliability; therefore small stimulus samples will typically produce more regression than large samples. Regressiveness is a pervasive feature of the entire research literature on frequency estimates (Fiedler 1991; Greene 1984).

Given this basic law, we begin to form a more refined picture of the specific cognitive algorithm that is triggered by split category retrieval cues. By splitting one category with a high or medium frequency into two subcategories of smaller frequency, there will be no underestimation of a large frequency but, instead, a relative overestimation of smaller subfrequencies. For example, given four stimulus categories with presentation frequencies 10, 6, 10, and 6, the basic regression effect will result in subjective frequencies of, say, 9, 7, 9, 7. However, when a 10-item category is split into two subcategories with frequency 5, the resulting stimulus distribution becomes (5, 5), 6, 10, 6, which might regress to (6, 6), 7, 9, 7. Pooling over the two split categories yields 12, 7, 9, 7. Apparently, the same 10-item category which is normally slightly underestimated (9) is now overestimated (12), due to regression after category split. Furthermore, the regression principle predicts that splitting a small 6-item category into even smaller 3-item subcategories will produce even stronger frequency illusions, because regressive overestimation will inflate these extremely infrequent subcategories even more. For example, a split might yield 10, (3, 3), 10, 6, which might regress to 9, (5, 5), 9, 6. Pooling over the subcategory split, the subjective distribution becomes 9, 10, 9, 6; splitting a small category produces even stronger overestimation.

In two controlled experiments, Fiedler and Armbruster (1994) used exactly these stimulus distributions. Participants saw a series of 32 geometric figures (Experiment 1) and 32 letters (Experiment 2), respectively, from four different categories (semicircle, triangle, square, pentagon; *d, f, n, r*; see Fig. 5.1). In the no stimulus-split condition, the stimulus series included 10 repetitions from two categories and 6 repetitions from the remaining two; the allocation of categories to frequency conditions was counterbalanced across participants. One participant of Experiment 1, for instance, could have seen 10 semicircles, 6 triangles, 10 squares, and 6 pentagons. In the stimulus-split condition, one

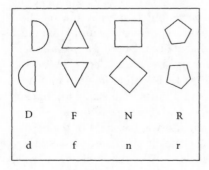

Figure 5.1 Geometric forms and letters used as stimulus materials in two experiments by Fiedler and Armbruster (1994).

of the four categories was split into two equal subcategories. For example, of the 10 semicircles, 5 showed the left half and 5 showed the right half of a circle. We also varied whether a large or a small category was split; thus, for other participants, the peak of 3 of the 6 triangles pointed upward and downward, respectively. Likewise, when the stimuli consisted of letters (Experiment 2), the 10 repetitions of, say, *d* were split into 5 lowercase *d* and 5 uppercase *D* letters; or the 6 *r* repetitions, were split into 3 lowercase *r* and 3 uppercase *R* stimuli.

In addition to the stimulus-split manipulation (whether two different subcategories actually occurred in the stimulus list) and the size of the split-category factor (large vs. small), we also manipulated split in retrieval cue. In the no retrieval-split condition, only the four basic category labels were used as judgement prompts, regardless of whether the stimulus list had contained one or two types of items. In contrast, in the retrieval-split condition, different judgements were prompted for both subcategories (e.g. upward and downward triangles) and the two judgements for the same category were pooled later on.

The results are summarized in Fig. 5.2. The ordinate represents *proportional deviations*, that is, the ratio of estimated frequencies divided by actually presented frequencies; values over 1 indicate overestimation whereas values under 1 indicate underestimation. The data support several implications of the regression analysis of category split. First, when there is no split, the basic regression effect is evident in overestimated 6-item categories and underestimated 10-item categories. Second, and central to the present argument, the strongest influence on frequency estimates is due to the retrieval-split manipulation. When separate judgements for two subtypes (e.g. two upward and downward triangles) were prompted and added up, the assessed frequencies were clearly enhanced, relative to the no retrieval-split condition (cf. shaded vs. black bars). This effect of retrieval split is fully confined to the stimulus split condition where two subtypes of one category were actually presented. This suggests that, at least in this task context, the retrieval–cue effect is contingent on actually encoded stimulus information in episodic memory. The findings summarized in Fig. 5.2 refer exclusively to the stimulus-split condition. Third, a three-way interaction indicates that the retrieval-split × stimulus-split interaction is stronger for small than for large categories. It is also noteworthy that the effect is stronger

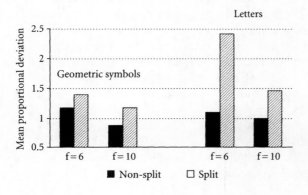

Figure 5.2 Mean proportional deviation scores for frequency judgements as a function of category split and actual presentation frequency (f = 6 vs. 10).

for the type of stimulus materials that led to generally less reliable judgements (i.e. for letters, as compared with geometric figures).

Altogether, the entire pattern of frequency judgements can be explained, in all detail, by the retrieval structure and the basic regressiveness of all frequency estimates. Splitting a category has two simultaneous effects that together elevate the subjectively experienced frequency. First, as the split results in smaller subcategories, the principle of regression produces overestimation of small frequencies. Second, increasing the number of retrieval cues warrants that information pertaining to the split categories is actually found in memory.

However, apart from some inevitable regression, frequency estimates reflect the *relative* frequencies of the four different stimulus categories quite well—if only the retrieval cues are comparable for all categories (i.e. in the no-split condition). If the prompting procedure is 'fair', for instance, by prompting frequency estimates of all categories with their own category labels, then the resulting judgements conserve the ordinal relations among frequencies. Regression-based inaccuracies only pertain to absolute judgements so that the relative order of judged frequencies remains accurate. Whenever different retrieval structures are used for different estimates, however, as in the category-split effect, frequency judgements may be severely biased and may violate the actual ordering.

In order to avoid the severe bias, judges would have to correct their estimates for the inflated regression effect resulting from the category split. A metacognitive device would be required that 'knows' about the regressiveness of judgements and that can quantify the differential regression to be expected when judging either *triangles* or *upward triangles + downward triangles*. As I will continue to argue and demonstrate empirically, human judges lack such a metacognitive device. They are not only unable to precisely quantify and correct for differential regression effects; they are almost totally insensitive to the regression effect that governs their quantitative estimates. And they are blind and insensitive to a number of other vicissitudes of judgement algorithms that use misleading retrieval cues.

The robust beauty of frequency memory revisited

The gradually developing two-sided picture of the assets and deficits of frequency memory is slightly different from the common one-sided notion of the robust beauty of quasi-automatic, remarkably accurate frequency judgements. Those studies that are commonly cited to highlight people's enormous capacity for frequency assessment typically pertain to relative frequencies, rather than absolute estimates, and comparable retrieval cues are used for all items to be judged. For example, people's noteworthy ability to judge word frequencies in the lexicon was demonstrated by prompting judgements of words with the word itself. Accuracy was demonstrated in terms of the relative order of judged word frequencies, rather than absolute numbers, thus mitigating the regression effect that provides the potential for serious biases in other experiments using more sensitive methods of analyses.

The relativity of the numerosity heuristic

Let us illustrate this revised view of the relative assets and deficits of frequency memory with reference to the so-called numerosity heuristic. Pelham, Sumarta, and Myaskovsky

(1994) made a strong case for numerosity as a cognitive primitive, a basic-level source of information that is encoded spontaneously and to which people resort universally, especially when cognitive load blocks the utilization of other, less natural quantities. Across several heterogeneous task situations, Pelham *et al.* (1994) pitted the influence of numerosity against the impact of value information. Paradigms included, for instance, decision tasks in which the number of risks was pitted against risk intensity, or size estimation tasks in which the number of pieces of a pie was pitted against the overall pie area. These authors were at pains to demonstrate repeatedly that numerosity overrides value information, which they considered to be less natural, or primitive.

For example, in one experiment, they showed that the same sum of numerical values is judged to be greater when distributed over 8 elements than when distributed over only 4 elements. Moreover, this numerosity effect was enhanced under cognitive load, when the numerical stimuli were presented very briefly. That is, when the arithmetic summation task was presented for only 2000 ms, rendering numerical encoding very hard, judges relied even more on the numerosity cue (i.e. the length of the vector of numbers to be summed). Based on such evidence, the authors attempted to establish numerosity, or number of elements in which some quantity is split, as a cognitive primitive that is more basic and encoded more quickly and effectively than other quantities (e.g. value, intensity, area size). Numerosity was shown to have a strong impact on subjective estimates of other quantities, whereas numerosity was assumed to be encoded prior to and independently of those other quantities.

However, in all experiments reported by Pelham and colleagues, numerosity was only used as an independent variable to be manipulated, whereas other quantities (area, value, intensity) served as the dependent measure that could reflect a judgement bias. Such a design guarantees that numerosity appears as the cause that induces biases in other quantitative dimensions. With this suspicion in mind, Fiedler and Koch (2001) conducted the following experiment that was modelled after the summation task with a different number of elements. The intention, however, was to allow for illusions to be manifested in both value judgements and frequency judgements. Using the cover story of a birthday party, a stimulus series was presented that consisted of 54 fictitious presents. Each item consisted in a brief verbal description of the present and a monetary equivalent, such as 'A book for 30 DEM', 'A computer game for 60 DEM' etc. All 54 items were drawn from four different categories: games (12), mechanical toys (15), clothing (12), and books (15). Items from the two more frequent categories (mechanical toys and books) were always denoted by the category name. However, items from the less frequent categories (games and clothing) were denoted by one of three subcategory names (computer games, sports games, party games; socks, underwear, towels). The three subcategories appeared with frequencies 3, 4, and 5 (in the indicated order). Thus, the less frequent categories were split whereas the more frequent categories were not split. Within both pairs of split and unsplit categories, one category (games and mechanical toys) was comprised of relatively expensive presents with a total value of 540 DEM, whereas the other category (clothing and books) included much cheaper presents with a value of only 180 DEM. Presentation order was randomized. This design allowed us to examine whether the numerosity phenomenon reflects some privileged status of frequency, as a cognitive primitive, or merely a normal effect of category split, or retrieval–cue structure. If frequency information is in fact a primitive, the average judgements of all four categories should correctly reflect the difference of large categories (15 items) and small categories (12 items). Moreover, judgements of item numerosity for each category should be generally more reliable than judgements of (monetary) value. Alternatively, given that the smaller categories are split into finer subcategories, split effects should override the original frequency differences between categories, and small categories should be markedly overestimated. Moreover, the split illusion should not only impact total value ratings (What was the summed value of all games, books etc.?) but frequency ratings as well (What was the total number of all games, books etc.?).

The main results are summarized in Fig. 5.3, again in proportional deviation scores. The pattern of results is consistent with the category-split interpretation. The average *value estimate* per category (after summing across subcategories in split conditions) reveals that split categories, though less numerous, receive dramatically higher value judgements than non-split categories of the same objective value. Moreover, the split effect is more pronounced for categories of objectively low value (180 DEM as opposed to 540 DEM), consistent with the above regression account. This pattern was obtained independently of an explicit attention manipulation, namely, whether judges were explicitly instructed to focus on either value or frequency information. Thus, regardless of whether the alleged priority of frequency information is implicit or reinforced by explicit instructions, value differences were overridden by strong category-split effects.

Looking at the *frequency estimates*, the findings corroborate the contention that quantitative judgements are mainly determined by the structure of retrieval cues. Split categories are generally overestimated, even though stimulus frequencies are higher in non-split categories control categories. As with the value judgements, this category-split effect tends to be stronger for small than for large categories. These data suggest that subjective frequency is equally dependent on the category structure as value judgements. Indeed, the proportional deviation scores exhibit even stronger overestimation resulting from category split for frequency than for value judgements (see right part of Fig. 5.3). Eighteen out of 20 participants show a stronger split effect for frequency than value estimates, suggesting that frequency may be a more labile subjective quantity than overall value.

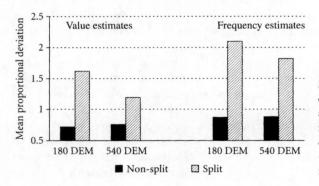

Figure 5.3 Mean proportional deviation scores for value estimates and frequency estimates as a function of category split and category value / numerosity (180 DEM, F = 15 vs. 540 DEM, f = 12).

In a second experiment, these findings were replicated, extended, and improved in several ways. Apart from more adequate counterbalancing of various task conditions, the confounded frequency and split manipulations were disentangled. Rather than constantly using the less frequent category for the split, the split manipulation was now applied separately to infrequent or frequent categories. This altered design allows for the simultaneous observation of numerosity (infrequent vs. frequent categories) and category split effects, as well as possible interactions. We also included a cognitive-load manipulation using a distracter task (responding with a keystroke to every value that includes the digit '4'). Category size (4, 8, and 12) was varied within participants such that three categories of presents (games, books, and clothing) appeared with frequencies 4, 8, and 12. One group of participants saw a stimulus series in which a small category (4 instances) was split (into 2+2) whereas a larger category (8 instances) was split (into 4+4) for the other group. The 4-item and the 8-item categories both included presents for 240 DEM altogether. Thus, the comparison of value judgements for these two categories resembles closely the equivalent 4-element and 8-element summation task used by Pelham *et al.* (1994). The largest category (12 instances, summing to 540 DEM) was never split and only served as a high frequency/value anchor. Both value and frequency estimates were again used as dependent measures.

The results, as summarized in Fig. 5.4, replicate and reinforce those already reported. The overall *value estimates* reflect a noticeable although not overwhelming split effect (i.e. higher ratings for the split than the no-split condition). As the summed value is constant (240 DEM) for the 4-element and the 8-element category, there is consequently no differential regression, and no differential split effect, for the rarest and the middle category. Thus, the basic numerosity effect is not replicated when numerosity and split are varied independently and a third, larger, category is included. That is, the same value sum is not judged higher when it is distributed over 8 than over 4 presents; if anything, pooling across all conditions, the proportional deviation scores for value estimates tend to be somewhat higher for the less numerous, 4-item category. Finally, the distracter task produces a general underestimation bias, suggesting that cognitive load interferes with the complete encoding of all presents.

The findings from the *frequency judgements* demonstrate once more that the basic regression effect as well as the category-split effect are even stronger than for the value

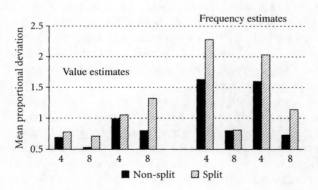

Figure 5.4 Mean proportional deviation scores for value estimates and frequency estimates as a function of category split, category frequency, and cognitive load.

judgements. Very rare categories and split categories are grossly overestimated, quite independent of cognitive load. As the proportional deviation scores for 4-item categories are almost twice as high as for 8-item categories, the regression effect fully overrides the actual numerosity of stimulus presentations.

Altogether, this review of category-split effect does not support the unqualified contention that frequency assessment is just a robust cognitive function that cannot be impaired and that is intrinsically superior to other aspects of subjective quantity. The often-noted accuracy and frugality of frequency assessment may be obtained under ideal prompting conditions, when comparable retrieval cues are used for all items and no complicating influences lead the judgement algorithm astray. However, when incomparable retrieval cues are used—as in category-split experiments—frequency judgements are at least as labile to misleading influences as other aspects of subjective quantity. Although judges ought to understand that doubling the number of cues for a category will inflate their judgements, they fail to apply such a metacognitive correction, just as they fail to correct for the universal regressiveness of judgements under uncertainty.

Spontaneous category split and illusory correlations

All experiments reviewed thus far have used explicit experimenter-provided category labels as retrieval cues. In reality, such artificial, explicit cues are rarely given but judgements have to rely on implicit, self-generated retrieval cues. Even when the experimenter does not provide explicit subcategory prompts, judges will spontaneously extract a category structure that can then be used as retrieval cues. This was shown in many recall experiments with categorized lists (Cohen 1966; Fiedler 1986). When the stimuli in a list fall into apparent subcategories (e.g. pieces of furniture, animals, hobbies), then those subcategories will be used effectively as self-generated retrieval cues, and participants will recall more items from the categorized subset than from the unstructured subset (cf. Fiedler & Stroehm 1986).

We (Fiedler & Armbruster 1994; Experiment 3) used the same rationale to modify illusory correlations via self-generated subcategory cues. Following Hamilton and Gifford's (1976) paradigm, participants were presented with positive and negative behaviour descriptions about two groups, a majority A and a minority group B. Altogether, the list included 18 positive and 8 negative items for Group A plus 9 positive and 4 negative behaviours of Group B. As the positive to negative ratio is the same for both groups (18:8 = 9:4), the correlation between groups and valence is zero and both groups should be judged equivalently. However, a well-established finding says that the minority will be judged less favourably (Fiedler 1991; Hamilton & Gifford 1976; Mullen & Johnson 1990). This illusory correlation between groups and desirability is evident in different dependent measures: frequency judgements, cued-recall assignments of behaviours to groups, and group impression ratings.

Although different factors may contribute to this disadvantage of minorities (Haslam *et al.* 1996), a minimal explanation for the bias against the minority can be based on differential regression alone. Although the same prevalence of positive behaviours holds for Group A and B, there are more observations about the majority; that is, participants have twice as many trials to learn the prevailing positivity in A than in B. As a consequence, B judgements are less reliable and therefore more regressive. Typically, the

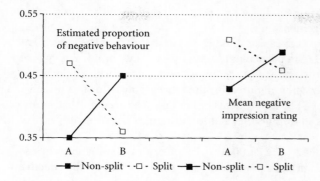

Figure 5.5 Mean frequency estimates and impression ratings of two groups, A and B as a function of two conditions in which 8 positive social behaviours are either split into two subsets of 4 positive behaviours vs. unsplit.

actual positivity rate $18/(18+8) = 9/(9+4) = 69\%$ regresses to something like 60% positive behaviours estimated for A but regresses more radically to hardly over 50% estimated for B. Thus, learning about the smaller group, with the smaller number of learning trials, is severely impaired.

Granting that differential regression produces this basic illusion, category split might be used to eliminate or reverse this effect. If the 8 positive behaviours of the minority group B are split into two subcategories (4+4 positive items), regression could be exploited to augment the impression of the minority. The resulting small subcategories should be overestimated, and the overall impression of B should benefit from this process. For an experimental demonstration, we used the following split manipulation. While all stimuli described social behaviours, the 8-item category of Group B was split into two subtypes of positive social behaviours (4 altruism and 4 sociability). Based on our own experience with categorized lists, we expected that participants would recognize the subcategory structure and spontaneously use the altruism and sociability split for additional retrieval cues. This should augment positive information retrieved about the minority. Figure 5.5 shows that this prediction was actually borne out. In the split condition, Group B received relatively more favourable judgements than in the no-split control conditions where Group A appeared more favourable. This finding demonstrates how category-split effects triggered by self-extracted retrieval cues relate to another prominent paradigm in social psychology.

Logical connectives in retrieval structures: conjunctive prompts

The notion of a retrieval structure cannot be reduced to the number of retrieval cues. In this section, I will briefly consider structural interactions of retrieval cues. For example, the frequency estimation algorithm may involve relations between retrieval cues, such as logical conjunction (AND) or disjunction (OR). An experimental task may either ask to estimate of the likelihood of people who are over 55 years AND who had a heart attack or, alternatively, people who are over 55 OR had a heart attack. The conjunctive prompt (AND) requires judges to scan memory for people who meet both features at the same time, whereas the disjunctive prompt (OR) requires them to focus on cases that fulfil one feature or the other or both. Logically, then, judgements elicited by OR-prompts should be systematically higher than judgements elicited by restrictive

AND-prompts. However, we found that conjunctive and disjunctive judgement prompts led to very similar judgements.

As in the category-split effect, it appears as if judges fail to understand the judgement algorithm they are using. They fail in particular to recognize the mnemonic and logical constraints of the connective retrieval structure. Severe biases or shortcomings occur because the metacognitive supervision required for the retrieval algortithm is missing. Just as participants in a category-split experiment do not take into account that multiple retrieval cues will inflate recall output, they fail to distinguish between conjunctive and disjunctive cues.

The following experiment (Fiedler *et al.* 2001a) demonstrates this inability to use logical constraints in a retrieval structure in the context of a task in which the effective encoding of frequency information is controlled. Participants were confronted with *Simpson's paradox*. The task was to assess, across an extended series of observations, how often direct and indirect mating strategies led to success. Each trial involved a male and a female person in a cartoon. The girl entered the computer screen from the right while the boy came from the left. The boy tried to date the girl, using either a direct strategy ('You are the cutest girl I have ever met') or an indirect strategy ('Can you show me the way to the railway station?'). The girl would either respond with approval, saying something nice and going with the boy, or with rejection, saying something nasty and going away. One strategy (the direct one) succeeded on 12 trials and failed on 4 trials. The other strategy (indirect) only succeeded 4 times and failed 12 times (75% vs. 25% success). However, there were two types of girls from different tribes, Pongals and Trisons. Girls from one tribe (say, the Pongals) were generally accepting and girls from the other tribe were generally rejecting. Upon closer inspection (see Fig. 5.6), it turns out that the advantage of the direct strategy is mainly due to the fact that most direct utterances were directed at Pongals, whereas most indirect strategies were directed at Trisons. When the influence of direct versus indirect strategies is considered separately for Pongals and Trisons, the different success rates (12/16 = 75% for strategies vs. 4/16 = 25% for indirect strategies) shrink considerably (to 100% vs. 80% for Pongals and 20% vs. 0% for Trisons). Thus, in Simpson's paradox, a marked difference is present (in the total set) and absent (within both subsets) at the same time.

After stimulus presentation, participants were asked to estimate the frequencies of success and failure with direct and indirect strategies when approaching Pongals and Trisons, using conjunctive prompts: Direct AND Pongals, Indirect AND Pongals, Direct AND Trisons, Indirect AND Trisons. As Fig. 5.7 reveals, these estimates were slightly

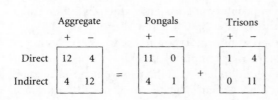

Figure 5.6 Stimulus distribution underlying Simpson's paradox. On aggregate, the relative number of positive (+) as compared with negative (−) reactions is much higher for direct than for indirect mating strategies (left part). However, within subsets of women (Pongals and Trisons), the relative advantage of direct over indirect strategies largely disappears.

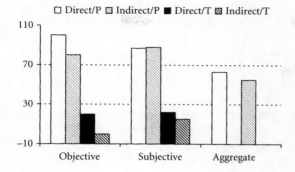

Figure 5.7 Percentage of positive (uniform) and negative (shaded) reactions elicited by Direct and Indirect mating strategies. The left part refers to objective stimulus frequencies, the middle part refers to subjective estimates, and the right part refers to overall judgements, pooling over Pongals and Trisons.

regressive, but generally sensitive to the actual frequencies. That is, judges had effectively encoded the stimulus frequencies at the highest level of resolution (the full $2 \times 2 \times 2$ contingency). However, when judges were asked to estimate the total success and failure rates for direct and indirect strategies, aggregating over Pongals and Trisons, the rather strong difference (12,4 vs. 4,12) was completely lost (see right part of Fig. 5.7). They no longer noticed any overall difference, obviously because they used their own preceding subset judgements as self-generated retrieval cues. Having just remembered, correctly, that the success rates for direct and indirect strategies was hardly different for either Pongals and Trisons, they inferred that strategies did not differ in general.

They were blind to the logical constraints of the retrieval prompts, failing to understand that memories solicited by conjunctive retrieval cues cannot be simply averaged to yield judgements at the aggregate level—another demonstration of the inability to compare algorithms that use structurally different cues.

Conditional retrieval structure and the baserate fallacy

An even more radical version of a frequency illusion comes from a recent investigation on the so-called base rate fallacy (Fiedler *et al.* 2000). To illustrate this point, consider the task to estimate the probability that a woman has breast cancer, given she has a positive mammogram (originally used by Eddy 1982). In one condition, judges are told that in the total population (of women over 40) the breast cancer base rate is 1%. Among those women who have breast cancer, the hit rate of a positive mammogram is 80%; among those who don't have breast cancer, the mammogram false alarm rate is 9.6%. Judges normally give rather inflated estimates of p(breast cancer/positive mammogram). Due to the low baserate, the normatively correct estimate would be only 7.8%.

In another condition, the identical problem is presented like this: Overall, there are 1000 women of which 10 have breast cancer and 990 have not (i.e. 1% base rate). Of those 10 women who have breast cancer, 8 have a positive mammogram (corresponding to 80% hit rate); of those 990 who do not have breast cancer, 95 have a positive mammogram (9.6% false alarm). In this condition, judgements of p(breast cancer/positive mammogram) are much more accurate. Judges realize correctly that among women with a positive mammogram the number of cases with breast cancer (8) is much lower than those without breast cancer (95).

According to Gigerenzer and Hoffrage (1995), the crucial difference between these two conditions lies in the use of *probability versus frequency format*. Whereas probability formats appeared rather late in evolution, frequency formats are well suited to the cognitive devices that human nature has evolved. However, although I tend to agree with this contention, our own evidence supports an explanation in terms of misleading retrieval or judgement prompts. In the frequency-format condition, judges are prompted to think of joint frequencies: women with a positive mammogram AND breast cancer, women with a positive mammogram AND no breast cancer, women with a negative mammogram AND breast cancer etc. All these prompts elicit memorized numbers (8, 2, 95, 895) that are conditional on the same total set (1000 women). Given such a common reference set, or scale, estimates are quite sensitive to the fact that $95 > 8$ and (apart from some regression) ratings should be quite accurate. However, as we have shown (Fiedler *et al.* 2000, Experiment 1), when there are no conflicting judgement prompts, a probability format leads to the same degree of accuracy. Thus, when judges learn that, relative to the totality of 100%, 0.8% have a positive mammogram AND breast cancer, 0.2% have a negative mammogram AND breast cancer, 9.5% have a positive mammogram AND no breast cancer etc., then they also find out that only a minority of women with a positive mammogram have breast cancer.

In contrast, in the other condition leading to severe fallacies, judges are prompted to think of the probability of positive mammogram conditional on breast cancer (small reference set of 10 women), which is 80%, and then to think of the probability of a positive mammogram conditional on no breast cancer (large reference set of 990 women), which is 9.6%. Because the two elements of the prompt are conditional on drastically different set sizes, the resulting quantities are extremely misleading. To be integrated into one judgement, the 80% quantity would have to be corrected downward (due to the small set size) and the 9.6% quantity would have to be corrected upward. However, judges lack the metacognitive device for this algorithm. In accordance with this interpretation, when the same incomparable conditionals are used in a frequency format (8 out of 10 women with breast cancer have a positive mammogram; 95 out of 990 women without breast cancer have a positive mammogram), judges also interpret 8 as high (relative to 10) and 95 as very low (relative to 990) and commit the same strong fallacy as in the original probability condition.

Now imagine that you have to estimate the likelihood of breast cancer given a positive mammogram from world knowledge stored in your memory, rather than given summary statistics. What kind of retrieval cues and what judgement algorithm would you use spontaneously? As the task is to judge breast cancer conditional on mammography, one ought to scan our memory for women with a positive mammogram. However, hardly anybody would do this presumably, simply because the information we encode about other people normally does not include artificial indicators like mammogram. Instead, most people, even physicians, would scan their memory for women with and without breast cancer. Moreover, within this active set of retrieved cases, almost all known positive cases with breast cancer will be included but only a small proportion of negative cases without breast cancer. Thus, positive cases will be strongly over-represented in the effective set triggered by the retrieval cues, as illustrated in Fig. 5.8. Although the breast cancer baserate is only 1% in the universe (left part), the effective

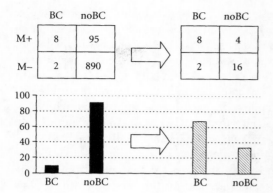

Figure 5.8 The impact of oversampling cases from a critical criterion category (women with breast cancer) on the sample estimate of the conditional probability p(breast cancer/positive mammogram) to be judged.

set will contain a much higher rate of breast cancer (right part). It is for this reason that p(breast cancer/positive mammogram) will be grossly overestimated. Within the effective set prompted by the retrieval cues, breast cancer *is* indeed very likely.

Other experiments (Fiedler *et al.* 2000, Experiments 2 and 3) confirm this interpretation. The self-determined information search process was made transparent using an external memory device. Index-card files (represented on the computer) were made available in which judges could search relevant information for the breast cancer task and various similar probability judgement tasks. In fact, two different index-card files were used for each problem. One file was organized by the diagnostic predictor, that is, there were different slots for women with positive and negative mammograms. Participants could draw index cards from one of the slots and see whether the chosen women had breast cancer or not. The other file was organized by the criterion, offering different slots for women with and without breast cancer, and feedback would then reveal whether the women had a positive or negative mammogram. Depending on the experimental condition, participants could search as many observations as they wanted from either index-card file. When they felt certain enough, they could stop information search and estimate p(breast cancer/positive mammogram).

To avoid overly long search processes, the base rates of the critical events were not as extreme as in previous studies (e.g. 1% breast cancer in Fig. 5.8). In this task, there were 125 cards in the file, five cases with breast cancer (4 with a positive and 1 with a negative mammogram) and 120 without breast cancer (20 with a positive and 100 with a negative mammogram). Thus, the breast cancer base rate was 4% and the correct value of p(breast cancer/positive mammogram) was 4/24 = 16.7%. The major independent variable was whether the search algorithm was conditionalized on the predictor or on the criterion. Participants were either randomly assigned a file type, or they could themselves choose the one file that they felt more appropriate for the judgement task. Task instruction clearly called for judgements of breast cancer conditionalized on mammography, not the other way around.

The results illustrate nicely how radically the conditional search direction (in an external memory store) influences the judgement outcome, and that strong biases originate in the judge's metacognitive blindness for the logical constraints of different search algorithms, or retrieval cues. On one hand, when searching by the predictor (i.e. drawing samples of women with positive and negative mammogram), judgements were rather accurate (mean estimate 21.8%), even when samples were small and the size of the two subsamples was unequal. Considering but a few cases of women with a positive mammogram was sufficient to recognize that breast cancer was rather rare within this group (see left side in Fig. 5.8). On the other hand, when searching information in the criterion file, the resulting judgements were highly inflated, in spite of the fact that the inductive search procedure guaranteed a perfectly natural frequency format. The mean estimate increased to 64.0%, dramatically higher than the original value of 16.7% in the index card file.

It is important to note where these inflated judgements came from. When information was searched by the criterion, almost as many cases were sampled from the rare breast cancer category as from the no-breast cancer category, thereby obscuring the low base rate of the critical event (see Fig. 5.8). In fact, the proportion of breast cancer given a positive mammogram *in the sample* prompted by the criterion search algorithm was as high as 85.12%. Subjective estimates followed this proportion in the sample rather accurately, for the breast cancer task and for a number of other tasks. Judges simply took the output of their search algorithm for granted, without ever understanding that oversampling from the breast cancer category had obscured the base rate and the conditional to be judged.

This demonstrates, once more, how insensitive human frequency judgements are, at the metacognitive level, to the constraints of their retrieval structure. When the outcome of information search can be taken for granted, as in the predictor search condition, they happen to produce accurate estimates. However, when the outcome of the algorithm needs to be supervised and corrected, the resulting judgements are severely biased. Judges do not understand that search processes that are conditional on the predictor versus the criterion are not equivalent (cf. Gavanski & Hui 1992; Sherman *et al.* 1992). This contention is corroborated by the challenging finding that when judges could themselves choose between the files organized by the predictor and the criterion, almost 40% freely chose the criterion file, which is the origin of the misleading retrieval structure.

Biases due to correlated retrieval cues

Finally, we will discuss findings from one more paradigm (Fiedler *et al.* 2001b) that demonstrate the amazing contrast between efficient frequency coding on the one hand and judgement biases due to unsupervised retrieval structures on the other hand. In a computer-simulated classroom setting, participants were asked to assess the performance of 16 students (8 boys and 8 girls) in up to eight different lessons or disciplines over an extended series of trials. Each trial involved (a) selecting a knowledge question, (b) observing which students raised their hands, (c) selecting one of the students, and (d) assessing whether the student's answer was correct or wrong. Differentiated parameter matrices determined for each discipline the probability with which a student raised his or her hand (motivation) and the probability of giving a correct answer (ability parameter). After each lesson, the teacher participants estimated all students' motivation (% of raising hands) and ability (% of correct answers).

In spite of the complexity of the multiple assessment task (assessing two parameters of 16 students in 8 disciplines; $2 \times 16 \times 8 = 256$), performance differences were correctly (though regressively) reflected in the judgements. After only 20 min of observation in each lesson, teachers were able to distinguish smart and poor students. These differences were not over-generalized but restricted to those lessons in which smart and poor students actually manifested their different ability. And judges were able to estimate motivation (participation rate) independently of ability (correct answers).

All these assets were restricted to situations in which comparable retrieval cues warranted a constant retrieval algorithm. If only the name of a student was prompted after a lesson, one at a time, and no other retrieval structure was there, the encoded information in memory was sufficient to echo valid judgements. However, when the retrieval algorithm was not constant and metacognitive corrections for different retrieval structures were called for, the resulting judgements were biased. In particular, teachers did not recognize, or correct for the fact, that of two equally smart students (i.e., with the same 80% ability parameter) the one who had answered more questions received more favourable ratings. Conversely, of two equally poor students, the one who gave more answers received more negative judgements. Thus, teachers did not correct for the differentially effective retrieval structures they had developed for well-attended and ill-attended students. Moreover, when motivation and ability were related, ability ratings could be misled by motivation cues, and teachers were metacognitively blind for the confounded retrieval structure, unable to correct for the bias. Thus, when raising hands (motivation) was regularly followed by correct answers, high-motivation students' ability was overestimated, whereas high-motivation students' ability was underestimated when raising hands was mostly followed by incorrect answers.

Summary and conclusions

The research reviewed in this chapter, referring to category-split experiments as well as other paradigms, suggest an answer to the puzzle of why frequency assessment is sometimes praised as a most efficient, almost automatic module of human adaptive behaviour and on other occasions identified as the origin of serious biases and shortcomings. The purpose here was to demonstrate that both tenets are at the same time correct and incorrect. On the one hand, subjective estimates will often conserve the order of actual event frequencies, but this does not mean that memory for frequency, as a cognitive primitive, warrants generally accurate frequency judgements under all conditions. Many frequentistic judgements have to be reconstructed in an active process that relies on specific retrieval algorithms. If this algorithm creates equal conditions for all estimates, the output (echo) of the retrieval algorithm will typically result in quite reliable judgements on an ordinal scale. However, if the algorithm uses variable retrieval structures, thus producing incomparable cues for different items, then the resulting judgements will be biased and judges will lack the metacognitive devices for understanding and correcting the impact of unequal retrieval structures.

References

Brown, N. R. (1995). Estimation strategies and the judgement of event frequencies. *Journal of Experimental Psychology: Learning, Memory, and Cognition, 21*:1539–1553.

Cohen, B. H. (1966). Some-or-none characteristics of coding behavior. *Journal of Verbal Learning and Verbal Behavior*, 5:182–187.

Combs, B. & Slovic, P. (1979). Newspaper coverage of causes of death. *Public Opinion Quarterly*, 56:837–843.

Eddy, D. M. (1982). Probabilistic reasoning in clinical medicine: Problems and opportunities. In D. Kahneman, P. Slovic & A. Tversky (eds) *Judgement under uncertainty: Heuristics and biases*. Cambridge: Cambridge University Press.

Fiedler, K. (1986). Person memory and person judgements based on categorically organized information. *Acta Psychologica*, 61:117–135.

Fiedler, K. (1991). The tricky nature of skewed frequency tables: An information loss account of distinctiveness-based illusory correlations. *Journal of Personality and Social Psychology*, 60:24–36.

Fiedler, K. & Armbruster, T. (1994). Two halfs may be more than one whole: Category-split effects on frequency illusions. *Journal of Personality and Social Psychology*, 66:633–645.

Fiedler, K., Brinkmann, B., Betsch, T. & Wild, B. (2000). A sampling approach to biases in conditional probability judgements: Beyond baserate neglect and statistical format. *Journal of Experimental Psychology: General*, 129:399–418.

Fiedler, K. & Koch, S. (2001). *The construction of subjective quantity: Type, value and number*. Unpublished research, University of Heidelberg.

Fiedler, K. & Stroehm, W. (1986). What kind of mood influences what kind of memory: The role of arousal and information structure. *Memory and Cognition*, 14:181–188.

Fiedler, K., Walther, E., Freytag, P. & Stryczek, E. (2002). Playing mating games in foreign cultures: A conceptual framework and an experimental paradigm for trivariate statistical inference. *Journal of Experimental Social Psychology*, 38:14–30.

Fiedler, K., Walther, E., Freytag, P. & Plessner, H. (in press). Judgement biases and pragmatic confusion in a simulated classroom—A cognitive-environmental approach. *Organizational Behavior and Human Decision Processes*.

Gavanski, I. & Hui, C. (1992). Natural sample spaces and uncertain belief. *Journal of Personality and Social Psychology*, 63:766–780.

Gigerenzer, G. (1991). From tools to theories: A heuristic of discovery in cognitive psychology. *Psychological Review*, 98:254–267.

Gigerenzer, G. & Hoffrage, U. (1995). How to improve Bayesian reasoning without instruction: Frequency formats. *Psychological Review*, 102:684–704.

Greene, R. L. (1984). Incidental learning of event frequencies. *Memory and Cognition*, 12:90–95.

Hamilton, D. L. & Gifford, R. K. (1976). Illusory correlation in interpersonal perception: a cognitive basis of stereotypic judgements. *Journal of Experimental Social Psychology*, 12:392–407.

Hamilton, D. L. & Sherman, J. W. (1994) Stereotypes. In R.S. Wyer, and T.K. Srull (eds) *Handbook of Social Cognition*, 2nd edn, vol. 1 (pp 1–68). Hillsdale, NJ: Erlbaum.

Hasher, L. & Zacks, R. T. (1984). Automatic processing of fundamental information: The case of frequency of occurrence. *American Psychologist*, 39:1372–1388.

Haslam, S. A., McGarty, C. & Brown, P. M. (1996). The search for differentiated meaning is a precursor to illusory correlation. *Personality and Social Psychology Bulletin*, 22:611–619.

Jenkins, H. M. & Ward, W. C. (1965). Judgement of contingency between responses and outcomes. *Psychological Monographs*, 79:(Whole No. 594).

Juslin, P. & Olsson, H. (1997). Thurstonian and Brunswikian origins of uncertainty in judgement: A sampling model of confidence in sensory discrimination. *Psychological Review*, 104:344–366.

Kahneman, D., Slovic, P. & Tversky, A. (eds) (1982). *Judgement under uncertainty: Heuristics and biases.* Cambridge: Cambridge University Press.

Klar, Y. & Giladi, E. E. (1997). No one in my group can be below the group's average: A robust positivity bias in favor of anonymous peers. *Journal of Personality and Social Psychology, 73*:885–901.

Mullen, B. & Johnson, C. (1990). Distinctiveness-based illusory correlations and stereotyping: A meta-analytic integration. *British Journal of Social Psychology, 29*:11–28.

Pelham, B. W., Sumarta, T. T. & Myaskovsky, L. (1994). The easy path from many to much: The numerosity heuristic. *Cognitive Psychology, 26*:103–133.

Price, P. C. (2001). A group size effect on personal risk judgements: Implications for unrealistic optimism. *Memory and Cognition, 29*:578–586.

Reber, A. S. (1989). Implicit learning and tacit knowledge. *Journal of Experimental Psychology: General, 118*:219–235.

Ross, M. & Sicoly, F. (1979). Egocentric biases in availability and attribution. *Journal of Personality and Social Psychology, 37*:322–336.

Schwarz, N., Bless, H., Strack, F., Klumpp, G., Rittenauer-Schatka, H. & Simons, A. (1991). Ease of retrieval as information: Another look at the availability heuristic. *Journal of Personality and Social Psychology, 61*:195–202.

Sedlmeier, P. (1999). *Improving statistical reasoning: Theoretical models and practical implications.* Mahwah, NJ: Lawrence Erlbaum Associates.

Sherman, S. J., McMullen, M. N. & Gavanski, I. (1992). Natural sample spaces and the inversion of conditional judgements. *Journal of Experimental Social Psychology, 28*:401–421.

Snyder, M. (1984). When belief creates reality. In L. Berkowitz (ed) *Advances in experimental social psychology*, vol. 18 (pp. 247–305). New York: Academic Press.

Tversky, A. & Kahneman, D. (1973). Availability: A heuristic for judging frequency and probability. *Cognitive Psychology, 4*:207–232.

Tversky, A. & Kahneman, D. (1983). Extensional versus intuitive reasoning: The conjunction fallacy in probability judgement. *Psychological Review, 90*:293–315.

Weinstein, N. D. (1980). Unrealistic optimism about future life events. *Journal of Personality and Social Psychology, 39*:806–820.

Zuckerman, M., Knee, C. R., Hodgins, H. S. & Miyake, K. (1995). Hypothesis confirmation: The joint effect of positive test strategy and acquiescence response set. *Journal of Personality and Social Psychology, 68*:52–60.

CHAPTER 6

EXPERIENTIAL AND CONTEXTUAL HEURISTICS IN FREQUENCY JUDGEMENT: EASE OF RECALL AND RESPONSE SCALES

NORBERT SCHWARZ AND MICHAELA WÄNKE

Abstract

Frequency estimates can be based on a number of different heuristic strategies, of which we address two. First, we revisit Tversky and Kahneman's (1973) availability heuristic. The reviewed findings indicate that individuals only rely on the subjective experience of ease of recall when its informational value is not called into question and their processing motivation is low. When the experience is not deemed informative, or processing motivation is high, individuals draw on recalled content. These diverging strategies result in markedly different frequency estimates and related judgements. Second, we address what respondents learn from frequency scales provided by the researcher. The reviewed findings indicate that respondents extract information about the assumed distribution and use this information in arriving at frequency estimates and related judgements. We highlight the consistency of these findings with general models of human judgement and conclude with a discussion of strategy selection.

As the contributions to this volume illustrate, individuals can arrive at judgements of frequency by relying on a variety of different strategies (see Brown, this volume, for a useful taxonomy), as holds true for any other complex task. Different strategies of frequency judgement have been addressed in separate, largely unconnected, research programmes, with diverging conclusions. On the one hand, laboratory studies based on relatively simple and distinct stimuli suggest that the frequency of stimulus occurrence is well encoded, resulting in relatively accurate frequency judgements (for a discussion see Zacks & Hasher, this volume). On the other hand, numerous naturalistic studies suggest otherwise. Most notably, when respondents are asked to report on the frequency of their own mundane behaviours, their reports are not only inaccurate, but also strongly influenced by contextual variables unrelated to the actual frequency of the respective behaviour (for reviews see Bradburn, Rips & Shevell 1987, and the contributions in Schwarz & Sudman 1994).

The latter observation reflects that respondents are unlikely to have detailed episodic representations of their own behaviour available in memory, unless the behaviour is rare and important. Instead, the individual instances of frequent behaviours blend into generic, knowledge-like representations that lack the time and space markers needed for episodic recall. Accordingly, respondents cannot follow an enumeration strategy based on recall and count procedures when asked to report on frequent mundane behaviours, but have to resort to estimation strategies to arrive at a plausible answer (for reviews see Bradburn, Rips & Shevell 1987; Schwarz 1990a; Schwarz & Sudman 1994; Strube 1987; Sudman, Bradburn & Schwarz 1996).

In this chapter, we address two of these estimation strategies (for a discussion of other strategies, see Brown, this volume). We first revisit Tversky and Kahneman's (1973) availability heuristic and highlight the role of experiential information in frequency estimation. Subsequently we address how the frequency scales provided by the researcher shape respondents' frequency estimates (e.g. Schwarz, Hippler, Deutsch & Strack 1985). These distinct, and unrelated, heuristic strategies draw on sources of information that are often neglected in cognitive psychology. In the case of the availability heuristic, respondents rely on the phenomenal experience of ease or difficulty of recall,[1] provided that the informational value of this experience is not called into question. In the case of frequency scales, respondents bring the tacit assumptions that underlie the conduct of conversation to the research situation and make systematic use of information that is, inadvertently, provided by the researcher.

In combination, research into these heuristics draws attention to two issues that figure prominently in social cognition research but have received limited attention in the domain of frequency judgement. First, human thought processes are accompanied by phenomenal experiences—like ease of retrieval, perceptual and conceptual fluency, or affective responses—that can serve as a source of information in their own right. Hence, understanding the interplay of declarative and experiential information is crucial for a full understanding of human judgement (for reviews see Schwarz & Clore 1996, and the contributions in Bless & Forgas 2000). Second, humans do much of their thinking in a social context and accounts of human judgement need to consider the interplay of cognitive and communicative processes (for reviews see Hilton 1995; Schwarz 1996). In addition, the heuristic strategies used in frequency estimation have far reaching implications for human judgement beyond the frequency domain, as will become apparent below.

Experiential information: the availability heuristic revisited

According to Tversky and Kahneman's availability heuristic, individuals estimate the frequency of an event, or the likelihood of its occurrence, 'by the ease with which instances or associations come to mind' (Tversky & Kahneman 1973, p. 208). Unfortunately, the

---

[1] Although the subjective experience of ease may also influence other judgements regarding the retrieved information as to its credibility, validity or convincingness (for a review see Wänke & Bless 2000) we restrict the present discussion to the impact of ease on frequency judgements.

classic studies are highly ambiguous with regard to the underlying process. For example, Tversky and Kahneman (1973, Experiment 3) observed that participants overestimated the number of words that begin with the letter r, but underestimated the number of words that have r as the third letter. These findings presumably reflect that words that begin with a certain letter can be brought to mind more easily than words that have a certain letter in the third position. Note, however, that this differential ease of recall may influence participants' frequency estimates in two different ways. On the one hand, participants may use the subjective experience of ease or difficulty of recall as a basis of judgement, as suggested by Tversky and Kahneman's (1973) description of the availability heuristic. If so, they would estimate a higher frequency if the recall task is experienced as easy rather than difficult, relying on a 'non-numerical' estimation strategy in terms of Brown's (this volume) taxonomy. On the other hand, they may recall as many words as they can within the time allotted to them and may base their judgement on the recalled sample, following an 'enumeration' strategy in terms of Brown's taxonomy. Given that the easier to recall words would be over-represented in the recalled sample, this strategy would again result in an estimate of higher frequency. Yet, in the latter case the estimate would be based on recalled *content* rather than on the *subjective experience* of ease of recall.

Similar ambiguities apply to other experimental procedures: throughout, manipulations intended to increase the subjective experience of *ease* of recall are also likely to affect the *amount* of recall (see Schwarz 1998 for a review). In most real world situations these two factors are naturally confounded, making it difficult to determine whether the obtained estimates of frequency, likelihood, or typicality are based on participants' phenomenal experiences or on a biased sample of recalled information. Although the latter possibility would render the availability heuristic rather trivial—after all, 'one's judgements are always based on *what* comes to mind' (Taylor 1982, p. 199, italics added)—it has become the standard textbook interpretation (e.g. Medin & Ross 1997, p. 522).

Ease of recall as information

Because experienced ease of recall and amount of recall are naturally confounded, targeted experimental manipulations are needed to isolate the role of subjective recall experiences. Three different strategies proved useful in this regard.

First, if subjective experiences of ease or difficulty of recall serve as a source of information in their own right, their impact should depend on the perceived diagnosticity of the experience, as has been observed for other types of experiential information, such as moods (Schwarz & Clore 1983) or arousal (Zillman 1978; see Schwarz & Clore 1996 for a comprehensive review). We can therefore test the role of subjective experiences by introducing (mis)attribution manipulations, which should give rise to augmentation and discounting effects, as described below.

Second, we may ask participants to recall either a few or many exemplars from a class of events or behaviours. Recalling few exemplars is experienced as easy, whereas recalling many exemplars is experienced as difficult. If participants rely on their subjective recall experience in forming a judgement, they should therefore arrive at a higher frequency estimate after recalling a few rather than many exemplars. In contrast, reliance on the amount of recall would result in higher frequency estimates after recalling many rather than a few exemplars. Hence, the experience-based and content-based strategies discussed above lead to opposite predictions under these conditions.

Third, we may keep the recall task constant and manipulate participants' subjective experience of ease or difficulty through unrelated manipulations, such as bodily feedback that induces experiences of ease or difficulty (Stepper & Strack 1993).

Converging findings from all three strategies consistently support the hypothesis that subjective recall experiences serve as a source of information in their own right, as a selective review may illustrate (see Schwarz 1998; Schwarz & Vaughn in press, for more comprehensive reviews).

Diagnosticity of the recall experience

To manipulate the perceived diagnosticity of ease of recall, Wänke, Schwarz, and Bless (1995) conducted a modified replication of Tversky and Kahneman's (1973, Experiment 3) letter experiment, described above. In the control condition, participants received a blank sheet of paper and were asked to first write down ten words that have t as the third letter, and subsequently ten words that begin with the letter t. Following this listing task, they rated the extent to which words beginning with a t are more or less frequent than words that have t as the third letter. As in Tversky and Kahneman's (1973) study, participants estimated that words that begin with a t are more frequent than words having a t in the third position, as shown in Table 6.1 ('control condition').

Note that participants had to recall the same number of words in each condition. To isolate the role of the subjective experience, the diagnosticity of the experience was manipulated in two experimental conditions. Specifically, participants had to record ten words that begin with a t on a sheet of paper that was imprinted with pale but visible rows of t's. Some participants were told that this background would make it easy to recall t-words ('facilitating condition'), whereas others were told that this background would interfere with the recall task ('inhibiting condition'). As expected, participants who could attribute the experienced ease of recall to the impact of their work sheet estimated that there are fewer t-words than did participants in the control condition. Conversely, participants who expected their work sheet to interfere with recall, but found recall easy nevertheless, estimated that there are more t-words than did participants in the control condition.

In combination, these discounting and augmentation effects indicate that participants did indeed base their frequency estimates on the *subjective experience* of ease of recall, rather than on the number of words they could bring to mind. Moreover, the misattribution effects obtained here have been replicated in other studies, to be addressed below. Hence, we may conclude that recall experiences are informative in their own

Table 6.1 Frequency of words having t as the first or third letter (alleged contextual influence)

Facilitating	Control	Inhibiting
3.8	5.4	6.1

Note: N is 9 or 10 per condition. Shown are mean ratings along an 8-point scale, with 1 = 'many more third letter words than first letter words', and 8 = 'many more first letter words than third letter words'. Adapted from Wänke, Schwarz, and Bless (1995).

right, much as has been observed for other types of experiential information (for reviews see Schwarz & Clore 1996; Strack 1992).

Differential implications of recall experience and amount of recall

To test the differential impact of recall experience and amount of recall, Schwarz et al. (1991b, Experiment 1) asked participants to report either 6 or 12 situations in which they behaved either assertively or unassertively. Following their reports, participants evaluated their own assertiveness. Although all participants could produce the required number, pretests had shown that recalling 6 examples was experienced as easy, whereas recalling 12 examples was experienced as difficult. Thus, a judgement strategy that is based on recalled amount would lead participants to infer higher assertiveness when the recall task brought 12 rather than 6 assertive examples to mind and vice versa for unassertive examples. In contrast, a strategy based on experienced ease or difficulty of recall would reverse this pattern: finding it difficult to recall 12 examples of assertive behaviours, participants may conclude that they can't be that assertive after all—or else bringing 12 examples to mind wouldn't be that difficult.

Table 6.2 shows that, as expected, participants reported higher assertiveness after recalling 6 examples of assertive, than after recalling 6 examples of unassertive behaviours. However, this difference did not increase as participants had to recall more examples. On the contrary, participants who had to recall assertive behaviours reported lower assertiveness after recalling 12 rather than 6 examples. Similarly, participants who had to recall unassertive behaviours reported higher assertiveness after recalling 12 rather than 6 examples. In fact they reported higher assertiveness after recalling 12 unassertive, rather than 12 assertive behaviours, in contrast to what one would expect on the basis of recalled content. Apparently, the experience that it was difficult to bring 12 examples to mind suggested to them that they can't be that assertive (or unassertive) after all. Thus, the experienced difficulty of recall induced participants to draw inferences opposite to the implications of recalled content.

This conclusion is further supported by a manipulation of the perceived diagnosticity of the recall experience. In one of their experiments, Schwarz et al. (1991b, Experiment 3) had participants listen to new age music played at half-speed while they worked on the recall task. Some participants were told that this music would facilitate the recall of

Table 6.2 Self-reports of assertiveness as a function of valence and number of recalled behaviours

	Type of behaviour	
	Assertive	Unassertive
Number of recalled examples		
Six	6.3	5.2
Twelve	5.2	6.2

Note: N is 9 or 10 per condition. Mean score of three questions is given; possible range is 1 to 10, with higher values reflecting higher assertiveness. Adapted from Schwarz, Bless, Strack, Klumpp, Rittenauer-Schatka, and Simons (1991, Experiment 1).

situations in which they behaved assertively and felt at ease, whereas others were told that it would facilitate the recall of situations in which they behaved unassertively and felt insecure. These manipulations render participants' recall experiences uninformative whenever the experience matches the alleged impact of the music—after all, it may simply be easy or difficult because of the music. In contrast, experiences that are opposite to the alleged impact of the music should be considered highly informative.

When the informational value of participants' experienced ease or difficulty of recall was *not* called into question, the previously obtained results replicated. Not so, however, when their subjective experiences of ease or difficulty of recall matched the alleged side-effects of the music. In this case, the meaning of the subjective experience is ambiguous and participants turned to a more informative source of information: the content they had just recalled. As a result, the previously obtained pattern reversed and participants reported higher assertiveness after recalling 12 rather than 6 examples of assertive behaviour, and lower assertiveness after recalling 12 rather than 6 examples of unassertive behaviour.

In combination with the Wänke *et al.* (1995) study, these results highlight that people do only rely on their phenomenal recall experiences when the informational value of the experience is not called into question. When the informational value of the experience is called into question, people turn to the content of recall, resulting in opposite patterns of judgement, as seen above.

Proprioceptive feedback

If accessibility experiences constitute a distinct source of information, *any* variable that leads people to experience a given recall task as easy or difficult should have the same effect as actual ease or difficulty of recall. A study by Stepper and Strack (1993, Experiment 2) supports this prediction. These authors asked all participants to recall 6 examples of assertive or unassertive behaviour, thus holding actual recall demands constant. To manipulate the subjective recall experiences, they induced participants to contract either their corrugator muscle or their zygomaticus muscle during the recall task. Contraction of the corrugator muscle produces a furrowed brow, an expression commonly associated with a feeling of effort. Contraction of the zygomaticus muscle produces a light smile, an expression commonly associated with a feeling of ease.

Reflecting the impact of this proprioceptive feedback, participants who recalled 6 examples of assertive behaviour while adopting a light smile judged themselves as more assertive than participants who adopted a furrowed brow. Conversely, participants who recalled 6 examples of unassertive behaviour while adopting a light smile judged themselves as less assertive than participants who adopted a furrowed brow. Apparently, the experience of difficulty conveyed by a furrowed brow led participants to conclude that they can't be that assertive (or unassertive) if it is so difficult to bring assertive (or unassertive, respectively) behaviours to mind.

Factors influencing the reliance on recall experiences

In combination, the reviewed studies indicate that individuals' subjective recall experiences are informative in their own right. Next, we address some of the factors that influence individuals' reliance on these experiences, namely their perceived expertise in the content domain, the personal relevance of the task, and the individual's mood at the time of judgement.

Perceived expertise

In the preceding studies, the informational value of recall experiences was manipulated by drawing participants' attention to an external source that allegedly facilitated or inhibited recall. The same logic applies to an individual difference variable, namely one's perceived expertise in the respective content domain. Suppose you are asked to list 12 famous Spanish matadors and find this task difficult. Chances are that you would not conclude that there aren't many famous Spanish matadors—instead, you would blame your own lack of expertise for the experienced difficulty, thus undermining its informational value. As this example illustrates, individuals may only rely on recall experiences when they consider themselves at least moderately knowledgeable in the relevant content domain. Experimental data support this conclusion.

For example, Biller, Bless and Schwarz (1992) had participants recall either 3 (easy) or 9 (difficult) examples of chronic diseases and subsequently asked them to estimate the percentage of Germans who suffer from chronic diseases. As expected, they estimated the prevalence of chronic diseases to be higher after recalling 3 ($M = 38.3\%$) rather than 9 ($M = 25.2\%$) examples, reflecting that they based their judgements on experienced ease of recall. To explore the role of perceived expertise, other participants were first asked to indicate how much they know about chronic diseases before they were asked to provide a prevalence estimate. The knowledge question was expected to draw their attention to their general lack of knowledge in this domain, thus undermining the informational value of their recall experiences. In this case, participants estimated the prevalence of chronic diseases to be lower after recalling 3 ($M = 23.1\%$) rather than 9 ($M = 33.0\%$) examples, reflecting that they based their judgements on the number of examples recalled. These findings suggest that people will only rely on their subjective recall experiences when they consider themselves knowledgeable, but will turn to recalled content when they do not (see also Schwarz & Schuman 1997).

Processing motivation

As a large body of research in social cognition indicates, individuals' judgement strategies depend, *ceteris paribus*, on the motivation they bring to the task. The more self-relevant and involving the task is, the more likely they are to adopt a systematic processing strategy, paying attention to the specific implications of the information that comes to mind. In contrast, heuristic processing strategies are preferred for less relevant and less involving tasks (e.g. the contributions in Chaiken & Trope 1999). If so, the self-relevance of the material addressed in the recall task may determine if individuals rely on a systematic evaluation of the implications of recalled content or draw on the experienced ease of recall as a heuristic strategy in forming a judgement.

To explore this possibility, Rothman and Schwarz (1998) asked male undergraduates to list either 3 (easy) or 8 (difficult) behaviours that they personally engage in, which may either increase or decrease their risk of heart disease. The personal relevance of the task was assessed via a background characteristic, namely whether participants had a family history of heart disease or not. Supposedly, assessing their own risk of heart disease is a more relevant task for males whose family history puts them at higher risk than for males without a family history of heart disease. Hence, participants with a family history of heart disease should be likely to adopt a systematic processing strategy, paying attention to the specific behaviours brought to mind by the recall task. In contrast, participants without a family history may rely on a heuristic strategy, drawing on the subjective experience of recall.

As shown in Table 6.3, the results supported these predictions (for a conceptual replication see Grayson & Schwarz 1999). The top panel shows participants' self-reported vulnerability to heart disease. As expected, men with a family history of heart disease reported higher vulnerability after recalling 8 rather than 3 risk-increasing behaviours, and lower vulnerability after recalling 8 rather than 3 risk-decreasing behaviours. In contrast, men without a family history of heart disease reported lower vulnerability after recalling 8 (difficult) rather than 3 (easy) risk-increasing behaviours, and higher vulnerability after recalling 8 rather than 3 risk-decreasing behaviours. In addition, participants' perceived need for behaviour change paralleled their vulnerability judgements, as shown in the bottom panel of Table 6.3.

Note that the observed impact of personal involvement on individuals' processing strategy contradicts the common assumption that reliance on the availability heuristic is independent of judges' motivation. In several studies, researchers offered participants incentives for arriving at the correct answer, yet such incentives rarely attenuated reliance on the availability heuristic (see Nisbett & Ross 1980; Payne, Bettman & Johnson 1993, for discussions). Unfortunately, these studies could not observe a change in processing strategy, even if it occurred. To see why, suppose that one asked participants to estimate whether words beginning with an r or which had r as a third letter were more frequent (Tversky & Kahneman 1973). Suppose further that some participants are offered an

Table 6.3 Vulnerability to heart disease as a function of type and number of recalled behaviours, and family history

	Type of Behaviour	
	Risk-increasing	Risk-decreasing
Vulnerability judgements		
With family history		
3 examples	4.6	5.8
8 examples	5.4	3.8
Without family history		
3 examples	3.9	3.1
8 examples	3.2	4.3
Need for behaviour change		
With family history		
3 examples	3.6	5.2
8 examples	6.3	4.7
Without family history		
3 examples	3.4	3.0
8 examples	2.8	5.6

Note: N is 8 to 12 per condition. Judgements of vulnerability and the need to change current behaviour were made on 9-point scales, with higher values indicating greater vulnerability and need to change, respectively. Adapted from Rothman and Schwarz (1998).

incentive to arrive at the correct estimate whereas others are not. Without an incentive, individuals may rely on a heuristic strategy, drawing on experienced ease of recall. This would lead them to conclude that there were more words beginning with an r. With a successful incentive, however, they may be motivated to invest more effort. If so, they may recall as many words as they can for both classes of words and may count the number of first letter words and that of third letter words in the recalled sample. Unfortunately, this systematic strategy would lead them to the same conclusion because the first letter words would be over-represented in the recalled sample. As this example illustrates, we can only distinguish between heuristic and systematic strategies when we introduce conditions under which both strategies lead to different outcomes. When this is done, we are likely to observe the expected impact of processing motivation, as the above findings illustrate.

Mood

Another factor influencing the use of heuristics is the individual's mood at the time of judgement. Numerous studies demonstrated that being in a happy mood fosters reliance on heuristic processing strategies, whereas being in a sad mood fosters reliance on systematic processing strategies (for a comprehensive review see Schwarz & Clore 1996; for theoretical accounts see Schwarz 1990b; Bless & Schwarz 1999; Bless 2000). Accordingly one may expect that happy individuals are more likely to draw on their subjective recall experience than sad individuals. Research by Bless and Ruder (1999) supports this assumption. They put participants in a happy or sad mood and asked them to generate either two or six reasons for or against different issues. Finally, participants reported their attitudes towards these issues. Participants in a happy mood replicated previous results of the moderating impact of ease of argument retrieval on attitudinal judgements (Wänke, Bless & Biller 1996). They reported more favourable attitudes after generating two rather than six favourable arguments, and more unfavourable attitudes after generating two rather than six unfavourable arguments. This pattern reflects that individuals relied on ease of retrieval rather than the number of arguments retrieved. In contrast, participants in a sad mood reported more favourable attitudes after retrieving six rather than two favourable arguments, and more unfavourable attitudes after retrieving six rather than two unfavourable arguments. Thus, happy individuals drew on their subjective recall experience, whereas sad individuals drew on the content of recall in arriving at an attitude judgement, consistent with the well documented influence of moods on processing style (Schwarz 1990b; Schwarz, Bless & Bohner 1991a).

Summary

In combination, the reviewed studies highlight that any recall task renders two distinct sources of information available: the recalled content and the ease or difficulty with which this content could be brought to mind. In most situations, these two sources of information are naturally confounded and the experience of ease of recall goes along with a greater amount of recall. This confound rendered many of the classic tests of the availability heuristic nondiagnostic. When this confound is disentangled, the available evidence supports the original formulation of the availability heuristic: Individuals estimate the frequency of an event, the likelihood of its occurrence, and its typicality 'by the ease with which instances or associations come to mind' (Tversky & Kahneman 1973, p. 208). However, they only rely on their subjective recall experience as a source of information when its informational value is not called into question. Finally, variables that are known to influence the likelihood of heuristic versus systematic processing strategies (see the contributions in Chaiken & Trope 1999) moderate individuals' reliance on subjective recall experiences versus recalled content, as theoretically expected.

The systematic nature of these effects, and their compatibility with general models of the interplay of experiential and declarative information in human judgement (see Schwarz & Clore 1996), is incompatible with the critique that the availability heuristic is a 'one-word explanation' (Gigerenzer 1996, p. 594) that lacks specific process assumptions and empirical support. It is similarly difficult to reconcile with the sweeping verdict that 'the availability approach is not a theory with any predictive power' (Betsch & Pohl this volume). Unfortunately, the critics have so far failed to provide *any* alternative account for the systematic, robust and predictable effects of the phenomenal experience of ease or difficulty of recall across a wide variety of judgements (for a review see Schwarz 1998). Instead, the critique focuses on a point that is well taken, but not germane to the availability heuristic per se: What determines the choice of *any* given strategy from the multitude of strategies available? We return to this issue in the final section of this chapter.

Frequency scales as a source of information

In this section, we turn to an estimation strategy that is distinctly different from the use of one's own subjective experiences, namely a strategy that relies on contextual information inadvertently provided by the researcher. In many studies, respondents are asked to report their behaviour by checking the appropriate alternative from a list of response alternatives of the type shown in Table 6.4. In this example, taken from Schwarz, Hippler, Deutsch, and Strack (1985), German respondents were asked how many hours they watch television on a typical day. To provide their answer, they were presented with a frequency scale that offered either high or low frequency response alternatives. While the selected response alternative is assumed to inform the researcher about the respondent's behaviour, it is frequently overlooked that a given set of response alternatives may also constitute a source of information for the respondent. Bringing the tacit assumptions underlying the conduct of conversation in daily life (Grice 1975) to the research situation, respondents assume that every contribution of the researcher is relevant to their task (for a review see Schwarz 1994, 1996). In research situations, these contributions include apparently formal aspects of the research instrument, like numeric values of rating scales (e.g., Schwarz, Knäuper, Hippler, Noelle-Neumann & Clark 1991) and frequency scales, as well as information regarding the purpose or target sample of the survey (Wänke 2000).

In the case of frequency scales, respondents assume that the researcher constructed a meaningful scale, based on his or her knowledge of, or expectations about, the distribution of the behaviour in the 'real world'. Accordingly, they assume that values in the middle range of the scale reflect the 'average' or 'usual' behavioural frequency, whereas the extremes of the scale correspond to the extremes of the distribution. These assumptions influence respondents' interpretation of the question, their behavioural estimates, and subsequent judgements. As expected on theoretical grounds, these influences are not

Table 6.4 Reported daily tv consumption as a function of response alternatives

Low frequency alternatives		High frequency alternatives	
Up to 1/2 h	7.4%	Up to 2 1/2 h	62.5%
1/2 h to 1 h	17.7%	2 1/2 h to 3 h	23.4%
1 h to 1 1/2 h	26.5%	3 h to 3 1/2 h	7.8%
1 1/2 h to 2 h	14.7%	3 1/2 h to 4 h	4.7%
2 h to 2 1/2 h	17.7%	4 h to 4 1/2 h	1.6%
More than 2 1/2 h	16.2%	More than 4 1/2 h	0.0%

Note: N = 132. Adapted from Schwarz, N., Hippler, H. J., Deutsch, B. & Strack, F. (1985). Response scales: Effects of category range on reported behavior and comparative judgements. *Public Opinion Quarterly, 49*:388–395. Reprinted by permission.

observed when respondents are aware that the scale may be uninformative for the task at hand, e.g., because it was taken from a study designed for a distinctly different population (Schwarz 1996).

Question interpretation

Suppose that respondents are asked how frequently they felt 'really irritated' recently. To provide an informative answer, respondents have to determine what the researcher means by 'really irritated'. Does this term refer to major or to minor annoyances? To identify the intended meaning of the question, they may consult the response alternatives provided by the researcher. If the response alternatives present low frequency categories, e.g., ranging from 'less than once a year' to 'more than once a month', they may conclude that the researcher has relatively rare events in mind. Hence, the question cannot refer to minor irritations, which are likely to occur more often, so the researcher is probably interested in more severe episodes of irritation. Several studies confirmed these predictions (see Schwarz 1996). For example, Schwarz, Strack, Müller and Chassein (1988) observed that respondents who had to report the frequency of irritating experiences on a low frequency scale assumed that the question referred to major annoyances, whereas respondents who had to give their report on a high frequency scale assumed that the question referred to minor annoyances. Thus, respondents identified different experiences as the target of the question, depending on the frequency range of the response alternatives provided to them. Similarly, Winkielman, Knäuper, and Schwarz (1998) observed that the length of the reference period, for example 'last week' versus 'last year', can profoundly affect question interpretation.

These findings have important implications for the comparison of concurrent and retrospective frequency reports. Consider, for example, the common observation that respondents report more intense emotions (Thomas & Diener 1990; Parkinson, Briner, Reynolds & Totterdell 1995), and more severe marital disagreements (McGonagle, Kessler & Schilling 1992), in retrospective than in concurrent reports. Findings of this type are typically attributed to the higher memorability of intense experiences. Yet, Winkielman et al.'s (1998) results suggest that discrepancies between concurrent and retrospective reports may in part be due to differential question interpretation: concurrent reports necessarily pertain to a short reference period, with one day typically being the upper limit, whereas retrospective reports cover more extended periods. Hence, the concurrent and retrospective nature of the report is inherently confounded with the length of the reference period. Accordingly, participants who provide a concurrent report may infer from the short reference period used that the researcher is interested in frequent events, whereas the long reference period used under retrospective conditions may suggest an interest in infrequent events. Hence, respondents may deliberately report on different experiences, rendering reports that pertain to different reference periods (Winkielman *et al.* 1998), or are provided on different frequency scales (Schwarz *et al.* 1988), non-comparable.

Frequency estimates

Given that respondents assume that the researcher's scale reflects the distribution of the behaviour in the population, they can draw on the range of the response alternatives as a frame of reference in estimating the frequency of their own behaviour. This results in higher frequency reports along high rather than low frequency scales, as shown in Table 6.4. In this study (Schwarz, Hippler, Deutsch & Strack 1985), only 16.2% of a sample of German respondents reported watching TV for more than $2\frac{1}{2}$ hours a day when the scale presented low frequency response alternatives, whereas 37.5% reported doing so when the scale presented high frequency response alternatives. Similar results have been obtained for a wide range of different behaviours, including sexual behaviours (Schwarz & Scheuring 1988; Tourangeau & Smith 1996), consumer behaviours (Menon, Rhagubir & Schwarz 1995), and reports of physical symptoms (Schwarz & Scheuring 1992; see Schwarz 1990a, 1996, for reviews).

For example, Schwarz and Scheuring (1992) asked 60 patients of a German psychosomatic clinic to report the frequency of 17 symptoms along a low frequency scale that ranged from 'never' to 'more than twice a month', or along a high frequency scale that ranged from 'twice a month or less' to 'several times a day'. Across 17 symptoms, 62% of the respondents reported average frequencies of more than twice a month when presented with the high frequency scale, whereas only 39% did so when presented with the low frequency scale, resulting in a mean difference of 23 percentage points. This impact of response alternatives was most pronounced for the ill-defined symptom of 'responsiveness to changes in the weather', where 75% of the patients reported a frequency of more than twice a month along the high frequency scale, whereas only 21% did so along the low frequency scale. Conversely, the influence of response alternatives was least pronounced for the better defined symptom 'excessive perspiration', with 50% vs. 42% of the respondents reporting a frequency of more than twice a month in the high and low frequency scale conditions, respectively.

As expected on theoretical grounds, the impact of response alternatives is more pronounced the more poorly the behaviour is represented in memory, thus forcing respondents to rely on an estimation strategy. When the behaviour is rare and important, and hence well represented in memory, the impact of response alternatives is small or eliminated, because respondents can rely on an enumeration strategy. The same is true when a respondent engages in the behaviour with high regularity (e.g. 'every Sunday'), in which case an estimate can be derived from rate-of-occurrence information (see Menon 1994; Menon *et al.* 1995, for a discussion). These observations are consistent with Brown's (this volume) multiple strategy perspective.

Subsequent judgements

Finally, respondents' placement on the scale provides them with information about their relative location in the distribution. For example, a TV consumption of '2½ h a day' constitutes a high response on the low frequency scale, but a low response on the high frequency scale shown in Table 6.4. A respondent who checks this alternative may therefore infer that her own TV consumption is above average in the former case, but below average in the latter. As a result, Schwarz *et al.* (1985) observed that respondents were less satisfied with the variety of things they do in their leisure time when the low frequency scale suggested that they watch more TV than most other people (see Schwarz 1990a for a review). Similarly, the psychosomatic patients in Schwarz and Scheuring's (1992) study reported higher health satisfaction when the high frequency scale suggested that their own symptom frequency is below average than when the low frequency scale suggested that it is above average. Note that this higher report of health satisfaction was obtained despite the fact that the former patients reported a higher symptom frequency in the first place, as seen above. Findings of this type reflect that respondents extract comparison information from their own placement on the scale and use this information in making subsequent comparative judgements.

However, not all judgements are comparative in nature. When asked how satisfied we are with our health, we may compare our own symptom frequency to that of others'. Yet, when asked how much our symptoms bother us, we may not engage in a social comparison but may instead draw on the absolute frequency of our symptoms. In this case, we may infer that our symptoms bother us more when a high frequency scale lead us to estimate a high symptom frequency. Accordingly, other patients who reported their symptom frequency on one of the above scales reported that their symptoms bother them more when they received a high rather than a low frequency scale (Schwarz 1999). Thus, the same high frequency scale elicited subsequent reports of higher health satisfaction (a comparative judgement) or of higher subjective suffering (a non-comparative judgement), depending on whether a comparative or a non-comparative judgement followed the symptom report.

Methodological implications

The reviewed findings have important methodological implications of the assessment of frequency reports (see Schwarz 1990a, for more detailed discussions).

Frequency scales

First, the numeric response alternatives presented as part of a frequency question may influence respondents' interpretation of what the question refers to. Hence, the same question stem in combination with different frequency alternatives may result in the assessment of somewhat different behaviours. This is more likely the less well defined the behaviour is.

Second, respondents' use of the frequency scale as a frame of reference influences the obtained behavioural reports. Aside from calling the interpretation of the absolute values into question, this also implies that reports of the same behaviour along different scales are not comparable, rendering comparisons between different studies difficult.

Third, the impact of response alternatives is more pronounced, the less respondents can recall relevant episodes from memory. This implies that reports of behaviours that are poorly represented in memory are more affected than reports of behaviours that are well represented (Menon *et al.* 1995). When behaviours of differential memorability are assessed, this may either exaggerate or reduce any actual differences in the relative frequency of the behaviours, depending on the specific frequency range of the scale.

Fourth, for the same reason, respondents with poorer memory for the behaviour under study are more likely to be influenced by response alternatives than respondents with better memory. Such a differential impact of response alternatives on the reports provided by different groups of respondents can result in misleading conclusions about actual group differences (e.g. Ji, Schwarz & Nisbett 2000).

Finally, the range of response alternatives may further influence subsequent comparative and non-comparative judgements. Hence, respondents' may arrive at evaluative judgements that are highly context dependent and may not reflect the assessments they would be likely to make in daily life.

To avoid these systematic influences of response alternatives, it is advisable to ask frequency questions in an open response format, such as, 'How many hours a day do you watch TV?——hours per day.' Note that such an open format needs to specify the relevant units of measurement, e.g. 'hours per day' to avoid answers like 'a few'.

Vague quantifiers

As another alternative, researchers are often tempted to use vague quantifiers, such as 'sometimes', 'frequently', and so on. This, however, is the worst possible choice (see Moxey & Sanford 1992; Pepper 1981 for reviews). To provide a meaningful answer, respondents have to anchor and calibrate the response scale. To do so, they rely on their subjective world knowledge and personal experiences. As a result, the same expression denotes different frequencies in different content domains and for different respondents (e.g. Schaeffer 1991; Wright, Gaskell & O'Muircheartaigh 1994), rendering comparisons across domains and respondents misleading.

The relationship between objective frequencies and vague quantifiers is further complicated by the fact that respondents may take the conversational context of the interview into account in selecting the proper term. For example, Wänke (2000) asked students how often they go to the cinema as part of an alleged survey of students or a survey of the general population. To the extent that respondents assume that students see more movies than other people do, they may adjust their use of vague quantifiers. The results supported this prediction. Although students reported the same absolute number of cinema visits in both surveys, they used lower vague quantifiers to describe this number when the survey was introduced as a survey of students rather than a survey of the general population. Findings of this type indicate that the use of vague quantifiers reflects objective frequencies relative to a respondent's subjective standard, rendering vague quantifiers inadequate for the assessment of actual frequencies.

For the same reason, vague quantifiers are more closely related to evaluative judgements than objective frequencies. For example, Wänke (1996) observed that the actual number of recalled pleasant or unpleasant consumer experiences was virtually unrelated to respondents' satisfaction judgements. In contrast, respondents' subjective impression of whether they had experienced 'many' or 'few' pleasant or unpleasant instances was a good predictor of consumer satisfaction. Note that the latter reports ('few' or 'many') already entail an evaluation of the actual frequency relative to a subjective standard ('less or more than usual'), whereas the evaluative implications of the objective frequency remain unknown in the absence of understanding the person's expectations.

Concluding remarks: strategy selection

This chapter reviewed two distinct strategies of frequency estimation, namely the use of subjective recall experiences postulated by Tversky and Kahneman's (1973) availability heuristic and reliance on the response scales provided by the researcher, first identified by Schwarz et al. (1985). Both strategies serve us well under many circumstances. Frequent events are indeed easier to recall than rare events, rendering reliance on ease of recall useful under many real world conditions. Moreover, people are sensitive to many of the conditions that may affect ease of recall and do not draw on their recall experiences when they have reason to doubt their informational value, as the reviewed studies illustrate (see Schwarz 1998 for a discussion). Similarly, the use of frequency scales as a frame of reference is defensible on the basis of the tacit assumptions that underlie the conduct of conversation in everyday life (Grice 1975). In fact, reliance on the scale would result in relatively accurate estimates for many respondents if values in the middle range of the scale did indeed reflect the average value in the population, as respondents assume. Yet these strategies are error prone under other conditions. Thus we may miss subtle influences that affect ease of recall or may miss that the researcher does not adhere to the norms of cooperative conversational conduct by providing a haphazard scale.

Although the two heuristics we addressed draw on different inputs and share few surface similarities, it is worth highlighting the underlying commonalities. Most important, respondents' use of experiential information (like ease of recall) as well as contextual information (like frequency scales) follows the same principles of applicability and diagnosticity as the use of any other information. Respondents only rely on the frequency scale presented by the researcher when it seems relevant to the task, but not when it is said to be taken from a study designed for a different population or explicitly presented as part of a pretest designed to test the appropriateness of the scale (see Schwarz 1996 for a review). Similarly, respondents only rely on their subjective recall experiences when these experiences seem relevant to the task at hand and their perceived diagnosticity is not undermined. Hence, they do not draw on their experiences when they perceive a lack of expertise in the content area (e.g., Biller, Bless & Schwarz 1992) or when the informational value of the experience is undermined through misattribution manipulations (e.g. Schwarz et al. 1991; Wänke et al. 1995). Moreover, their reliance on these as well as other estimation strategies depends on the accessibility of relevant behavioural information in memory. Hence, the impact of response scale decreases when relevant episodic information or rate-of-occurrence information is available (Menon et al. 1995; Ji et al. 2000), but increases as the required judgement becomes more demanding (Bless, Bohner, Hild & Schwarz 1992). Finally, respondents' strategy choice is, in part, a function of their processing motivation. Thus, respondents turn to the content of recall, rather than to their subjective experiences, under conditions known to foster reliance on systematic rather than heuristic strategies of judgement (Rothman & Schwarz 1998; Grayson & Schwarz 1999).

None of these observations is surprising and all are consistent with general models of human judgement that conceptualize the interplay of experiential and declarative information (see Schwarz & Clore 1996) and the role of conversational processes in judgement (see Hilton 1995; Schwarz 1994, 1996). In light of these consistencies, sweeping verdicts like 'the availability approach is not a theory with any predictive power' (Betsch & Pohl this volume) are overstated, at best, in particular in light of the failure to provide an alternative theoretical account for the available data. Nevertheless, we agree that the available research has primarily explored the parameters that determine reliance on a given strategy, and the relative size of the observed impact, rather than the variables that determine the choice between strategies. We address this issue in the final section.

Strategy selection

Our take on strategy selection is compatible with Brown's (this volume) multiple strategies perspective (see Menon *et al.* 1995, for a related discussion). When the behaviour that respondents are asked to report on is rare and important, they can access some episodic representations and follow an enumeration strategy, which may require extrapolation, depending on the reference period. Most behaviours researchers are interested in, however, are frequent and mundane and hence do not meet these fortunate conditions (see the contributions in Schwarz & Sudman 1994). When such behaviours are engaged in with high regularity ('daily' or 'every Sunday'), respondents can draw on rate-of-occurrence information, which usually leads to a correct inference (for a discussion see Menon 1994; Menon *et al.* 1995). Yet most behaviours lack this regularity and respondents have to rely on other estimation strategies to determine a plausible answer. In most cases, a number of different strategies will be applicable to the task, raising the question which one respondents are likely to select.

In our reading, there is little mystery to the basic principle of this selection process. Consistent with a large body of knowledge accessibility research (for a review see Higgins 1996), we propose that respondents first select the strategy that is applicable to the information that is most accessible at that point in time. Hence, a respondent who is presented with a frequency scale is likely to rely on this salient piece of information, instead of engaging in a memory search that may produce a subjective recall experience. Conversely, a respondent who has just completed a recall task is likely to draw on the information rendered highly accessible by this task, namely the subjective recall experience or the recalled content, depending on the conditions discussed earlier. If the most accessible information is considered nondiagnostic for one of the reasons discussed above, the respondent will search for other accessible information that may bear on the task. The nature of this alternative accessible input will determine the next strategy the respondent is likely to test.

From this perspective, strategy selection is largely determined by the *accessibility of applicable inputs* that are perceived as diagnostic for the task at hand. Accordingly, strategy selection is not only a function of the representation variables addressed by Brown (this volume), but also a function of contextual influences that determine the relative accessibility of different inputs as well as their perceived diagnosticity. As in other domains of judgement, we can not understand frequency estimation without taking such contextual influences into account. Unfortunately, this has unwelcome methodological implications, none of which is unique to frequency judgement.

First, any frequency question will necessarily be presented in some context. This context, as well as the characteristics of the behaviour or event of interest, will influence respondents' strategy selection by rendering some inputs more accessible than others.

Second, experimental procedures that allow for focused analyses of the operation of any one strategy render the selection of other strategies less likely. Once we manipulate variables relevant to Strategy A, variables that may serve as inputs into Strategy B will be relatively less accessible, hence favouring Strategy A over Strategy B. An exception are tasks that provide accessible inputs for different strategies, as was the case for recalled content and recall experiences in the studies reviewed above. Yet even in this case, the accessible inputs restrict the choice of strategies to the ones that are applicable to one of these inputs.

Third, this implies that our methodological tools are better suited for understanding the operation of a given strategy than for understanding the conditions under which this strategy is spontaneously selected over another one. That the participants in our studies, for example, drew on the subjective recall experiences evoked by the experimental task does not tell us under which conditions they would spontaneously try to *generate* such experiences when asked to answer a frequency question. Nevertheless, there is much that can be learned by focusing on a given strategy, as the above identification of the interplay of recalled content, subjective recall experiences, perceived diagnosticity and processing motivation illustrates.

Finally, the proposal that strategy selection is driven by the accessibility of applicable inputs entails that there is little hope for a comprehensive model that consistently predicts strategy selection under natural conditions. Although we can derive predictions about the accessibility of some inputs from analyses of memory representations (see Brown, this volume), these representations are only one of the numerous determinants of what happens to come to mind at any given point in time. Hence, we are likely to obtain consistent results in experiments that render one or the other input accessible, but equally likely to face numerous surprises under natural conditions, where the accessibility of different inputs is subject to haphazard influences of uncontrolled contextual variables.

Acknowledgement

Preparation of this chapter was supported through a fellowship from the Center for Advanced Study in the Behavioural Sciences, Stanford, CA, to Norbert Schwarz, which is gratefully acknowledged.

References

Biller, B., Bless, H. & Schwarz, N. (1992, April). *Die Leichtigkeit der Erinnerung als Information in der Urteilsbildung: Der Einfluß der Fragenreihenfolge* (Ease of recall as information: The impact of question order). Paper presented at Tagung experimentell arbeitender Psychologen, Osnabrück, FRG.

Bless, H. & Forges, J. (eds) (2000). *The message within: The role of subjective experience in social cognition and behavior*. Philadelphia, PA: Psychology Press.

Bless, H. & Ruder, M. (1999). *Reliance on the availability heuristic. A question of individuals' mood*. Paper presented at Biennial conference on subjective probability utility and decision making, SPUDEM 17, Mannheim, FRG.

Bless, H. & Schwarz, N. (1999). Sufficient and necessary conditions in dual process models: The case of mood and information processing. In S. Chaiken & Y. Trope (eds) *Dual process theories in social psychology* (pp. 423–440). New York: Guilford.

Bless, H. (2000). The mediating role of general knowledge structures. In J. P. Forges (ed.) *Feeling and thinking: The role of affect in social cognition* (pp. 201–222). Cambridge, UK: Cambridge University Press.

Bless, H., Bohner, G., Hild, T. & Schwarz, N. (1992). Asking difficult questions: Task complexity increases the impact of response alternatives. *European Journal of Social Psychology, 22*:309–312.

Bradburn, N. M., Rips, L. J. & Shevell, S. K. (1987). Answering autobiographical questions: The impact of memory and inference on surveys. *Science, 236*:157–161.

Chaiken, S. & Trope, Y. (eds) (1999). *Dual-process theories in social psychology.* New York: Guilford.

Gigerenzer, G. (1996). On narrow norms and vague heuristics: A reply to Kahneman and Tversky (1996). *Psychological Review, 103*:592–596.

Grayson, C. E. & Schwarz, N. (1999). Beliefs influence information processing strategies: Declarative and experiential information in risk assessment. *Social Cognition, 17*:1–18.

Grice, H. P. (1975). Logic and conversation. In P. Cole, & J. L. Morgan (eds) *Syntax and semantics, vol. 3: Speech acts* (pp. 41–58). New York: Academic Press.

Higgins, E. T. (1996). Knowledge activation: Accessibility, applicability, and salience. In E. T. Higgins & A. Kruglanski (eds) *Social psychology: Handbook of basic principles* (pp. 133–168). New York: Guilford Press.

Hilton, D. J. (1995). The social context of reasoning: Conversational inference and rational judgement. *Psychological Bulletin, 118*:248–271.

Ji, L., Schwarz, N. & Nisbett, R. E. (2000). Culture, autobiographical memory, and behavioural frequency reports: Measurement issues in cross-cultural studies. *Personality and Social Psychology Bulletin, 26*:586–594.

McGonagle, K. A., Kessler, R. C. & Schilling, E. A. (1992). The frequency and determinants of marital disagreements in a community sample. *Journal of Social & Personal Relationships, 9*:507–524.

Medin, D. L. & Ross, B. H. (1997). *Cognitive psychology* 2nd edn. Fort Worth: Harcourt Brace.

Menon, G. (1994). Judgements of behavioural frequencies: Memory search and retrieval strategies. In N. Schwarz & S. Sudman (eds) *Autobiographical memory and the validity of retrospective reports* (pp. 161–172). New York: Springer Verlag.

Menon, G., Raghubir, P. & Schwarz, N. (1995). Behavioural frequency judgements: An accessibility-diagnosticity framework. *Journal of Consumer Research, 22*:212–228.

Moxey, L. M. & Sanford, A. J. (1992). Context effects and the communicative functions of quantifiers: Implications for their use in attitude research. In N. Schwarz & S. Sudman (eds) *Context effects in social and psychological research* (pp. 279–296). New York: Springer Verlag.

Nisbett, R. E. & Ross, L. (1980). *Human inference: Strategies and shortcomings of social judgement.* Englewood Cliffs: Prentice Hall.

Parkinson, B., Briner, R. B., Reynolds, S. & Totterdell, P. (1995). Time frames for mood: Relations between momentary and generalized ratings of affect. *Personality and Social Psychology Bulletin, 21*:331–339.

Payne, J. W., Bettman, J. R. & Johnson, E. J. (1993). *The adaptive decision maker.* New York: Cambridge University Press.

Pepper, S. C. (1981). Problems in the quantification of frequency expressions. In D. W. Fiske (ed.) *Problems with language imprecision* (New Directions for Methodology of Social and Behavioural Science, vol. 9). San Francisco: Jossey-Bass.

Rothman, A. J. & Schwarz, N. (1998). Constructing perceptions of vulnerability: Personal relevance and the use of experiential information in health judgements. *Personality and Social Psychology Bulletin, 24*:1053–1064.

Schaeffer, N. (1991). Hardly ever or constantly? Group comparisons using vague quantifiers. *Public Opinion Quarterly, 55*:395–423.

Schwarz, N. (1990a). Assessing frequency reports of mundane behaviours: Contributions of cognitive psychology to questionnaire construction. In C. Hendrick & M. S. Clark (eds) *Research methods in personality and social psychology*. Review of Personality and Social Psychology, vol. 11 (pp. 98–119). Beverly Hills, CA: Sage.

Schwarz, N. (1990b). Feelings as information: Informational and motivational functions of affective states. In E. T. Higgins & R. M. Sorrentino (eds) *Handbook of motivation and cognition: Foundations of social behaviour* vol. 2 (pp. 527–561). New York, NY: Guilford Press.

Schwarz, N. (1994). Judgement in a social context: Biases, shortcomings, and the logic of conversation. In M. Zanna (ed.) *Advances in experimental social psychology* vol. 26. San Diego, CA: Academic Press.

Schwarz, N. (1996). *Cognition and communication: Judgemental biases, research methods, and the logic of conversation*. Hillsdale, NJ: Erlbaum.

Schwarz, N. (1998). Accessible content and accessibility experiences: The interplay of declarative and experiential information in judgement. *Personality and Social Psychology Review*, 2:87–99.

Schwarz, N. (1999). Frequency reports of physical symptoms and health behaviours: How the questionnaire determines the results. In Park, D.C., Morrell, R.W. & Shifren, K. (eds) *Processing medical information in aging patients: Cognitive and human factors perspectives* (pp. 93–108). Mahaw, NJ: Erlbaum.

Schwarz, N., Bless, H. & Bohner, G. (1991a). Mood and persuasion: Affective states influence the processing of persuasive communications. *Advances in Experimental Social Psychology*, 24:161–199.

Schwarz, N., Bless, H., Strack, F., Klumpp, G., Rittenauer-Schatka, H. & Simons, A. (1991b). Ease of retrieval as information: Another look at the availability heuristic. *Journal of Personality and Social Psychology*, 61:195–202.

Schwarz, N. & Clore, G. L. (1983). Mood, misattribution, and judgements of well-being: Informative and directive functions of affective states. *Journal of Personality and Social Psychology*, 45:513–523.

Schwarz, N. & Clore, G. L. (1996). Feelings and phenomenal experiences. In E.T. Higgins & A. Kruglanski (eds) *Social psychology: A handbook of basic principles* (pp. 433–465). New York: Guilford.

Schwarz, N., Hippler, H. J., Deutsch, B. & Strack, F. (1985). Response categories: Effects on behavioural reports and comparative judgements. *Public Opinion Quarterly*, 49:388–395.

Schwarz, N., Knäuper, B., Hippler, H. J., Noelle-Neumann, E. & Clark, F. (1991). Rating scales: Numeric values may change the meaning of scale labels. *Public Opinion Quarterly*, 55:570–582.

Schwarz, N. & Scheuring, B. (1988). Judgements of relationship satisfaction: Inter- and intraindividual comparison strategies as a function of questionnaire structure. *European Journal of Social Psychology*, 18:485–496.

Schwarz, N. & Scheuring, B. (1992). Selbstberichtete Verhaltens-und Symptomhäufigkeiten: Was Befragte aus Anwortvorgaben des Fragebogens lernen. (Frequency-reports of psychosomatic symptoms: What respondents learn from response alternatives.) *Zeitschrift für Klinische Psychologie*, 22:197–208.

Schwarz, N. & Schuman, H. (1997). Political knowledge, attribution, and inferred political interest: The operation of buffer items. *International Journal of Public Opinion Research*, 9:191–195.

Schwarz, N., Strack, F., Müller, G. & Chassein, B. (1988). The range of response alternatives may determine the meaning of the question: Further evidence on informative functions of response alternatives. *Social Cognition*, 6:107–117.

Schwarz, N. & Sudman, S. (1994). *Autobiographical memory and the validity of retrospective reports*. New York: Springer Verlag.

Schwarz, N. & Vaughn, L. A. (in press). The availability heuristic revisited: Recalled content and ease of recall as information. In T. Gilovich, D. Griffin & D. Kahneman (eds) *The psychology of intuitive judgment: Heuristics and biases*. Cambridge: Cambridge University Press.

Stepper, S. & Strack, F. (1993). Proprioceptive determinants of emotional and nonemotional feelings. *Journal of Personality and Social Psychology, 64*:211–220.

Strack, F. (1992). The different routes to social judgement: Experiential versus informational strategies. In L. L. Martin & A. Tesser (eds) *The construction of social judgements* (pp. 249–276). Hillsdale, NJ: Erlbaum.

Strube, G. (1987). Answering survey questions: The role of memory. In H.J. Hippler, N. Schwarz, & S. Sudman (eds) *Social information processing and survey methodology* (pp. 86–101). New York: Springer Verlag.

Sudman, S., Bradburn, N. M. & Schwarz, N. (1996). *Thinking about answers: The application of cognitive processes to survey methodology*. San Francisco, CA: Jossey-Bass.

Taylor, S. E. (1982). The availability bias in social perception and interaction. In D. Kahneman, P. Slovic & A. Tversky (eds) *Judgement under uncertainty: Heuristics and biases* (pp. 190–200). Cambridge: Cambridge University Press.

Thomas, D. L. & Diener, E. (1990). Memory accuracy in the recall of emotions. *Journal of Personality and Social Psychology, 59*:291–297.

Tourangeau, R. & Smith, T. W. (1996). Asking sensitive questions. The impact of data collection, mode, question format, and question context. *Public Opinion Quarterly, 60*:275–304.

Tversky, A. & Kahneman, D. (1973). Availability: A heuristic for judging frequency and probability. *Cognitive Psychology, 5*:207–232.

Wänke, M. (1996). *Determinanten der Informationsnutzung bei Einstellungsurteilen*. Unpublished Habilitation Thesis. Universität Heidelberg.

Wänke, M. (2000). Conversational norms and the interpretation of vague quantifiers. Manuscript under review.

Wänke, M. & Bless. H. (2000). How ease of retrieval may affect attitude judgements. In H. Bless & J. Forgas (eds) *The role of subjective states in social cognition and behaviour* (pp. 143–161). Psychology Press.

Wänke, M., Bless, H. & Biller, B. (1996). Subjective experience versus content of information in the construction of attitude judgements. *Personality and Social Psychology Bulletin, 22*: 1105–1115.

Wänke, M., Schwarz, N. & Bless, H. (1995). The availability heuristic revisited: Experienced ease of retrieval in mundane frequency estimates. *Acta Psychologica, 89*:83–90.

Winkielman, P., Knäuper, B. & Schwarz, N. (1998). Looking back at anger: Reference periods change the interpretation of (emotion) frequency questions. *Journal of Personality and Social Psychology, 75*:719–728.

Wright, D. B., Gaskell, G. D. & O'Muircheartaigh, C. A. (1994). How much is 'quite a bit'? Mapping absolute values onto vague quantifiers. *Applied Cognitive Psychology, 8*:479–496.

Zillman, D. (1978). Attribution and misattribution of excitatory reactions. In J.H. Harvey, W.I. Ickes & R. F. Kidd (eds) *New directions in attribution research* (Vol. 2). Hillsdale, NJ: Erlbaum.

CHAPTER 7

TVERSKY AND KAHNEMAN'S AVAILABILITY APPROACH TO FREQUENCY JUDGEMENT: A CRITICAL ANALYSIS

TILMANN BETSCH AND DEVIKA POHL

Abstract

This chapter examines the predictive power of Tversky and Kahneman's availability approach to frequency estimation. It is shown that the original formulation of the availability approach does not provide a falsifiable theory that would allow us to derive a priori predictions about the psychological mechanisms by which people arrive at frequency judgements. Rather, availability is one among other possibilities that people might rely on when being confronted with a frequency estimation task. A brief review of the empirical evidence reveals that the availability approach indeed describes one possible estimation strategy. However, the review also shows that a considerable amount of evidence cannot be accounted for in terms of availability. Altogether, the evidence indicates that people make use of a variety of strategies when judging frequency. We conclude that it is not helpful to accumulate evidence in favour of either mechanism. Rather, one should attempt to develop theories of frequency estimation that spell out the conditions under which a particular strategy of judgement is likely to be used.

Introduction

As a common denominator, a wide array of cognitive theories share the assumption that judgements are made on the basis of subsets of information which are available at the time of judgement (Wyer & Srull 1989; Higgins *et al.* 1977). The emphasis of availability as a guiding principal in judgement was strongly promoted by Tversky and Kahneman's seminal article on the availability heuristic (Tversky & Kahneman 1973). This paper had a tremendous impact on the judgement and decision literature when it was put forward almost thirty years ago. During the subsequent decades, numerous studies have been conducted to test the viability of Tversky and Kahneman's availability approach. At present, this approach still provides the theoretical background for hypotheses generation, especially in social psychology (Schwarz 1998, for an overview), although its viability as a theoretical model has often been cast into doubt (Brown 1995; Brown & Siegler 1993; Fiedler 1983; Sedlmeier 1999; Wallsten 1983). This chapter aims to provide an evaluation of the *original* formulation of the availability approach. Moreover, it will sketch out some trajectories to future theory development.

Tversky and Kahneman's availability approach has survived over roughly thirty years of research due, at least in part, to an overestimation of its predictive power. There is an apparent and consequential discrepancy between the original statement of the availability approach and the manner in which it has been interpreted as a theoretical background for hypotheses generation by many researchers. Often, these interpretations take the following form: *Individuals estimate the frequency of an event by the ease with which instances or associations come to mind.*[1]

This judgement mechanism relying on the ease with which instances can be retrieved has been termed the 'availability heuristic'. The form of this statement suggests that subjective frequency estimation is a function of availability. According to such an interpretation of the availability approach, one would predict that people generally use the availability heuristic when confronted with a frequency estimation task. Such an interpretation in turn, would justify research that tests people's propensity to employ the availability heuristic. However, such a view is based on a misinterpretation. The availability approach, as we will show, is not designed as a predictive theory, which would allow any falsifiable prediction of what people do when confronted with a frequency estimation task. We will further show that the lack of predictive power inheres in the original formulation which avoids spelling out conditions under which the availability heuristic will be used. The next section provides an analysis which works out the limitations of the availability approach.

The predictive power of the availability approach

We henceforth refer to the first statement of the availability approach (Tversky & Kahneman 1973). All quotations in this section, with their page numbers in brackets, refer to this article. The availability approach aims to describe the psychological mechanisms by which individuals arrive at frequency or probability judgements (p. 207). Therefore, it has to face two basic problems, (1) the selection of judgemental heuristics or strategies, and (2) the employment of a particular one, namely the availability heuristic.

Under which conditions are people likely to employ the availability heuristic?

Although Tversky and Kahneman are primarily concerned with the functioning of the availability heuristic, one can detect various assumptions spread throughout their article that refer to the initial step of strategy identification and selection. The first assumption of this kind appears early in the introduction and shows that the availability heuristic is not the only way people can arrive at quantitative judgements:

We propose that when faced with the difficult task of judging probability or frequency, people employ a limited number of heuristics which reduce these judgements to simpler ones. (p. 207)

---

[1] Similar interpretations can be found for example in Bruce, Hockley and Craik (1991, p. 301), Lewandowsky and Smith (1983, p. 347), Sedlmeier (1999, p. 35), Wänke, Schwarz and Bless (1995, p. 83), and Williams and Durso (1986, p. 388).

One implication of this assumption is that the availability heuristic will not always be applied in frequency judgement. Rather, the availability heuristic is one among other strategies to be considered for solving a frequency estimation task. In line with this presumption, Tversky and Kahneman restrict themselves to the existential statement that one *may* estimate frequency by assessing availability (see e.g. p. 208). This is a modest claim, but the most intriguing question remains: When do people select a particular heuristic and employ it in judgement? Tversky and Kahneman consider difficulty of the task as a precondition for choosing the availability heuristic (pp. 207–208). This simply means that people are assumed to rely on shortcut strategies if they cannot construct and enumerate all instances of the sample to be judged in terms of frequency (p. 211). Some conditions are mentioned that hinder a decision to use the recall-based version of the availability heuristic:

There are situations, however, in which occurrences cannot be retrieved, e.g., when the total number of items is large, when their distinctiveness is low, or when the retention interval is long. In these situations, subjects may resort to a different method of judging frequency. (p. 221)

As we will see later, however, application of the availability heuristic is not assumed to generally require the retrieval of instances. Hence, the above statement does not allow us to predict when people are especially reluctant to employ the availability heuristic.

Nevertheless, the authors make some statements on the conditions concerning the use of either the availability or the representativeness heuristic[2] when they conclude on the results of a study on relative frequency estimation:

. . . the frequency of a class is likely to be judged by availability if the individual instances are emphasized and by representativeness if generic features are made salient. (p. 220)

If availability and representativeness were the only methods by which people estimate frequencies, this assumption would make up a strong prediction indeed. However, the authors emphasize that the availability and the representativeness heuristic are not the only methods by which frequencies can be estimated (p. 222). They enlarge the heuristic toolbox in the remainder of the paper by referring to a couple of other methods. For example, they also allow for the possibility that people can rely on cognitive 'counters' (Underwood 1969), which represent online-established memory records of event frequency. Another possibility mentioned is to infer probability or rate of occurrence from schemas (Howell 1970). One can easily think of other strategies, if one takes related work by the same authors into account (e.g. the simulation heuristic). Even if emphasizing of individual instances serves as (is) a facilitating condition for choosing an availability based strategy, a prediction cannot be made until the conditions are specified leading people *not to rely* on cognitive counters, schemas or other bases of knowledge.

---

[2] When employing the representativeness heuristic, judges are assumed to base their estimates on the degree that a target element is typical or representative of a class or category.

Summary

Tversky and Kahneman suggest that the availability heuristic is just one among other strategies for frequency estimation. They do not specify the absolute number of distinct strategies. Most importantly, they do not specify under which conditions the availability heuristic is likely to be employed. Therefore, it is not possible to derive any prediction from the original formulation of the availability approach when people rely on the availability heuristic in frequency estimation.

How does the availability heuristic work: mechanism of assessing associative strength

The second, and most well known, part of the availability approach is concerned with the psychological processes characterizing the employment of the availability heuristic. The central tenet is that 'A person is said to employ the availability heuristic whenever he estimates frequency or probability by the ease with which instances or associations come to mind' (p. 208).

Ease of retrieval reflects the associative strength between a prompt and a target. Since associative bonds are strengthened by repetition, the strength of a bond may often provide quite a valid cue for frequency, although it may be affected by other factors irrelevant to actual frequency.

What are the cognitive strategies by which ease of retrieval could be assessed? Tversky and Kahneman (1973, p. 220) suggest at least two methods. The first one is implied in the definition of availability itself. Accordingly, the individual

> ... attempts to recall some instances and judges frequency by availability, i.e., by the ease with which instances come to mind. As a consequence, classes whose instances are readily recalled will be judged more numerous than classes of the same size whose instances are less available.

Aside from assessing availability via recall, they suggest another possibility:

> To assess availability it is not necessary to perform the actual operations of retrieval or construction. It suffices to assess the ease with which these operations could be performed, such as the difficulty of a puzzle or mathematical problem can be assessed without considering specific solutions. (p. 208)

Unfortunately, the functioning of this alternative method is not spelled out. The examples in the quotation invite the interpretation that assessments of ease or difficulty might also be inferred from domain-specific knowledge stored in long-term memory. For instance, most people know that the ease by which a puzzle can be assembled depends on the number of pieces. Consequently, one might judge a fifty-pieces puzzle easier to be assembled than one consisting of five hundred pieces. If such processes were also considered as being based on an assessment of associative strength or availability, the distinction between availability and alternative methods, such as the reliance on schemata, is blurred. The opening of the availability approach to processes, other than retrieval-based assessment of ease, entails serious problems. Most importantly, it undermines the major motivation behind the model, which is to provide an account for systematic biases in frequency judgement. If the availability heuristic were confined to retrieval of instances, one could make the strong prediction that all factors affecting associative strength and retrieval process would impact subsequent judgements, given that the availability heuristic is applied (Sherman & Corty 1984). This assumption, indeed, has often been considered to be the model's major contribution and has motivated a considerable amount of research effort in the past thirty years (Bruce et al. 1991; Gabrielcik & Fazio 1984; Greene 1989; Lewandowsky & Smith 1983; Manis et al. 1993; Schwarz et al. 1991; Sedlmeier 1999; Tversky & Kahneman 1973; Wänke et al. 1995; Williams & Durso 1986). However, the original formulation of the availability account does not justify this prediction, unless it were to be extended by explicit assumptions, under which conditions availability assessments would be based on the experienced ease of retrieval.

Summary

Tversky and Kahneman subsume different psychological processes under the concept of availability. They do not predict, for example, when people retrieve instances from memory, or when they employ recall-free techniques to assess availability.

Conclusion: the theoretical status of the availability approach

The original availability approach put forward by Tversky and Kahneman can neither predict when people use the availability heuristic for judging frequencies, nor the processes used for assessing availability. Hence, it does not even help to treat the availability approach as a conditional statement, spelling out what happens *after* the individual has decided to employ the availability heuristic. Since the model does not assume that assessments of availability require a memory search for instances, multiple processes can come into play. As a consequence, the availability approach fails to predict the psychological processes of how people arrive at frequency judgements. Therefore, the predictive power of this approach is very low. It merely makes assumptions regarding the *existence* of processes. However, a theory requires more than the making of such existential claims. Existential claims do not allow (to make) any falsifiable predictions to be made. Thus, the availability approach in its original formulation does not count as a scientific theory (Popper 1961).

Do people (sometimes) base their frequency judgements on availability?

A great deal of effort has been devoted to testing for the prevalence of the availability heuristic during the past three decades. Some tests of the availability approach used *confirmatory tests*, i.e. they focused on one version of the availability heuristic. Specifically, confirmatory tests investigated whether judgements reflect the ease by which relevant instances can be retrieved from memory. Alternative strategies, such as recall free versions of the availability heuristic or other judgemental heuristics, were not considered. In accordance with the original demonstrations of the availability heuristic (e.g. the 'famous-names' experiment by Tversky & Kahneman 1973), a couple of studies assessed availability by recall measures (Bruce et al. 1991; Lewandowsky & Smith 1983). However, there are also attempts to operationalize ease of retrieval independent from recall content in order to avoid the confound between amount of recalled instances and ease of retrieval (Gabrielcik & Fazio 1984; Rothman & Schwarz 1998; Schwarz et al. 1991; Wänke et al. 1995; see also Reber & Zupanek, this volume; Schwarz & Wänke, this volume). Regardless what kind of measure of ease of retrieval has been used, all of the above cited studies found evidence supporting the notion that people *can* employ the availability heuristic. Yet even these confirmatory tests of the availability approach indicate that the employment of the availability heuristics depends on context conditions. We will return to this issue below.

Another class of studies employed *competitive* designs. This means that assumptions of the availability approach were tested against predictions from other theories of frequency processing and judgement. The evidence obtained from such competitive tests is rather mixed. Results from a few studies suggest that ease of retrieval is the predominant psychological process underlying frequency estimation (Greene 1989; Williams & Durso 1986). Results from other studies support the view discussed earlier that people may base their judgements on availability if certain context conditions are met (Betsch et al. 1999; Manis et al. 1993). Finally, there is a considerable number of studies indicating that people do not rely on availability when judging frequencies. Watkins and LeCompte (1991) and Sedlmeier and colleagues (1998), for example, tested for various versions of availability in a series of studies, so that the proposed mechanism, if as pervasive as often assumed, had a fair chance of being detected. In these studies the influence of availability on frequency judgement was marginal at best. Frequency estimates in the studies of Brooks (1985) and Howell (1973), to give some further examples, were not also affected by availability but revealed a remarkable degree of relative accuracy. Concluding on their findings, these authors interpreted their results as evidence for the automatic encoding model (Hasher & Zacks 1984; Zacks & Hasher, this volume).

The automatic encoding model posits that frequencies are automatically encoded and stored in memory. Later, individuals can directly rely on such memory structures in their frequency judgements, even if concrete memories about the events to be judged are no longer accessible. As a consequence, frequency judgements should not reflect the transient state of memory (e.g. availability) at the time of judgement, but rather the frequencies of previously sampled information. The automatic encoding model has been supported by number of studies which are mostly hosted in the experimental literature (Alba et al. 1980; Freund & Hasher 1989; Howell 1973; Jonides & Jones 1992; Zacks, Hasher & Sanft 1982; but see Birnbaum & Taylor 1987; Hanson & Hirst 1988 for counter evidence).

Summary

Tversky and Kahneman described the availability heuristic—the assessment of the ease by which instances can be brought to mind—as one possible strategy laypeople use to form frequency judgements. The evidence from a variety of studies indicates that indeed people do *sometimes* rely on availability in frequency judgement. In convergence with the availability approach, the evidence clearly shows that the availability heuristic is just one among other strategies. Although this brief review does not allow us to specify people's general propensity to rely on availability, one might doubt that availability is the dominant strategy in frequency estimation.

When do people base their frequency judgements on availability?

Little is known to date about the dynamics of strategy choice in frequency judgement. Only a few moderating variables have been studied thus far, and occasionally the evidence is mixed.

Informational value of availability

There is some evidence, however, indicating that individuals seem to be sensitive to factors which favour or disfavour the informational value of availability. Often researchers take recall measures before they assess frequency judgement. If the recall measure is prone to render differences in availability salient, then participants are likely to rely on recall content in subsequent judgement (Betsch et al. 1999). Paper and pencil methods, for example, bear the risk that judges become aware of the differences in the number of recalled category instances while reviewing their own protocols. When they are later asked to judge the frequencies of the corresponding categories, they are prone to rely on the proportions generated in the recall protocol. If the recall procedure is altered, so that participants cannot review their own protocols, differential availability no longer impacts frequency judgements (Betsch et al. 1999, Experiments 5 and 6). The assumption that the recall tests may invite participants to rely on availability is also corroborated by studies that vary the order of recall test and frequency measure. Bruce and his colleagues (1991), for example, found that the impact of availability on judgement decreases if the recall test was administered *after* the frequency estimation task. Other factors may explicitly call the informational value of availability into question, and consequently can prevent people from relying on availability. In a couple of studies from different domains of judgement, Schwarz and colleagues accumulated evidence indicating that when availability information is considered as non-diagnostic or when its informational value is explicitly discredited by a misattribution manipulation, participants are likely not to rely on ease of retrieval (Schwarz, Bless et al. 1991; Wänke et al. 1995). Perceived diagnosticity of recall experiences might also moderate which variant of the availability heuristic is employed. In the case that ease of recall is rendered non-diagnostic, individuals may be more likely to rely on recall content (for overviews see Schwarz 1998; Schwarz & Wänke, this volume).

Encoding task

The nature of the encoding task can also affect subsequent use of the availability heuristic. When exemplars of the category are encoded and the focus of attention is placed on the category to be judged later, subsequent judgements are unlikely to be biased by availability (Betsch et al. 1999). Focusing on a category during encoding might enhance the likelihood that frequency records can be established online for the appropriate category (Barsalou & Ross 1986). Subsequent judgements can be based on the resulting memory structure without necessitating an assessment of availability (see Haberstroh & Betsch, this volume). In accordance with this notion, availability effects decrease in magnitude or vanish if the experimental procedure ensures that exemplars are completely encoded with reference to the category to be judged for frequency later. For example, Greene (1989, Experiment 2) enhanced the salience of the appropriate encoding category by presenting items of the same category in a consecutive order. This procedure led to a remarkable decrease of the impact of availability on judgement. Bruce and colleagues (1991) found no evidence for availability, when participants named the categories to which items belonged while encoding the stimuli. For further discussion of the role of encoding factors in frequency judgement, see also the contribution by Brown (this volume).

Judgement domain

Manis and colleagues (1993) proposed that the use of the availability heuristic might depend on the judgement domain. They provided evidence that individuals employ the availability heuristic to judge category frequency (set-size judgement). In contrast, when judging repetition frequency of the same event (frequency-of-occurrence judgements), people seemed to rely on unitary memory structures, which were established automatically during encoding. However, subsequent research showed that this pattern of results can be reversed when controlling for category focus during encoding and category salience at the time of judgement (Betsch *et al.* 1999).

Time

Time also seems to be an important factor. Bruce and colleagues (1991) showed that a long time delay between encoding and judgement may increase the tendency to rely on availability. The time period allocated to make a judgement also moderates the use of the availability heuristic. If participants make their judgements spontaneously without allotting much time for deliberation, frequency judgements are unbiased by availability (see Haberstroh, Betsch, Pohl & Aarts 2000).

The evidence reviewed so far merely provides a first step towards a deeper understanding of the conditions promoting the employment of the availability heuristic. To date we are not able to provide a taxonomy of conditions or even a general mechanism of strategy selection in frequency judgements.

Directions of future research and theorizing

What is the scientific progress we have achieved since the original outline of the availability approach by Tversky and Kahneman back in 1973? Today we know that people may sometimes rely on availability and sometimes on other strategies. It was not until the 1990s that researchers eventually began to take moderating factors into account in order to understand *when* people rely on the availability heuristic. Although such endeavours are promising and have already yielded first insights into strategy choice, we still know little about the underlying mechanisms of strategy selection in quantitative estimation. Taken together, we are not far beyond the may-and-might-be, which characterized the original formulation of the availability approach. Why does progress take so long?

One reason is that the availability approach has often been treated as a true theory which would allow us to predict the psychological mechanisms of frequency judgement. We showed that the original version of the availability approach is not a theory with predictive power. Although this has been lamented before (e.g., Slovic, Fischhoff & Lichtenstein 1977; Sherman & Corty 1984), the implications of this critique did not seem to precipitate hypothesis generation and research for a long time. The importance of the availability heuristic in frequency judgement was overemphasized. It was rarely treated as just one among a couple of other strategies. Rather, it was often treated as a candidate for the dominant mechanism, as evident from the numerous studies which tested whether people employ the availability heuristic or not. However, this view led researchers to fight battles at the wrong fronts. The premises of the availability approach cannot be falsified, because it is impossible to derive predictions that could fail. Any evidence showing that people make use of (other) strategies other than the availability heuristic cannot threaten the availability approach, because it does not prohibit anything. Logically, it cannot be falsified but only verified, since it simply assumes the existence of a strategy but does not make predictions when it is applied (see Popper 1961, for a detailed discussion of this problem).

We do not need further demonstrations or existential proofs that the availability heuristic or any other mechanism is applied for frequency estimation. Future research should rather focus on investigating the factors which govern the reliance on a particular strategy. Most of all we need theoretical frameworks that guide research interests in a systematic fashion. A few movements towards this direction have recently been made. For example, Brown and Siegler suggested a theoretical framework integrating three approaches to quantitative estimation: heuristics, domain-specific knowledge, and intuitive statistics (Brown & Siegler 1993). In another paper, Norman Brown (1995) provided a taxonomy to cluster estimation strategies. On a broad level, he distinguishes between enumerative and non-enumerative strategies. The entire taxonomy, consisting of various subcategories, allows clustering of almost all of the strategies discovered so far (see also Brown, this volume).

Elsewhere in this book the reader will find a couple of other attempts to achieve theoretical process, both on the subordinate level concerning moderating variables of strategy choice, and on the superordinate level of universal mechanisms of frequency estimation and strategy choice. The book will demonstrate that the field is on its way to overcoming the period of stagnancy in theory development which characterized research on frequency estimation since the availability approach has been put forward.

References

Alba, J. W., Chromiak, W., Hasher, L. & Attig, M. S. (1980). Automatic encoding of category size information. *Journal of Experimental Psychology: Human Learning and Memory,* 6:370–378.

Barsalou, L. W. & Ross, B. H. (1986). The roles of automatic and strategic processing in sensitivity to superordinate and property frequency. *Journal of Experimental Psychology: Learning, Memory and Cognition,* 12:116–134.

Betsch, T., Siebler, F., Marz, P., Hormuth, S. & Dickenberger, D. (1999). The moderating role of category salience and category focus in judgements of set size and frequency of occurrence. *Personality and Social Psychology Bulletin,* 25:463–481.

Birnbaum, I. M. & Taylor, T. H. (1987). Is event frequency encoded automatically? The case of alcohol intoxication. *Journal of Experimental Psychology,* 13:251–258.

Brooks, J. E. (1985). Judgements of category frequency. *American Journal of Psychology,* 98:363–372.

Brown, N. R. (1995). Estimation strategies and the judgement of event frequency. *Journal of Experimental Psychology: Learning, Memory and Cognition,* 21:1539–1553.

Brown, R. B. & Siegler, R. S. (1993). Metrics and mappings: A framework for understanding real-world quantitative estimation. *Psychological Review, 100*:511–534.

Bruce, D., Hockley, W. E. & Craik, F. I. M. (1991). Availability and category frequency estimation. *Memory & Cognition, 19*:301–312.

Fiedler, K. (1983). On the testability of the availability heuristic. In R.W. Scholz (ed.) *Decision making under uncertainty* (pp. 109–119). Amsterdam: Elsevier, North-Holland.

Freund, J. S. & Hasher, L. (1989). Judgements of category size: Now you have them now you don't. *American Journal of Psychology, 102*:333–352.

Gabrielcik, A. & Fazio, R. H. (1984). Priming and frequency estimation: A strict test of the availability heuristic. *Personality and Social Psychology Bulletin, 10*:85–89.

Greene, R. L. (1989). On the relationship between categorical frequency estimation and cued recall. *Memory and Cognition, 17*:235–239.

Haberstroh, S., Betsch, T., Pohl, D. & Aarts, H. (2000). When guessing is better than thinking: The strategy application model for frequency judgement. Manuscript submitted for publication.

Hanson, C. & Hirst, W. (1988). Frequency encoding of token and type information. *Journal of Experimental Psychology: Learning, Memory and Cognition, 14*:289–297.

Hasher, L. & Zacks, R. T. (1984). Automatic processing of fundamental information: The case of frequency of occurrence. *American Psychologist, 12*:1372–1388.

Higgins, E. T., Rholes, W. S. & Jones, C. R. (1977). Category accessibility and impression formation. *Journal of Experimental Social Psychology, 13*:141–154.

Howell, W. C. (1970). Intuitive 'counting' and 'tagging' in memory. *Journal of Experimental Psychology, 85*:210–215.

Howell, W. C. (1973). Storage of events and event frequency: A comparison of two paradigms in memory. *Journal of Experimental Psychology, 98*:260–263.

Jonides, J. & Jones, C. (1992). Direct coding for frequency of occurrence. *Journal of Experimental Psychology: Learning, Memory and Cognition, 18*:107–128.

Lewandowsky, S. & Smith, P. W. (1983). The effect of increasing the memorability of category instances on estimates of category size. *Memory & Cognition, 11*:347–350.

Manis, M., Shedler, J., Jonides, J. & Nelson, T. E. (1993). Availability heuristic in judgements of set-size and frequency of occurrence. *Journal of Personality and Social Psychology, 65*:448–457.

Popper, K. R. (1961). *The logic of scientific discovery*. New York: Science eds.

Rothman, A. & Schwarz, N. (1998). Constructing perceptions of vulnerability: Personal relevance and the use of experiential information in health judgements. *Personality and Social Psychology Bulletin, 24*:1053–1064.

Schwarz, N. (1998). Accessible content and accessibility experiences: The interplay of declarative and experiential information in judgement. *Personality and Social Psychology Review, 2*:87–99.

Schwarz, N., Bless, H., Strack, F., Klumpp, G., Rittenauer-Schatka, H. & Simons, A. (1991). Ease of retrieval as information: Another look at the availability heuristic. *Journal of Personality and Social Psychology, 61*:195–202.

Sedlmeier, P. (1999). *Improving statistical reasoning: Theoretical models and practical implications*. Mahwah, NJ: Lawrence Erlbaum Publishers.

Sedlmeier, P., Hertwig, R. & Gigerenzer, G. (1998). Are judgements of the positional frequencies of letters systematically biased due to availability? *Journal of Experimental Psychology: Learning, Memory and Cognition, 24*:754–786.

Sherman, S. J. & Corty, E. (1984). Cognitive heuristics. In R. S. Wyer & T. K. Srull (eds) *Handbook of social cognition* vol. 1, (pp. 189–286). Hillsdale, NJ: Lawrence Erlbaum Associates, Inc.

Slovic, P., Fischhoff, B. & Lichtenstein, S. (1977). Behavioral decision theory. *Annual Review of Psychology, 28*:1–39.

Tversky, A. & Kahneman, D. (1973). Availability: A heuristic for judging frequency and probability. *Cognitive Psychology, 5*:207–232.

Underwood, B. J. (1969). Attributes of memory. *Psychological Review, 76*:559–573.

Wallsten, T.S. (1983). The theoretical status of judgemental heuristics. In R.W. Scholz (ed.) *Decision making under uncertainty* (pp. 21–39). Amsterdam: Elsevier.

Wänke, M., Schwarz, N. & Bless, H. (1995). The availability heuristic revisited: Experienced ease of retrieval in mundane frequency estimates. *Acta Psychologica, 89*:83–90.

Watkins, M. J. & LeCompte, D. C. (1991). Inadequacy of recall as a basis for frequency knowledge. *Journal of Experimental Psychology: Learning, Memory and Cognition, 17*:1161–1176.

Williams, K. W. & Durso, F. T. (1986). Judging category frequency: Automaticity or availability. *Journal of Experimental Psychology: Learning, Memory, and Cognition, 12*:387–396.

Wyer, R. S. & Srull, T. K. (1989). *Memory and cognition in its social context.* Hillsdale, NJ: Lawrence Erlbaum Associates.

Zacks, R. T., Hasher, L. & Sanft, H. (1982). Automatic encoding of event frequency: Further findings. *Journal of Experimental Psychology: Learning, Memory and Cognition, 8*:106–166.

CHAPTER 8

A MEMORY MODELS APPROACH TO FREQUENCY AND PROBABILITY JUDGEMENT: APPLICATIONS OF MINERVA 2 AND MINERVA DM

MICHAEL R. P. DOUGHERTY AND ANA M. FRANCO-WATKINS

Abstract

A fundamental question in decision making has been how people make judgements of frequency and probability. In this chapter, we describe a new theory, Minerva DM (MDM) that ties many of the heuristics and biases together within the context of a multiple-trace memory model for frequency judgements. We provide a brief tutorial on how Minerva 2 and MDM simulates non-conditional and conditional frequency judgements, and then demonstrate how the models can be used to describe two possible accounts of the availability heuristic. We end by discussing where MDM fits within the context of other models of frequency judgement.

Introduction

A fundamental question in decision making has been how people make judgements of frequency and probability. This question has been at the heart of over four decades of research, numerous theories, and countless debates (Gigerenzer 1996; Kahneman & Tversky 1996). However, despite the attention it has received, few overarching theories of probability and frequency judgement have emerged. Instead, many of the empirical phenomenon discovered in the lab have remained a dissociated grab-bag of cognitive heuristics and their associated biases. In this chapter, we describe a new theory, Minerva DM, that ties many of the heuristics and biases together within the context of a multiple-trace memory model.

Historically, frequency-estimation processes have been investigated from two distinct, and often dissociated, paradigms. There is a long-standing history of examining frequency processing in the memory literature. From within the memory paradigm, frequency judgements have often been treated as a special case of recognition memory (Hintzman

1988), and much of this research has focused on: (a) examining the memory factors underlying frequency judgements, and (b) using the frequency estimation task as a means to test models of recognition memory. This research has revealed that frequency judgements are often affected by the same factors that affect recognition memory, including item similarity (Hintzman 1988; Hintzman et al. 1992), encoding (Greene 1988) and context effects (Brown 1995). Theoretically, frequency tasks have been used to distinguish between different models of memory, such as whether frequency judgements rely solely on a recognition process (Hintzman 1988) or a recall plus recognition process (Hintzman et al. 1992), and whether source monitoring affects memory and judgement (Hockley and Christi 1996).

In contrast to the memory paradigm, frequency judgement research in the decision making paradigm has emphasized the accuracy (or inaccuracy) of judgement, with much less regard for the cognitive (or memory) processes underlying judgement. Early research on frequency judgements was heavily influenced by the work of Tversky and Kahneman (Tversky & Kahneman 1973, 1974). They proposed that frequency judgements were made by selecting a heuristic from a tool-kit assortment and using that heuristic to make a quick and dirty judgement. For example, work on the availability heuristic suggested that frequency judgements were based on the ease with which instances could be brought to mind (Tversky & Kahneman 1973), and work on the representative heuristic suggested that judgements were based on the similarity of a target event to a prototype stored in memory (Tversky & Kahneman 1982). Application of these heuristics to explain extant data was widespread. For example, Lichtenstein et al. (1978) used the availability heuristic as a mechanism for describing how people made assessments of risk: People tended to overestimate the frequency of rare, but catastrophic events relative to more frequent, but less catastrophic events. Additionally, the availability heuristic was used to explain category judgements and judgements of set-size (Manis et al. 1993; Williams & Durso 1986).

Conceptually, there is considerable overlap between the memory processes and the decision making approaches to frequency judgement. However, despite this conceptual overlap, there has been relatively little cross-fertilization between the two domains. Many of the advances made in the memory literature have gone unnoticed by decision researchers, and conversely, memory researchers have not adequately addressed many of the phenomena in the decision literature. We argue that integrating these two approaches will lead to more advances as well as a deeper understanding of both memory and decision-making processes.

In this chapter, we present a relatively new theoretical account of frequency and probability judgement, Minerva DM (MDM) (Dougherty 2001; Dougherty et al. 1999) that draws on recent advances in memory theory. MDM is a modified version of Hintzman's (1988) Minerva 2 model, and as such retains all of Minerva 2's original capabilities as a memory model (the general term Minerva is used when describing isomorphic properties of the two models). Thus, the revised model can simultaneously account for a variety of memory and decision-making phenomena, including availability biases, conservatism, overconfidence, and base rate neglect among others, as well as make new predictions regarding the memory factors that affect frequency and probability judgement. In what follows, we present an overview of both Minerva 2 and MDM and provide sample calculations to describe how the models simulate non-conditional and conditional frequency judgements. Following this brief tutorial, we will demonstrate how the models can be used to describe two possible accounts of the availability heuristic. We will end by discussing where MDM fits within the context of other models of frequency judgement.

Model descriptions

Minerva 2 and MDM belong to the class of models known as multiple-trace memory models. Multiple-trace models assume that memory consists of a database of instances representing a person's past experiences. Accordingly, each new event that is experienced in the environment is assumed to create a *new* trace in memory. If an event is experienced multiple times, it is assumed to be represented separately, and independently, for each experienced event—rather than updating a single trace with new details each time the event occurs. Thus, a participant who studies a word list in which the word *apple* appears three times is assumed to store three copies (or traces) of the word *apple* in memory.

Retrieval is achieved by computing the similarity between a probe vector and each trace in memory. The probe vector is a retrieval cue that is assumed to be gleaned from the to-be-judged stimulus. For example, imagine that one is asked to judge the frequency of deaths that result from murder in the USA. In this case, the participant would construct a memory probe consisting of features of murders and match that probe against all traces in memory. The resulting familiarity value that arises from probing memory is assumed to be the basis of both frequency and probability judgements.

For mathematical convenience, traces and probes are represented as vectors of features, with the features taking on a tertiary code of $+1, -1$, and 0. One can think of $+1$'s and -1's as representing features that are excitatory and inhibitory respectively, and 0's as representing missing features (Hintzman 1988). Following Hintzman (1988), features in an environmental event vector are represented by a $+1$, -1, or 0 with equal probability. When environmental events are experienced, a copy of the event (the trace) is encoded in memory. Because of factors that affect encoding, such as perceptual acuity or divided attention, traces are degraded copies of the stimulus events that created them. Encoding is modelled by a learning rate parameter, L, which determines how well individual traces are stored in memory. L ranges from 0 and 1: with low values of L, few features in the original event are transferred into the memory trace. With probability $1 - L$, a feature in the original stimulus event *fails* to be copied to the memory trace and takes on the value of 0 (which corresponds to a missing feature). How well an event is stored in memory will affect the degree to which specific traces are activated when memory is probed.

Minerva and non-conditional frequency judgements

Frequency judgements are modeled using a non-conditional computational process described by three equations. Figure 8.1 illustrates a non-conditional computation between a probe vector and one trace; we refer to this illustration in our description of the equations that follow.

The first equation computes the similarity (S) between the probe with each trace in memory. Similarity is based on the extent to which the probe and trace share common

Feature j =	1	2	3	4	5	6	7	8	9
Probe	−1	+1	−1	0	+1	+1	0	−1	+1
Trace	−1	0	−1	0	+1	−1	0	−1	0
$P_j \bullet T_{i,j}$	+1	0	+1	0	+1	−1	0	+1	0
N_i	1	1	1	0	1	1	0	1	1

$$S_i = \frac{\sum_{j=1}^{N} P_j T_{i,j}}{N_i} = \frac{(+1)+(0)+(+1)+(0)+(+1)+(-1)+(0)+(+1)+(0)}{1+1+1+0+1+1+0+1+1} = \frac{3}{7} = 0.429$$

$$A_i = S_i^3 = 0.429^3 = 0.079$$

$$I = \sum_{i=1}^{M} A_i = 0.079$$

Figure 8.1 Computational example of the non-conditional process in Minerva with one trace stored in memory.

features. As can be seen in Figure 8.1, the trace and probe share several features (represented by the same feature values). The formula for computing similarity is:

$$S_i = \frac{\sum_{j=1}^{N} P_j T_{ij}}{N_i} \qquad (1)$$

where P_j is feature j in the probe, T_{ij} is feature j in trace i, and N_i is the number of corresponding non-zero features in the probe and trace. In words, to get the numerator, each feature of the probe is multiplied by the corresponding feature of the trace and then summed over all features. In Figure 8.1, the product of the first feature of the probe and the first feature of the trace is $-1 \times -1 = 1$, the product of the second feature of the probe and the second feature of the trace is $+1 \times 0 = 0$, and so forth. The sum of the trace and probe feature calculations in the example is 3. The denominator corresponds to the number of non-zero features in either the trace or the probe. Because there are 7 such non-zero features, the similarity between the probe and trace in this example is 3/7 = 0.429. Note that in this example only one trace was stored in memory. Typically there would be many traces stored in memory, one for each event (word, picture, etc.) studied in an experiment, in which case a separate S value would be computed for each trace.

The original Minerva 2 model uses a cubing function on the similarity values. This cubing function allows those traces that most closely match the probe to be weighted more heavily, while preserving the sign. There are an infinite number of functions that could be applied to the similarity value that would serve the same purpose, however, because the cubing function in Minerva 2 has enabled reasonably good data fits in memory experiments, it is retained here. Thus, the activation, A, of each trace i is given by:

$$A_i = S_i^3 \qquad (2)$$

In the example, the $A_i = S_i^3 = 0.429^3 = 0.079$.

An important assumption of Minerva is that all traces in memory contribute to the output of the model. Thus, the overall echo intensity, I, is given by the third equation:

$$I = \sum_{i=1}^{M} A_i \qquad (3)$$

I is the sum of the activation of all traces, M, stored in memory. Because the cubing function gives more weight to highly similar traces, similar traces will tend to dominate the overall echo intensity. Echo intensity in our example is equal to 0.079, as we have only one trace stored in memory. I is the output of the model, and is assumed to be the basis of both old/new (yes/no) recognition decisions and frequency judgements. In the case of frequency judgements, multiple criteria are assumed to be set to represent different subjective frequencies. For example, for a judgement scale ranging from 0 to 5, five criteria are set to discriminate between the different levels of frequency. Recognition decisions are treated as a special case of frequency judgements (frequency 1 or 0), with a single criterion. In the standard recognition memory experiment, an *old* (yes the word was on the study list) decision would be made if I were above the criterion and a *new* (no the word was not on the study list) decision would be made if I were below the criterion. The echo intensity value depends on the number of traces in memory that are similar to the probe, as well as how well traces were encoded.

Figure 8.2 Panel A characterizes how echo intensity is calculated with several traces. In this example, *apple* serves as the probe and is matched against all traces in memory. Similarity and activation are computed for each trace. Note that the three *apple* traces have different S values. This is because the learning rate parameter, L, has degraded some of the traces, and has lessened the degree to which they match the probe. Note also that this has affected the overall echo intensity, which is 2.57. As L approaches 1.0, echo intensity in this example will approach 3.0 (because there are three traces in memory, and with $L = 1.0$, Minerva becomes an event counter that includes random error from the non-matching traces). But because L is less than 1.0 in our example, the frequency output is less than 3.0. Note that the two non-matching traces contributed little to the overall echo intensity.

MINERVA DM and conditional probability and frequency judgements

Frequency judgements often involve a conditional decision that requires a more elaborative process than simple frequency judgements. At this point, the distinction between Minerva 2 and MDM is emphasized by MDM's ability to model conditional frequency and probability judgements. The conditional process involves a two-stage memory match. (Although we describe the process in terms of stages, it is assumed to be a relatively automatic process that proceeds in parallel.) Suppose a decision maker is asked to estimate 'Of males, how many are golfers?' The question of interest pertains to the frequency of golfers in the subset of males (i.e. f[golfer | male]). The first stage of the conditional process involves focusing on the condition while ignoring the conditional. The condition is used to discriminate between instances of males and females in memory, since we are only interested in instances of males. The discrimination process requires

Panel A

										S_i	A_i
P Apple	0	+1	-1	0	-1	+1	0	-1	+1		
				M							
T_1 Apple	0	+1	-1	0	-1	+1	0	-1	+1	1.0	1.0
T_2 Apple	0	+1	-1	0	-1	+1	0	-1	0	0.83	0.571
T_3 Basket	+1	+1	0	-1	0	0	-1	+1	0	0.0	0.0
T_4 Paper	0	0	+1	-1	0	+1	-1	0	+1	0.125	0.002
T_5 Apple	0	+1	-1	0	-1	+1	0	-1	+1	1.0	1.0

$$f(\text{Apple}) = I = \sum_{i=1}^{M} A_i = 1.0 + 0.571 + 0.0 + 0.002 + 1.0 = 2.57$$

Panel B

	Hypothesis mini-vector									Data mini-vector									D	H	K
P	+1	-1	0	0	-1	+1	+1	0	-1	-1	0	+1	0	+1	-1	-1	0	+1	S_i	A_i	
									M												
T_1	+1	-1	0	0	-1	+1	+1	0	-1	-1	0	+1	0	+1	0	-1	0	+1	0.83	1.0	+
T_2	+1	-1	0	0	0	+1	+1	0	-1	-1	0	+1	0	+1	-1	-1	0	+1	1.0	0.57	+
T_3	+1	0	0	0	-1	+1	+1	0	-1	-1	0	0	0	+1	-1	0	0	+1	0.66		
T_4	+1	-1	0	0	-1	+1	+1	0	0	-1	0	+1	0	+1	0	-1	0	+1	0.83	0.57	+
T_5	-1	+1	0	+1	0	+1	0	0	-1	+1	-1	0	0	+1	-1	+1	0	+1	0.14		
T_6	+1	-1	0	0	-1	+1	+1	0	-1	-1	0	+1	0	0	-1	-1	0	+1	0.83	0.57	+

$K = 4$

$$P(H \mid D) = I_c = \frac{I_{S_i \geq S_c}}{K} = \frac{1.0 + 0.57 + 0.57 + 0.57}{4} = \frac{2.71}{4} = 0.678$$

Figure 8.2 Sample calculations for non-conditional (Panel A) and conditional (Panel B) judgements. The conditional judgement assumes $S_c = 0.70$. $K = 4$ because four traces have passed the criterion value. The non-conditional frequency is given by summing the activation values for the 5 traces. The conditional likelihood is given by averaging A_i for the 4 traces that passed the S_c. Note: M refers to set of traces in memory, T_i refers to trace i, and P refers to the probe used to access traces in memory.

establishing a criterion value for assessing whether a condition has been met, and in MDM, this criterion value is the parameter S_c. Using (golfer | male), the first stage requires discriminating between instances of 'males' and 'females' in order to determine the subset of instances in memory corresponding to 'males'. The second stage involves probing the subset with the conditional statement to determine the likelihood or frequency of golfers in the subset of male instances.

In applying MDM to the golfer|male example, we assume that traces consist of multiple components, with each component corresponding to a portion of the trace (vector). Thus, traces are assumed to consist of a sex component (male vs. female) as well as a component corresponding to different athletic pastimes (golf, baseball, soccer, basketball). For generality, we can think of 'golfer' as a hypothesis (we want to assess the frequency or probability of the hypothesis that a person is a golfer) and 'male' as the data (we want to use sex as the conditioning variable for the judgement). Thus, this type of judgement corresponds to a $P(H|D)$ judgement where H is the golfer hypothesis and D is the data or information used to make the inference. Although our example contains only 'male' as the data, we could just as easily consider multiple cues to make the judgement (e.g. does the person own golf clubs, do they watch golf tournaments on television, do they belong to a golf club). A judgement based on multiple cues would be modelled with multiple data vectors (e.g., $P[H|D_1D_2D_3\ldots]$ where each D_i is a different portion of the memory vector).

Figure 8.2 Panel B illustrates the conditional process for a $P(H|D)$ judgement. The left side contains the information about H (golfer vs. non-golfer) and the right side information about D (male vs. female). In the first step, each trace in memory is compared to a probe that contains only the D information, and an S value is computed for each trace. S values for the D minivectors are shown to the right of each trace in Fig. 8.2B. S is then compared to the S_c criterion value to determine whether the condition was met. The condition is met if:

$$S_i \geq S_c \qquad (4)$$

where S_c is a criterion similarity, and S_i is the similarity between the D portion of trace i and the probe. In the example, $S_c = 0.70$, so only those traces that meet or exceed this value will be used in the second stage; traces where $S_i < S_c$ are eliminated from the process and are not used in the second stage. In our example, only 4 traces met the criterion, therefore only these will be used in the subsequent steps of the conditional process.

In the second step, the H portion of the probe is matched against the H portion of the traces in the subset. S_i is computed for each trace in the subset and then cubed to get an activation for each trace (using Equation 2). Note that in the example, the cubed similarity function is shown only for the traces that passed S_c (T_1, T_2, T_4, T_6). The *conditional* echo intensity is given by summing the activations of all the traces in the subset and dividing by the number of traces in the subset. The equation for *conditional* echo intensity is:

$$I_c = \frac{I_{S_i \geq S_c}}{K} \qquad (5)$$

where I_c is the conditional intensity and K is the number of traces that passed the criterion value. Note that equation 5 computes a mean I_c. Thus, I_c will decrease as the number of dissimilar traces passing the criterion increase. In Figure 8.2, I_c is equal to 0.678.

In the situation where the decision maker evaluates two or more hypotheses, judged probability is assumed to be equal to the ratio of the I_c for the focal hypothesis to the sum of the I_c's for all alternative hypotheses considered:

$$P(H|D) = I_c(focal) : \Sigma I_{C's}(alternatives) \qquad (6)$$

This ensures that the sum of all explicitly considered hypotheses sums to 1.0.

What factors affect conditional echo intensity? One factor is encoding quality. In general, as L approaches 1.0, the probability of a relevant trace (e.g. male) being included in the subset increases (hit rate increases). Thus, I_c will tend to be larger as L increases, because more matching traces will pass S_c. As L approaches 0.0, the probability that a relevant trace will *fail* to pass the criterion increases (miss rate increases). A second factor affecting conditional echo intensity is criterion setting. As S_c increases, only those traces with a high degree of similarity to the probe condition will pass the criterion; hence, hit rate and false-alarm rates both decrease as S_c increases. Decreasing S_c tends to lower the overall I_c because the number of dissimilar traces passing S_c increases. Because each dissimilar trace that passes S_c contributes little, if any, to the numerator of Equation 4, but contributes 1.0 to the denominator, dissimilar traces tend to decrease I_c. Both S_c and L are fundamental to MDM's predictions and affect how closely the model's estimates track Bayes' theorem in Bayesian inference problems and the degree of overconfidence in overconfidence-type problems (Dougherty 2001; Dougherty *et al.* 1999).

Simulations of frequency and conditional probability judgements

Simple frequency judgement task

Minerva 2 can be used to simulate a number of findings regarding frequency judgements. In this section, Minerva is used to simulate the results of a simple list-learning frequency-estimation experiment collected by Greene (1988). Participants in Greene's experiment demonstrated sensitivity to absolute frequencies as well as a generation effect for frequency judgements. The generation effect is the tendency for retrieval to be better for information that is created or produced by the person than for information presented externally (Slamecka & Graf 1978). In Greene's Experiment 6, words on a list appeared one, two, three, or four times. Participants had to either perform a copy or generation task. The copy task required participants to copy words to paper and required minimal elaborative encoding. In contrast, the generation task required participants to generate words from an anagram. This type of generation task generally is believed to improve encoding quality (because it involves elaborative rehearsal), thus the functional difference between the copy and generation conditions is the level of encoding: encoding was higher for the generate condition. Greene's results are represented on the left of Figure 8.3. Two results are worth noting: (1) the generation task produced higher frequency estimates than the copy task, and (2) the frequency estimates are amazingly accurate, showing little evidence of systematic bias. Can Minerva account for these results?

In Minerva, encoding is modelled with the L parameter, so tasks that improve encoding are modeled with higher values of L. Greene's results were simulated using $L = 0.97$

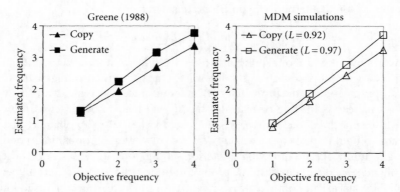

Figure 8.3 Results of Greene (1988) and Minerva 2 simulation of Greene's results. Simulations used values of $L = 0.92$ and $L = 0.97$ for the copy and generate conditions respectively.

for the generation task and $L = 0.92$ for the copy task. These values were chosen because they closely approximated Greene's data, though we did not use formal data fitting procedures. The larger L for the generation task reflects the fact that this task enhances encoding compared to the copy task. Four words were stored in memory with frequencies of 1, 2, 3, or 4 for a total of 10 instances in memory. For example, one can think of the 10 instances consisting of *apple* stored once, *chair* stored twice, *envelope* stored thrice, and *diamond* stored four times. In order to ensure stable results, 1,000 participants were used in the simulation.

The results of the simulation are presented on the right of Figure 8.3. The predictions reflect the model's ability to account for both the sensitivity to trace frequency and the generation effect. Sensitivity to trace frequency is shown by the increase in predicted frequency as a result of increased word repetitions. The generation effect is accounted for by the fact that increasing L led to increased predicted frequency.

Conditional probability judgements and the conservatism effect

MDM's conditional process can be used to model the conservatism effect. Conservatism is the tendency to underestimate the extremity of probabilities predicted by Bayes' theorem. The conservatism effect is typified by an S-shaped curve, where people are conservative for extreme posterior probabilities but nearly optimal for objective probabilities between 10:1 to 1:10. Explanations for this effect have ranged from misaggregation of true probabilities, misperception of probabilities, and response bias (DuCharme 1970), to regression effects due to random error variance (Erev *et al.* 1994). MDM's explanation departs from these previous accounts by modelling conservatism as a function of criterion setting. Specifically, as S_c decreases, likelihood estimates systematically deviate from Bayes' theorem showing the S-shaped pattern. Recall that S_c determines the number of relevant and irrelevant traces that are included in the subset and subsequently used in the conditional process. Because I_c is a mean, decreasing S_c allows for more irrelevant traces to be included in the subset, which tends to lower the overall I_c. This is most clearly seen by inspection of Equation 5: irrelevant traces that pass the criterion increment the numerator of Equation 5 very little (since the average similarity will tend towards 0.0 for non-matching traces), but increment the denominator by 1.

Figure 8.4 plots the results from two sets of simulations: one in which S_c is manipulated and the other in which L is manipulated. In both simulations, we stored 1506 instances in memory representing the heights of males and females in a hypothetical population. The exact frequencies of males and females at each height are given in Table 8.1. We used MDM's conditional process to estimate P(male|height) and P(female|height) for all heights. L and S_c were each manipulated independently to illustrate the effects of encoding and criterion setting. The left graph illustrates the effect of decreasing S_c while holding L constant at 0.75. The right graph illustrates the effect of decreasing L while holding S_c constant at 0.75. The objective probabilities, which can be computed using Bayes' theorem, fall on the identity line. In the limit, MDM predicts Bayesian performance when both S_c and L equal 1.0. As can be seen in Fig. 8.4, Bayes' theorem is approximated even with parameters less than 1.0, but these predictions become increasingly more conservative as S_c decreases. When S_c is high, few irrelevant traces pass the criterion, however, as S_c decreases, a greater number of irrelevant traces pass

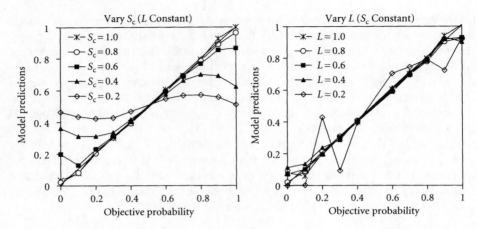

Figure 8.4 Manipulation of S_c produces the familiar S-shaped curve indicative of the conservatism effect (top). Manipulating L exaggerates variability (bottom).

Table 8.1 Distribution of hypothetical male and female heights used in conservatism simulations

Height (cm)	Males	Females	P(males|Height)
193.0	25	0	1.0
188.0	54	4	0.9
182.9	96	24	0.8
177.8	175	75	0.7
172.7	180	120	0.6
167.6	120	180	0.4
162.6	75	175	0.3
157.5	24	96	0.2
152.4	4	54	0.1
147.3	0	25	0.0
Totals	753	753	

through the criterion. This increases conservatism because mismatching (or irrelevant) traces contribute to the denominator of Equation 5, but not the numerator. L's influence on predicted probability is to increase or decrease variability. When L is high, more relevant traces pass S_c so the plots are fairly stable. However, when L is low, fewer relevant traces pass the criterion, which exaggerates the amount of variability in the plots.

In sum, our explanation of conservatism departs from previous accounts. We propose that conservatism is the result of fallible memory retrieval processes—the inability to completely discriminate between relevant and irrelevant traces. When discrimination is high (i.e. S_c is high), MDM shows little conservatism, and in fact tracks Bayes' theorem quite closely. However, MDM predicts conservatism when discrimination is moderate to low, with the amount of conservatism increasing monotonically with decreases in S_c.

Conditional and non-conditional accounts of availability biases

MDM's conditional process allows us to model a variety of judgemental biases, including biases that can be conceptualized as arising from the use of the availability heuristic. The availability heuristic is a mental shortcut that people presumably use to judge the frequency or probability of events. According to Tversky and Kahneman (1973), people make these estimates by assessing the *ease* with which instances can be brought to mind. In their original treatment, Tversky and Kahneman did not specify a process model for assessing ease, though they did speculate as to how ease *might* be assessed (see Schwarz et al. 1991). In this section, we propose two new mechanisms for availability that can account for the common biases associated with the use of the availability heuristic—we call these availability biases. In what follows, we intentionally replace the term 'availability heuristic' with the term 'availability bias' in an attempt to distance ourselves from the vagueness inherent in the original notion of the availability heuristic. Our accounts of availability are intended to provide plausible process models that predict when bias in frequency judgement might occur.

A paradigm case example of an availability bias is illustrated by the now classic famous-names study (Tversky & Kahneman 1973). In the famous name study, participants were presented with a recording of 39 names consisting of 19 names of famous men (e.g. John Wayne) and 20 names of less famous women (e.g. Lana Turner). The experiment was also done using famous women and less-famous males. Participants were asked to judge which had occurred more frequently: males or females. Participants consistently rated the less frequent, but more famous names as having occurred more often. This illustrated a bias in frequency judgement whereby the less frequent event was erroneously rated as more frequent. Subsequent analyses revealed that subjects in these experiments were able to recall more of the famous (12.3) than the less-famous names (8.4), illustrating a positive correlation between number recalled and judged frequency.

Several alternative explanations have been proposed to account for availability biases, including number of instances recalled, feeling of knowing (Tversky & Kahneman 1973), and subjective ease of retrieval (Schwarz et al. 1991) to name a few. Our explanations involve the use of Minerva's echo intensity and biases that are built into the memory system as a result of the differential encoding of events in the environment and/or differential experience of events learned in different contexts. Certainly, there is abundant evidence that frequency judgements can be accurate, as illustrated by Greene's (1988) data presented earlier and our simulations of Greene. This point was also demonstrated by Sedlmeier et al. (1998) in their research using a letter frequency-judgement task. However, under what conditions are frequency judgements inaccurate, and what memorial factors underlie these inaccuracies?

Our explanations of availability biases assume that frequency and probability judgements are based on the relative familiarity of events in memory as assessed by MDM's echo intensity calculations, without regard to overt recall. Echo intensity forms the basis of frequency and probability judgements in MDM; thus, any process that affects echo intensity also affects judgement.

One mechanism that affects echo intensity is encoding: Events that receive higher levels of encoding contribute relatively more to the output of the model. Thus, given two events that occur equally often, and have the same number of traces in memory, if one event has received better encoding it will be estimated as more frequent than the event that received poorer encoding. We refer to this as the differential-encoding hypothesis of availability biases.

Differential encoding arises from the tendency to devote more attentional resources to some (perhaps more important) environmental events while devoting less attention to other (perhaps less important) events. Differential encoding as a mechanism may very well be adaptive. For example, it is in a deer's best interest to carefully note threatening stimuli, such as a mountain lion, and remember the location of these threatening stimuli. It is much less important that the deer attends to and remembers where non-threatening stimuli reside. Everything else being equal, the differential encoding of threatening stimuli relative to non-threatening stimuli will likely lead the deer to choose the pasture where the fewest threatening stimuli have been seen. We suggest that differential encoding is a general principle of learning and memory where exciting, more interesting, or more important events are given more attention at encoding, with greater attention resulting in better encoding (Murnane and Phelps 1995), and applies to a range of stimuli including the famous names study (the assumption being that famous people are more interesting and so attract more attention). In sum, our differential-encoding hypothesis places the locus of availability biases at the encoding stage of learning, with built-in encoding biases leading to biased frequency judgements.

A second mechanism for explaining availability biases is the inability to completely discriminate between events learned in different contexts. Research on context effects in recognition memory has shown that context information aids in discriminating items learned in different contexts, but that discrimination based on context is fallible (Dougherty & Franco-Watkins 2001; Fernandez & Glenberg 1985; Thomson 1988). Again, take the famous-names study as a simple example. It is likely that participants had extensive exposure to the famous names on the list prior to the experiment. For example, a person may have experienced John Wayne in various contexts aside from the experimental context (e.g. movies, advertisements, biographies etc.). It also is likely that participants were imperfect at discriminating between the names learned prior to the experiment and those learned during the experiment, and that these pre-experimental traces influenced people's judgements of the number of times the name occurred on the list. We refer to this mechanism as the context–discrimination account of availability biases because the degree to which events learned in alternative (and irrelevant) contexts influence the judged frequency of events learned in the target context depends on how easily the two contexts can be discriminated. Again, we propose the context–discrimination hypothesis as a general principle of learning and memory that affects how well people discriminate between items learned in different contexts (Murnane & Phelps 1995; Murnane et al. 1999) which in turn affects frequency judgement (Dougherty & Franco-Watkins 2001; Hockley & Christi 1996).

The differential-encoding and context-discrimination accounts of availability biases can be demonstrated using MDM. The first example focuses on how differential encoding affects frequency judgements in the famous name experiment. A simulation was run using different values of L to model the encoding of the famous and less famous names in Tversky and Kahneman (1973). Thirty-nine instances were stored in memory: 19 famous, 20 less famous. L was estimated on the basis of the recall data in Tversky and

Kahneman, and set to the proportion of names recalled in each condition. $L = 0.65$ for famous names (12.3 out of 19 names recalled) and $L = 0.42$ for the less famous names (8.4 out of 20 names recalled). If L were identical for both lists, the less-famous names would be predicted to be more frequent because there are more less-famous names stored in memory. However, when differential encoding is introduced, the famous names are predicted to be judged more likely than the less-famous names. In our simulations, the relative echo intensities (after being normalized to 1.0) were 0.74 and 0.26 for the famous and less-famous names respectively. Converting these values to choice probabilities using Luce's Choice Axiom yields predictions consistent with the results of Tversky and Kahneman.

The second example simulates the effect of pre-experimental traces on frequency judgements, and provides a second mechanism to explain availability biases. Although Dougherty et al. (1999) did not model the context-discrimination account using the conditional-matching process, one can readily be applied. Figure 8.5 presents the results of three simulations demonstrating the effect of pre-experimental traces (traces learned in a pre-experimental context) on frequency judgement. Each simulation used 39 traces (19 famous and 20 less famous names) with the context component corresponding to the experimental context. In addition to these 39 traces, we stored 0, 26, or 46 pre-experimental context traces for each famous trace and 0, 1, or 1 pre-experimental context traces for each less-famous trace. These numbers were chosen on the basis of citation frequencies in the media of famous and less-famous names (e.g. famous male actors were cited 46 times more often than less-famous male actors). The 39 names studied in the experiment had identical experimental context components. The pre-experimental traces had pre-experimental context components. L was held constant at

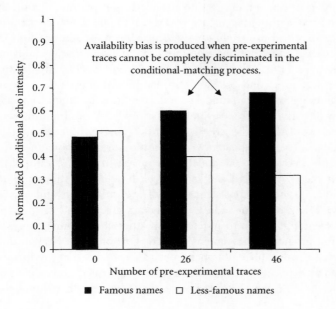

Figure 8.5 Famous names study.

0.75, and S_c was arbitrarily set to 0.50 for all simulations (though it should be noted that similar results are obtained with a variety of values of S_c, with lower values producing greater bias, and higher values producing less bias). MDM's conditional-matching mechanism was used to simulate the context-discrimination hypothesis, using the context as the condition (e.g., L[Males|experimental context]). As can be seen in the left-most pair of bars in Fig. 8.5, MDM predicts a slight advantage (higher echo intensity) for the less-famous names when no extra-experimental traces are present in memory. In contrast, the echo intensity for famous names increases as the number of pre-experimental traces increases. In the latter cases, the model predicts that the famous names will be judged more frequent, even though they actually occurred less often on the list. This result also is consistent with the results of Tversky and Kahneman (1973). Note that decreasing S_c increases the impact of the pre-experimental traces.

Both the differential (biased) encoding and context-discrimination hypotheses provide plausible mechanisms for an availability-type mechanism. We don't view these mechanisms as competing, but rather as complementary. Sometimes differential encoding might be responsible for biased judgement and other times context-discrimination might be responsible, and still other times the two may operate in combination. Certainly, these two mechanisms, as instantiated in MDM, do not provide the *only* basis of availability biases but rather provide plausible process models that explain the possible source of bias found in some frequency judgements.

Conclusions

In this chapter, we introduced MDM as a memory-based account of several decision-making phenomena, and showed how it could account for two of these phenomena: biases due to availability and the conservatism effect. Whereas MDM has been successful at simulating a number of phenomena, by no means does it provide a complete account of frequency and probability judgements. MDM can model only how memory processes influence frequency and probability judgements. It cannot account for how subjective feelings of retrieval and emotional cues are used to assess frequency, as has been shown by Schwarz *et al.* (1991) and Aarts and Dijksterhuis (1999). In addition, MDM is silent with respect to second-order confidence judgements, in contrast to Sedlmeier's (1999) PASS model which can model people's confidence in their frequency judgements. Finally, while MDM shares several properties and assumptions with Brunswikian models such as BIAS (Fiedler 1996) and PMM (Gigerenzer *et al.* 1991), it departs from these models in several important ways. Most important among these differences is that MDM is first and foremost a model of memory, whereas BIAS and PMM are primarily models of the environment. While issues of how information is represented in the environment are entirely consistent with MDM (and can be built into it's multiple-trace memory), MDM does not address these issues directly.

Our programme of research has been to systematically study the memory processes underlying frequency and probability judgement with the goal of developing an integrative theory that is able to both explain old phenomena and provide new predictions. While our focus has been on explicating the processes underlying concepts such as availability and representativeness and their resulting biases, our theorizing has shifted to a lower level of analysis in that we have focused on developing well defined and falsifiable process models. We are by no means satisfied with the verbal descriptions afforded by labels such as 'availability' and 'representativeness'. However, we believe that many of the findings born during the heuristics and biases programme are important and robust, and should be accounted for by any model of judgement and decision making. Our work with MDM has enabled us to go beyond these verbal descriptions to describe the memory processes underlying judgement and decision making.

References

Aarts, H. & Dijksterhuis, A. (1999). How often did I do it? Experienced ease of retrieval and frequency estimates of past behavior. *Acta Psycholgica, 103*:77–89.

Brown, N. R. (1995). Estimation strategies and the judgement of event frequency. *Journal of Experimental Psychology: Learning, Memory, and Cognition, 21*:1539–1553.

Dougherty, M. R. P. (2001). Integration of the ecological and error models of overconfidence using a multiple-trace memory model. *Journal of Experimental Psychology: General, 130*:579–599.

Dougherty, M. R. P. & Franco-Watkins, A. (2001). Context discrimination and source monitoring in judgements of frequency. Manuscript submitted for publication.

Dougherty, M. R. P., Gettys, C. F. & Ogden, E. (1999). Minerva DM: A memory process model for judgements of likelihood. *Psychological Review, 106*:180–209.

DuCharme, W. (1970). Response bias explanations of conservative human inference. *Journal of Experimental Psychology, 85*:66–74.

Erev, I., Wallsten, T. S. & Budescu, D. V. (1994). Simultaneous over- and underconfidence: The role of error in judgment processes. *Psychological Review, 101*:519–527.

Fernandez, A. & Glenberg, A. M. (1985). Changing environmental context does not reliably affect memory. *Memory and Cognition, 13*:333–345.

Fiedler, K. (1996). Explaining and simulating judgment biases as an aggregation phenomenon in probabilistic, multiple-cue environments. *Psychological Review, 103*:193–214.

Gigerenzer, G. (1996). On narrow norms and vague heuristics: A reply to Kahneman and Tversky (1996). *Psychological Review, 103*:592–596.

Gigerenzer, G., Hoffrage, U. & Kleinbolting, H. K. (1991). Probabilistic mental models: A Brunswikian theory of confidence. *Psychological Review, 98*:506–528.

Greene, R. L. (1988). Generation effects in frequency judgment. *Journal of Experimental Psychology: Learning, Memory, and Cognition, 14*:298–304.

Hintzman, D. L. (1988). Judgments of frequency and recognition memory in a multiple trace model. *Psychological Review, 95*:528–551.

Hintzman, D. L., Curran, T. & Oppy, B. (1992). Effects of similarity and repetition on memory: Registration without learning? *Journal of Experimental Psychology: Learning, Memory, and Cognition, 18*:667–680.

Hockley, W. E. & Christi, C. (1996). Tests of the separate retrieval of item and associative information using a frequency-judgment task. *Memory and Cognition, 24*:796–811.

Kahneman, D. & Tversky, A. (1996). On the reality of cognitive illusions. *Psychological Review, 103*:582–591.

Lichtenstein, S., Slovic. P., Fischhoff, B., Layman, M. & Combs, B. (1978). Judged frequency of lethal events. *Journal of Experimental Psychology: Human Learning and Memory, 4*:551–578.

Manis, M., Shedler, J., Jonides, J. & Nelson, T. (1993). Availability heuristic in judgments of set size and frequency of occurrence. *Journal of Personality and Social Psychology, 65*:448–457.

Murnane, K. & Phelps, M. P. (1995). The effects of changes in relative cue strength on context-dependent recognition. *Journal of Experimental Psychology: Learning, Memory, and Cognition, 21*:158–172.

Murnane, K., Phelps, M. P. & Malmberg, K. (1999). Context-dependent recognition memory: The ICE theory. *Journal of Experimental Psychology: General, 128*:403–415.

Schwarz, N., Bless, H., Strack, F., Klumpp, G., Rittenauer-Schatka, H. & Simons, A. (1991). Ease of retrieval as information: Another look at the availability heuristic. *Journal of Personality and Social Psychology, 61*:195–202.

Sedlmeier, P. (1999). *Improving statistical reasoning: Theoretical models and practical implications.* Mahwah, NJ: Lawrence Erlbaum Associates.

Sedlmeier, P., Hertwig, R. & Gigerenzer, G. (1998). Are judgments of the positional frequency of letters systematically biased due to availability? *Journal of Experimental Psychology: Learning, Memory, and Cognition, 24*:754–770.

Slamecka, N. J. & Graf, P. (1978). The generation effect: Delineation of a phenomenon. *Journal of Experimental Psychology: Human Learning and Memory, 4*:592–604.

Thomson, D. M. (1988). Context and false recognition. In G. M. Davies and D. M. Thomson (eds) *Memory in context: Context in memory* (pp.285–304). New York: John Wiley and Sons.

Tversky, A. & Kahneman, D. (1973). Availability: A heuristic for judging frequency and probability. *Cognitive Psychology, 5*:207–232.

Tversky, A. & Kahneman, D. (1974, September). Judgment under uncertainty: Heuristics and biases. *Science, 185*:1124–1130.

Tversky, A. & Kahneman, D. (1982). Judgments of and by representativeness. In D. Kahneman, P. Slovic & A. Tversky (eds) *Judgements under certainty: Heuristics and biases* (pp. 84–98). New York: Cambridge University Press.

Williams, K. W. & Durso, F. T. (1986). Judging category frequency: Automaticity or availability? *Journal of Experimental Psychology: Learning, Memory, and Cognition, 12*:387–396.

CHAPTER 9

ASSOCIATIVE LEARNING AND FREQUENCY JUDGEMENTS: THE PASS MODEL

PETER SEDLMEIER

Abstract

When events are successively encoded, people usually make adequate judgements of relative frequencies and probabilities, although high relative frequencies are generally under- and low relative frequencies overestimated. Moreover, if relative frequencies change over time the most recently encountered proportions are given more weight. The sensitivity for frequencies goes even beyond simple estimates: with increasing sample size, the accuracy of judgements and the confidence therein increases as well. It is argued that associative learning is the obvious mechanism to simulate judgements of relative frequency. PASS (Probability ASSociator), the specific associationist model proposed, consists of two parts, FEN (Frequency Encoding Network), a neural network, and the CA (Cognitive Algorithms)-module, which operates on the output of the neural network. FEN encodes events, including their contexts, by their featural description and builds up a representation of the frequency with which features co-occur. The CA-module consists of only two algorithms that suffice to model the results usually found in studies on relative frequency estimates as well as on confidence judgements about such estimates. Several extensions of PASS that allow judgements of absolute frequencies and the simulation of biased estimates are suggested, and PASS is compared to competing models that have been used to simulate relative frequency judgements.

Associative learning is based on the frequency with which events occur and co-occur—the higher the frequency of co-occurrence the higher the association between events (Slamecka 1985). Therefore, if judgements about relative frequencies or probabilities are to be made it seems straightforward to use the results of associative learning. There are associationist models of probability judgement but these are restricted to contingency learning, that is, the learning of action–outcome and cue–outcome relationships. Although such relationships are an important topic in psychological research, this chapter will *not* be concerned with contingency learning, at least not in the usual sense. It will be argued that most kinds of frequency judgements in everyday life as well as those studied in the literature are based on the encoding of events or objects in the absence of explicit contingencies. Thus, a model of contingency learning would not be adequate and therefore another kind of model is called for.

The issue of how such an associative-learning model of relative frequency judgements should look is discussed in the next paragraph. A model for relative frequency estimates has to simulate known facts about judgements of relative frequency; these are briefly reviewed. After that, a family of associative models of relative frequency judgement that obey the constraints posed by associative learning and the constraints due to empirical findings are introduced, and their simulation results are compared with the empirical data. I also explore how the models might account for biased estimates and for estimates of probabilities and absolute frequencies. Finally, associationist models are compared with exemplar models that represent each encountered item as a trace on its own and that have already been successfully used to simulate relative frequency judgements.

Associative learning as the basis for frequency estimates

Before we see how associative learning can be the basis for frequency estimates, we must clarify what associative learning is. I will argue that there are restrictions to the kind of associative learning to be used in a model that simulates judgements of relative frequency.

Two kinds of associative learning

Typically, an associative learning situation is seen as one in which there exists a contingent relationship between events (arranged by the environment or by an experimenter) that allows us to predict one or more of the events from the presence of one or more others. Predictive events can be external cues or a person's own actions and a predictive relationship can be either causal or structural (Shanks 1995). In causal relationships, one event or set of events is followed, after a certain amount of time, by another that may be used as corrective feedback. For example, there is usually a causal relationship between turning the steering wheel to the right and the car's movement to the right. In structural relationships, events co-occur without being causally related to each other, as, for instance, the sight and the smell of a steak on the grill.

Although causal relationships between events are the topic mostly studied in associative learning experiments, they do not play a dominant role in everyday estimates of frequency and probability. There, one is usually interested in the frequencies and probabilities of events without considering other co-occurring events. If there is only one event, what kind of associations are learned in this case? The view taken here, which is the view taken in most contemporary models of associative learning, that is, in connectionist models, is that objects or events can be described by their elements or features: the features present and their associations determine what kind of object or event one experiences. Learning consists in modifying the strengths of the associations between features following the exposure to a given object or event. When concentrating on features instead of events, structurally contingent relationships can be treated as complex events, that is, instead of taking into account the features of only one event, one deals with the features of all (simple) events in question. This, however, does not work with causally related events, because of the strict linear ordering (cause before effect). The family of models introduced below conforms to the second kind of associative learning: associations between features of an event or a complex event (a structural relationship).

From associative learning to frequency estimates

The PASS (*Probability ASSociator*) model (Sedlmeier 1999) regards the ability to come up with frequency and probability estimates as a by-product of associative learning. As a result of the associative learning process, concepts and categories are created in memory (Elman, Bates, Johnson, Karmiloff-Smith, Parisi & Plunkett 1996; Rumelhart & McClelland 1985). These concepts develop gradually as, for instance, when a child experiences many dogs and eventually comes up with a category for 'dog'. The strength of the concepts depends on the frequency of objects encountered. So if a child has seen only a few dogs, and she hears the word 'dog' or sees a new exemplar, the neural representation of 'dog' is only mildly activated. The more dogs she has seen, the higher the activation of the neural representation will be. This is the general mechanism postulated in PASS: The amount of activation elicited by a stimulus or prompt such as something one observes, thinks of, or is asked about is the basis for frequency and probability estimates of all sorts. The exact mechanism will be explained below.

What should be modelled?

Actual frequency estimates

'Ask about the relative numbers of many kinds of events, and you are likely to get answers that reflect the actual relative frequencies of the events with great fidelity' (Jonides & Jones 1992). Even before the beginning of school, children show this sensitivity to relative frequencies in both choice and estimation tasks (Fischbein 1975; Huber 1993; Kuzmak & Gelman 1986; Reyna & Brainerd 1994). People's sensitivity to relative frequencies, however, does not mirror actual relative frequency exactly (for a more comprehensive overview see Sedlmeier 1999). Estimates of relative frequencies or proportions usually show systematic deviations from actual relative frequencies: first, they are usually 'regressed', that is, high relative frequencies are underestimated and low relative frequencies are overestimated (Fiedler 1991; Hintzman 1969; Sedlmeier, Hertwig & Gigerenzer 1998). Second, if relative frequencies change over time the most recently encountered proportions are given more weight (Chapman 1991; Lopez, Shanks, Almaraz & Fernandez 1998; Sedlmeier & Hertwig unpublished manuscript). And third, sample size influences judgements about relative frequencies in that with increasing sample size, the judgements become more exact and are given higher confidence ratings (DuCharme & Peterson 1969; Erlick 1964; Irwin, Smith & Mayfield 1956; Sedlmeier 1998). A model of relative frequency estimates has to account for these empirical results.

What about the biases in frequency estimates?

Evidence for the sensitivity to relative frequencies has been found in studies with children and in a large number of studies with adults as well (Hasher & Zacks 1984; Hock, Malcus & Hasher 1986; Jonides & Jones 1992; Watkins & LeCompte 1991). However, there also have been contradictory findings in studies with adults. For instance, Lichtenstein,

Slovic, Fischhoff, Layman and Combs (1978) found that people gave inexact estimates of the frequency of lethal causes. For some causes such as tornadoes, cancer, and murder, estimates were exaggerated, whereas for other causes such as diabetes, tuberculosis, and asthma, estimates were too low. Is this evidence that people are not sensitive to frequencies? It turns out that people's frequency estimates are closely tied to the frequencies with which such events are reported in the news. Whereas deaths due to a tornado are reported in every newspaper, news about the death of a diabetic will seldom be published. Thus, although participants' frequency judgements did not mirror the actual state of affairs, they seem to have reflected the frequency with which these causes of death had been presented to them.

Such an explanation, however, does not hold for the results in one of the best-known demonstrations of biased estimates of relative frequency, Tversky and Kahneman's (1973) 'letter study', which is cited in almost every cognitive psychology textbook as evidence of faulty estimates of relative frequency. In that study, participants had to judge whether a certain letter (e.g. the letter 'r') appears more often in the first or the third position in English words and what the ratio of the letter appearing in the two positions would be. Tversky and Kahneman (1973) found that about two-thirds of their participants erroneously judged the majority of the letters used (which all occurred more often in the third position) to occur more often in the first position. There seems to be only one published but largely ignored replication of this study, a single page article that was unable to duplicate the original result (White 1991). Recently, Sedlmeier et al. (1998) replicated Tversky and Kahneman's (1973) study several times and consistently found that participants' estimates of the relative frequency of letters conformed well with actual relative frequencies. What can explain the different findings? One explanation for Tversky and Kahneman's original findings may be that, due to the fact that there are more letters in first than in third position in the English language (because the one- and two-letter words don't have a third letter), letters in the first position may be generally over-represented in memory (see Sedlmeier et al. 1998).

In a third well-known study (Tversky & Kahneman 1973), participants had to judge whether a list of 39 names contained more males or more females. The list included either 19 names of famous men and 20 names of less-famous women or 19 names of famous women and 20 names of less-famous men. The names of the famous sex were generally judged to be more frequent in the list. A possible explanation for this effect could be that participants were not able to totally separate the information received in the experimental task from their pre-existing knowledge. Famous people are usually famous because one reads and hears their names very often. Participants might, to a certain extent, have used their pre-experimental knowledge about frequencies in their frequency judgements during the experiment.

Taken together, these three studies, which are probably the most often cited ones for arguing that frequencies are not adequately processed, do not provide convincing evidence for the view that adults are bad frequency estimators. Sometimes, people's relative frequency estimates might rather accord with the frequencies with which *descriptions* of events occurred (e.g. in the news) than with actual frequencies; and sometimes people might not be able to separate the sources of frequency information cleanly.

The PASS family of models

To sum up the constraints for an associative model that adequately simulates relative frequency judgements: it must rely on the co-occurrence of features of successive events as the sole input and it has to learn without any corrective feedback. Its estimates of relative frequencies should roughly conform to actual relative frequencies, but the model should produce a regression effect, be responsive to a change in relative frequency over time, become more exact with increasing sample size, and put increasing confidence in judgements as sample size goes up. The PASS model attempts to obey all these constraints. PASS is not really one single model because the constraints outlined above do not suffice to narrow the possible candidate models down to a single one—rather it is a family of models. The variants differ in respect to the kind of neural network used. The crucial property of all these models is that learning must work without corrective feedback.

Architectures and learning rules

In our simulations, we have used several quite different architectures and associative learning rules, among others feed-forward models that use competitive learning (Grossberg 1976; Rumelhart & Zipser 1985), and both feedforward and recurrent architectures that use Hebbian learning (see Sedlmeier 1999). All the simulations produced roughly identical results (Sedlmeier & Köhlers, unpublished data). Therefore, only one variant is now described in more detail.

Figure 9.1 shows the architecture of this variant, which consists of a CA-module (cognitive algorithms module) that interprets the activation of FEN (frequency encoding network; the specific version of FEN described here corresponds with FEN II in Sedlmeier [1999]).

FEN learns by modifying the strength of the associations or weights between its units in response to successively encountered objects or stimuli. It consists of only one layer of units with recurrent connections (Fig. 9.2, bottom) and uses a variant of the Hebbian learning rule adapted from stimulus sampling theory (Bower 1994; Estes 1950). The learning rule that governs Δw_{ij}, that is, the change in the associative strength or the weight connecting units i and j, in a particular trial is

$$\Delta w_{ij} = \begin{cases} \theta_1(1 - w_{ij}), \text{ if both units } i \text{ and } j \text{ are active} \\ -\theta_2 w_{ij}, \text{ if either unit } i \text{ or unit } j \text{ is active} \\ -\theta_3 w_{ij}, \text{ if neither unit } i \text{ or unit } j \text{ is active} \end{cases} \quad (1)$$

The θ_1, θ_2, and θ_3 in equation 1 stand for learning, interference, and decay rates, respectively (Schwartz & Reisberg 1991). This learning rule allows weights to vary between 0 and 1. When both units are active in the input pattern, θ_1 determines the amount of change in the weight connecting these units. Interference occurs when one unit is active and the other is not, which results in a reduction of the weight connecting these two units. The amount of interference is set by θ_2. If neither of two connected

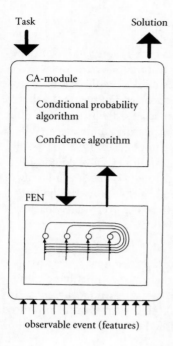

Figure 9.1 Architecture of the PASS (Probability Associator) model, consisting of the FEN (Frequency Encoding) and CA (Cognitive Algorithms) modules.

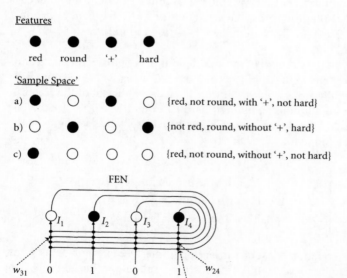

Figure 9.2 A variant of the FEN module (bottom) and some simple patterns consisting of four features each (top) which could be used as input to FEN.

units is active, the weight connecting these two units decays by a proportion of the current weight specified by the decay rate θ_3.

Consider as an example the situation depicted in Figure 9.2 (bottom) where input units I_2 and I_4 are active and the other two are not, and the weights are set arbitrarily. Assuming w_{24}, the weight connecting unit I_2 to unit I_4 being 0, and $\theta_1 = 0.1$, the updated weight after presentation of pattern S_1 would be $w_{24\ (\text{new})} = w_{24\ (\text{old})} + \Delta w_{24} = 0 + 0.1(1 - 0) = 0.1$. As another example, if w_{14} prior to the presentation of S_1 were 0.2 and $\theta_2 = 0.05$, then $w_{14(\text{new})} = 0.2 - 0.05 * 0.2 = 0.19$. And finally, if w_{13} was 0.5, and $\theta_3 = 0.02$, then $w_{13(\text{new})} = 0.5 - 0.02 * 0.5 = 0.49$. Note that the weights of FEN need not to be initialized randomly; the learning process also can start from all weights being zero (the difference in long-term learning results is negligible).

Simulation of empirical results

At all stages of learning, that is, after any number of pattern presentations (which result in successive updating of weights), PASS can be probed to make judgements of relative frequency. To every probe PASS gives a *response* that is taken to be the sum of the activations of all units to that probe. These responses are the basis for judgements of relative frequency. Take as an example the simplified sample space shown in Fig. 9.2 (top). Here, a given event is defined by four features that can be either present or absent. The sample space consists of only three events. Imagine that these three events have been repeatedly presented to the network. If we now want PASS to estimate the relative frequency of things marked with a '+' given that they are red, PASS would use the relative-frequency algorithm (see Table 9.1). In Step 1, the response of PASS to pattern (a) in Fig. 9.2 will be determined. In Step 2, the result from Step 1 and the response to pattern (c) will be summed up. Finally, the result from Step 1 is divided by that from Step 2 and this quotient gives the final estimate of the relative frequency.

Table 9.1 The two algorithms of the CA-module

Relative-frequency algorithm	Confidence algorithm
IF	IF
Task is to estimate the relative frequency of X given Y	Task is to decide in which of two estimates of the relative frequency of X given Y that differ only in sample size, $n_1 \neq n_2$, one should be more confident
THEN	THEN
1 Determine responses to all (different) patterns that contain the conjunction of patterns X and Y and sum up these responses.	1 Assess variance in FENS's unit activations elicited by pattern X after n_1 pattern presentations.
2 Determine responses to all (different) patterns that contain pattern Y as a part and sum up these responses.	2 Assess variance in FENS's unit activations elicited by pattern X after n_2 pattern presentations.
3 Divide the summed response from 1 by summed response from 2.	3 Be more confident in estimate that is associated with larger variance.
4 Use the result from 3 as your estimate.	

Regression

In PASS, the regression effect, that is, the overestimation of low and the underestimation of high relative frequencies usually found in empirical data, arises from two sources: (i) partial overlap of input patterns and (ii) a plausible amount of interference. In everyday life, almost all patterns we encode show some degree of overlap, be it in colour, size, shape, or other dimensions. For instance, all members of a category such as 'rose' or 'car' have many features in common—they show a considerable amount of overlap. The second source of regressed estimates in PASS stems from the ratio of θ_2, the interference parameter to θ_1, the learning rate (θ_3 is assumed to be usually much smaller than both θ_1 and θ_2). It is assumed that θ_2 reaches its maximum when it is equal to θ_1 but usually, interference is weaker than learning. Figure 9.3 shows the effects of feature overlap and amount of interference on PASS's relative frequency estimates of a pattern that was presented in 50% (50 out of 100) of all presentations. The three patterns used in this simulation were either '1 0 0 0', '0 1 0 0', and '0 0 1 0' (no overlap) or '1 0 0 1', '0 1 0 1', and '0 0 1 1' (25% overlap), which were presented 50, 30, and 20 times, respectively, in random order. Consider, for example, the results for the pattern that was presented in 50% of all cases. Figure 9.3 shows PASS's estimates for this pattern after every presentation, averaged over 100 runs (the sequence of 100 patterns was presented 100 times). In the beginning, PASS's mean estimate for this 50% pattern is 33% because there are three patterns and the weights are either randomly distributed or all set to zero (which makes no systematic difference). After about 40 pattern presentations PASS already estimates a relative frequency of 50%, but only if there is no overlap and if interference is at its maximum, that is, $\theta_1 = \theta_2$ (Fig. 9.3, uppermost curve). This is, however, an unrealistic state of affairs. If either features overlap or a plausible amount of interference ($\theta_1 > \theta_2$) is

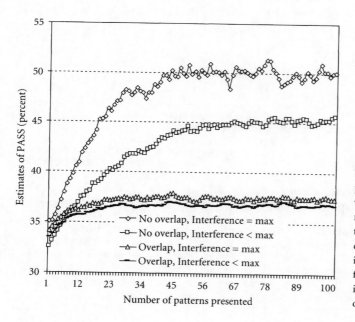

Figure 9.3 Illustration of PASS's relative-frequency and probability estimates. Mean values (averaged over 100 runs) are shown for a pattern that was presented in 50% of all cases. The figure also illustrates the impact of feature overlap and interference on the amount of regression.

used, estimates are regressed and the highest amount of regression results if both factors are combined (Fig. 9.3, lowermost curve). The parameters used in the simulation were $\theta_1 = \theta_2 = \theta_3 = 0.1$ (interference = max) and $\theta_1 = 0.1$ and $\theta_2 = \theta_3 = 0.05$ (interference < max). The results are analogous for the 30% and the 20% pattern (not shown). For these patterns, regression means, of course, that PASS's relative frequency estimates are higher than the actual values. The basic pattern of results does not depend on the absolute size of the parameters.

Changing proportions

Human relative frequency estimates show a strong recency effect if proportions change over time. If asked to judge the relative frequency of a pattern that was presented 100 times in a sequence of 200 presentations, it would, for instance, make a big difference whether the pattern is presented 10 times in the first 100 presentations and 90 times in the second 100 presentations or in the reverse order. In the first case, the relative frequency estimate could be expected to be usually larger than 50% and in the second to be considerably smaller. In several experiments using visually presented patterns, we found that participants were sensitive to a change of proportions over time and their results could be simulated well by PASS (Sedlmeier 1999; Sedlmeier & Hertwig, unpublished manuscript).

Impact of sample size

With increasing sample size, human relative frequency estimates usually become more exact and the confidence in the estimates increases. Figure 9.3 shows how PASS models the first finding: with increasing sample size, that is, a growing number of patterns presented, estimates become increasingly more exact (until they reach an asymptotic level due to the regression effect). How does PASS model the second finding, the increasing confidence? For that, it uses the confidence algorithm of the CA-module (Table 9.1). In the course of learning, FEN's weights should be increasingly well tuned to the input patterns. Recall that in the beginning all weights are either randomly distributed or all set to 0. Therefore, if a given pattern is presented more and more often, the weights can be expected to specialize on that and on the other patterns, and the activation values of the units will have an increasingly higher variance. This is why the variance in FEN's unit activations is taken as an indicator of confidence. Figure 9.4 shows how the variance of the three patterns '1 0 0', '0 1 0', and '0 0 1', which were presented to PASS 50, 30, and 20 times, respectively, in random order, increases with the number of pattern presented. Results are averaged over 100 runs and the parameters in this example were set to $\theta_1 = \theta_2 = \theta_3 = 0.1$. Again, the basic result, that is, variances increasing with sample size, is not dependent on specific patterns or parameters.

Simulation of biased estimates

One could argue that the usually regressed nature of relative frequency estimates should be regarded as a bias. If so, PASS has already been shown to be able to simulate this phenomenon. What about other biases? How could PASS simulate the overestimation of famous names (Tversky & Kahneman 1973) or of salient causes of death (Fischhoff, Slovic & Lichtenstein 1979)? To achieve this, PASS assumes that it is not possible to totally separate learning outside the experimental context from learning during the experiment. For instance, if one has read the name John Wayne some hundred times in

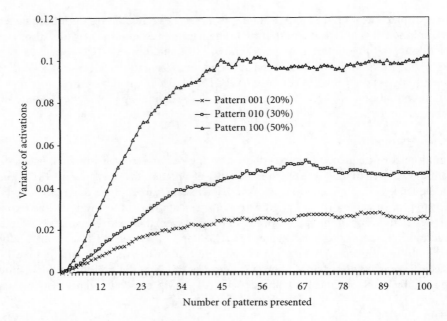

Figure 9.4 Illustration of the impact of sample size (number of pattern presentations) on PASS's confidence judgements (measured by the variance in units' activations elicited in response to a given pattern).

magazines or newspapers, the result of this pre-experimental associative learning process may have an impact on the estimate about the relative frequency of men in the experiment. If pre-experimentally learned associations are considered in addition to the associations learned during the experiment, PASS also arrives at an overestimation of the relative frequency of famous names (Renkewitz, Sedlmeier & Wettler, unpublished data). A similar explanation holds for the overestimation of salient death rates.

There are, however, results that cannot be explained by reference to pre-experimental associative knowledge. For instance, Lewandowsky and Smith (1983) found that more salient stimuli were judged to be more frequent. How can PASS account for results like these? Here, the basic assumption is that stimuli to which stronger attention is diverted are encoded with a higher learning rate θ_1. An increased learning rate leads to higher associations among the features included, which in turn elicits higher relative frequency estimates for such stimuli (Renkewitz et al., unpublished data).

Biased relative frequency estimates may arise if an inadequate prompt is used or if the frequency estimate is made on a biased sample (Fiedler 2000; Fiedler, Brinkmann, Betsch & Wild 2000). PASS will also exhibit biased estimates if it uses an inadequate prompt. If, for instance, the system would have encountered many exemplars of atypical dogs (such as dachshunds and chows) it would give a biased estimate of the relative number of dogs (say, relative to dogs and cats) if the prompt was a more typical dog such as a German shepherd.

Finally, there are demonstrations of biased estimates that would require additional mechanisms not included in PASS. Recall that PASS uses a model of implicit memory.

If in addition to relying on implicit memory, explicit deliberations determine frequency judgements (e.g. Schwarz & Wänke, this volume), PASS, in its present version, cannot yet simulate the results.

Extensions of the basic model

This paper argues that associative learning is the most plausible candidate mechanism for a model that simulates relative frequency judgements. Such a model has to learn without corrective feedback and it has to simulate several empirical results found in studies on relative frequency judgements. The PASS model introduced here learns without feedback and mimics empirical results well. In the remainder of the paper, I describe how and why PASS can be used to simulate probability estimates and how it can be used to make estimates of absolute frequency. Then I consider how the model deals with small frequencies and rich environments and finally, it is evaluated against competing models of frequency judgement.

Probability estimates

PASS was initially devised to simulate probability estimates—hence its name (*probability associator*). So far, however, we have only been talking about judgements of relative frequency. Can probability judgements be treated analogously? Not differentiating between the two can be justified on the grounds that in the widely accepted frequentist interpretation of probability, the input on which such judgements are based is the same: frequencies of events. In their review of developmental studies on probability judgement, Reyna and Brainerd (1994) concluded that children indeed use frequency information to make probability judgements. But there is also recent evidence for the equivalence of judgements about relative frequencies and probabilities (Sedlmeier 1999; Sedlmeier & Köhlers, unpublished data). In these studies, participants were seated in front of a computer screen and different visual patterns were presented, one at a time. One group of participants was asked for relative frequency and the other for probability judgements. In the relative frequency condition, participants had to estimate in what percentage of presentations each of the symbols had appeared over the whole sequence. In the probability condition, the question was about the probability with which each of the symbols would occur in the immediately following presentation (participants did not know that there would be no more such presentations). Note that these two questions differed both in perspective (past vs. future) and in whether they referred to relative frequencies or single event probabilities. Despite these differences, participants' responses did not differ in any systematic way. These results imply that a model for simple probability estimates does not have to differentiate between estimates of relative frequency and probability.

What about the numerous studies showing that the representational format, that is, probabilities versus frequencies, yields remarkably different judgements (Fiedler 1988; Gigerenzer & Hoffrage 1995; Tversky & Kahneman 1983)? On closer inspection the apparent contradiction resolves: in these studies, the difference concerned the numerical *input* used in tasks. If this input was frequencies, judgements were generally much better than if it was probabilities or percentages. Consistent with that finding, the results of training studies (Sedlmeier 1997, 1999; Sedlmeier & Gigerenzer 2001) showed that if participants learned how to transform input given in probabilities or percentages into frequencies, they were able to solve statistical tasks even when the response was required in the form of probabilities. Thus, a restriction applies to the equivalence between estimates of relative frequencies and probabilities: it only holds in cases in which the input consists of serially encountered frequencies of events. This is the sole input to PASS.

Absolute frequency estimates

PASS has not yet been used to simulate judgements of absolute frequency. This could, however, be achieved by assuming that PASS knows at least one correspondence between relative and absolute frequencies beyond the starting point zero where relative and absolute frequencies are corresponding to each other anyway. For instance, if PASS knows the total absolute frequency (e.g. the total number of pattern presentations) all frequencies in between zero and maximum can be easily determined by adding a third rule to the CA-module: If the task is to judge the absolute frequency of X, (i) determine the relative frequency of X and (ii) multiply the relative frequency of X by the total frequency. This gives the absolute frequency estimate. If, for instance, the absolute frequency was 50 and PASS's relative frequency estimate was 0.4, its absolute frequency estimate would be $0.4 * 50 = 0.2$. There is evidence that humans do it similarly, that is, start with relative frequencies and use some kind of scale or range to arrive at absolute frequency estimates (Brown & Siegler 1993; Hertwig, Hoffrage & Martignon 1999). The value for the maximum frequency to be used by PASS, however, has to be supplied from the outside. It could, for a given task, be taken from participants' additional estimates about the maximum number of patterns presented.[1] Note that with this postulated estimation process, no systematic regression effects can be expected. This result is consistent with the corresponding empirical findings (e.g. Brown, this volume).

Small frequencies and enumeration

Before the first presentation, PASS predicts that the average relative frequency estimate for any of n events is $1/n$ and then gets increasingly accurate as the number of presentations increases (see Fig. 9.3). Although one can, for instance, expect a relative frequency estimate of 1/3 if a participant knows that there are only three possible events and has not yet encountered any instances of those events (and knows nothing else about the events in question), this changes with encountering more and more events. However, the average change in estimates is not always as smooth as depicted in Fig. 9.3. Sometimes people count events explicitly or are able to enumerate them from memory. If they do so, there is a sudden increase in accuracy, followed by less accurate estimates when counting or enumeration breaks down as the number of presentations increases beyond a certain limit; after that, estimates again become increasingly accurate (Erlick 1964). In its current version, PASS only considers the implicit processes involved in frequency estimates.

---

[1] Usually frequency estimates follow a negatively accelerated function of actual frequencies (for overview see Hasher & Zacks 1984). This also holds for other kinds of estimates that can be assumed to be based on frequency information (Anderson & Schooler 1991; Landauer & Dumais 1997). The absolute frequency estimates of PASS also exhibit such a diminishing increase due to the same mechanisms that cause regression.

Detailed assumptions about when people tend to count events explicitly or enumerate them from memory are needed as a basis for an extension of the model.

Rich environments

The examples in this chapter used very 'poor' set-ups (Fig. 9.2). How does PASS behave in 'rich' environments, that is, in environments that include everyday life target events embedded in their context? Basically, PASS behaves in the same way as demonstrated above and does not need any extensions. Because it encodes an event by its features, it does not matter whether the event is simple or complex. Technically, though, the size of the input vector has to be made sufficiently large to represent all the features in question. When PASS is to make a frequency estimate of some sort for a given event, irrelevant features in the prompt, that is, features that are not part of that event, are just given a zero value.

Related models

The core of PASS is an associative memory model that integrates every newly encoded event as a change in its weights.[2] Although similar models—often termed *prototype models* (Smith & Medin 1981)—have frequently been used to simulate judgements of causal contingency, PASS seems to be the only one of this kind to be used for modelling the judgement of relative frequencies. There is, however, another class of models, often called *exemplar models* (Estes 1986), that store every newly presented event or exemplar in an ever growing matrix of memory traces. Frequency estimates are derived by calculating some measure of similarity between a prompt and all traces in memory. One of these models that has been used to mimic frequency estimates is MINERVA 2 (Hintzman 1988). MINERVA has recently been extended by Dougherty and colleagues (Dougherty, Gettys & Ogden 1999; Dougherty & Franco-Watkins, this volume) in MINERVA-DM, which allows one to model adequate as well as biased estimates. Despite clear differences in architecture and the way estimates are derived, the assumed causes for biased estimates are in part similar to those postulated in PASS. For instance, one explanation for the famous names effect relies on the same assumption used in PASS, that is, famous names are encountered more often and participants are not able to separate pre-experimentally from experimentally encountered stimuli (Dougherty & Franco-Watkins, this volume). Another model that relies on a similar representation but might not be strictly classifiable as an exemplar model because of its different underlying theoretical assumptions is Fiedler's (1996) BIAS (Brunswikian Induction Algorithm for Social cognition) model. BIAS has been successfully used to simulate many kinds of biased judgements in a social context.

Although exemplar models arrive at basically the same simulation results as PASS for many kinds of tasks, they do not (yet) have mechanisms to simulate relative frequency judgements when proportions change over time. Moreover, to date they do not include ways to model the growth in confidence with increasing sample size.

---

[2] This does not mean that only general information can be represented in such models (e.g. Rumelhart & McClelland 1985).

Conclusion

Obviously, PASS does not cover all possible kinds of frequency and probability judgements. It can be applied to simulate frequency and probability judgements if the events in question have been successively encoded either in reality or in the imagination. It also does not yet cover cases in which frequency estimates are codetermined by explicit deliberations, although it might well integrate in a more comprehensive theoretical framework that includes a model about such deliberations. These limitations aside, PASS simulates many basic results described in the literature on frequency and probability estimates (see also Sedlmeier 1999). At the moment, there seems to be no convincing evidence to make preference judgements between prototype and exemplar models and it may transpire that these two classes of models can be treated analogously at a more abstract level. However, a strong advantage of PASS in comparison to other models lies in the fact that it is firmly embedded in one of the oldest and most important theories in psychology: the theory of associative learning.

Acknowledgement

This research was supported by the German Science Foundation (SE 686/6–1).

References

Anderson, J. R. & Schooler, L. J. (1991). Reflections of the environment in memory. *Psychological Science, 2*:396–408.

Bower, G. H. (1994). A turning point in mathematical learning theory. *Psychological Review, 101*:290–300.

Brown, N. R. & Siegler, R. S. (1993). Metrics and mappings: A framework for understanding real-world quantitative estimation. *Psychological Review, 100*:511–534.

Chapman, G. B. (1991). Trial order affects cue interaction in contingency judgement. *Journal of Experimental Psychology: Learning, Memory and Cognition, 17*:837–854.

Dougherty, M. R. P., Gettys, C. F. & Ogden, E. (1999). Minerva-DM: A memory process model for judgements of likelihood. *Psychological Review, 106*:180–209.

DuCharme, W. M. & Peterson, C. R. (1969). Proportion estimation as a function of proportion and sample size. *Journal of Experimental Psychology, 81*:536–541.

Elman, J. L., Bates, E. A., Johnson, M. H., Karmiloff-Smith, A., Parisi, D. & Plunkett, K. (1996). *Rethinking innateness: A connectionist perspective on development.* Cambridge, MA: MIT Press.

Erlick, D. E. (1964). Absolute judgements of discrete quantities randomly distributed over time. *Journal of Experimental Psychology, 57*:475–482.

Estes, W. K. (1950). Toward a statistical theory of learning. *Psychological Review, 57*:94–107.

Estes, W. K. (1986). Array models for category learning. *Cognitive Psychology, 18*:500–549.

Fiedler, K. (1988). The dependence of the conjunction fallacy on subtle linguistic factors. *Psychological Research, 50*:123–129.

Fiedler, K. (1991). The tricky nature of skewed frequency tables: An information loss account of distinctiveness-based illusory correlations. *Journal of Personality and Social Psychology, 60*:24–36.

Fiedler, K. (1996). Explaining and simulating judgement biases as an aggregation phenomenon in probabilistic, multiple-cue environments. *Psychological Review, 103*:193–214.

Fiedler, K. (2000). Beware of samples! A cognitive ecological sampling approach to judgement biases. *Psychological Review, 107*:659–676.

Fiedler, K., Brinkmann, B., Betsch, T. & Wild, B. (2000). A sampling approach to biases in conditional probability judgements: Beyond base rate neglect and statistical format. *Journal of Experimental Psychology: General, 129*:399–418.

Fischbein, E. (1975). *The intuitive sources of probabilistic thinking in children.* Dordrecht, The Netherlands: D. Reidel.

Fischhoff, B., Slovic, P. & Lichtenstein, S. (1979). Subjective sensitivity analysis. *Organizational Behavior and Human Performance, 23*:339–359.

Gigerenzer, G. & Hoffrage, U. (1995). How to improve Bayesian reasoning without instruction: Frequency formats. *Psychological Review, 102*:684–704.

Grossberg, S. (1976). Adaptive pattern classification and universal recoding, I: Parallel development and coding of neural feature detectors. *Biological Cybernetics, 23*:121–134.

Hasher, L. & Zacks, R. T. (1984). Automatic processing of fundamental information: The case of frequency of occurrence. *American Psychologist, 39*:1372–1388.

Hertwig, R., Hoffrage, U. & Martignon, L. (1999). Quick estimation: Letting the environment do the work. In G. Gigerenzer, P. M. Todd, & the ABC Research Group (eds) *Simple heuristics that make us smart* (pp. 209–234). New York: Oxford University Press.

Hintzman, D. L. (1969). Apparent frequency as a function of frequency and the spacing of repetitions. *Journal of Experimental Psychology, 80*:139–145.

Hintzman, D. L. (1988). Judgements of frequency and recognition memory in a multiple-trace memory model. *Psychological Review, 95*:528–551.

Hock, H. S., Malcus, L. & Hasher, L. (1986). Frequency discrimination: Assessing global-level and element-level units in memory. *Journal of Experimental Psychology: Learning, Memory, and Cognition, 12*:232–240.

Huber, O. (1993). The development of the probability concept: Some reflections. *Archives de Psychologie, 61*:187–195.

Irwin, F. W., Smith, W. A. S. & Mayfield, J. F. (1956). Tests of two theories of decision in an 'expanded judgements' situation. *Journal of Experimental Psychology, 51*:261–268.

Jonides, J. & Jones, C. M. (1992). Direct coding for frequency of occurrence. *Journal of Experimental Psychology: Learning, Memory, and Cognition, 18*:368–378.

Kuzmak, S. D. & Gelman, R. (1986). Young children's understanding of random phenomena. *Child Developmentelopmentelopment, 57*:559–566.

Landauer, T. K. & Dumais, S. T. (1997). A solution to Plato's problem: The latent semantic analysis theory of acquisition, induction, and representation of knowledge. *Psychological Review, 104*:211–240.

Lewandowsky, S. & Smith, P. W. (1983). The effect of increasing the memorability of category instances on estimates of category size. *Memory & Cognition, 11*:347–350.

Lichtenstein, S., Slovic, P., Fischhoff, B., Layman, M. & Combs, B. (1978). Judged frequency of lethal events. *Journal of Experimental Psychology: Human Learning and Memory, 4*:551–581.

Lopez, F. J., Shanks, D. R., Almaraz, J. & Fernandez, P. (1998). Effects of trial order on contingency judgements: A comparison of associative and probabilistic contrast accounts. *Journal of Experimental Psychology: Learning, Memory, and Cognition, 24*:672–694.

Reyna, V. R. & Brainerd, C. J. (1994). The origins of probability judgement: A review of data and theories. In G. Wright & P. Ayton (eds) *Subjective probability* (pp. 239–272). Chicester, England: Wiley.

Rumelhart, D. E. & McClelland, J. L. (1985). Distributed memory and the representation of general and specific information. *Journal of Experimental Psychology: General*, 114:159–188.

Rumelhart, D. E. & Zipser, D. (1985). Feature discovery by competitive learning. *Cognitive Science*, 9:75–112.

Schwartz, B. & Reisberg, D. (1991). *Learning and Memory*. New York: Norton.

Sedlmeier, P. (1997). BasicBayes: A tutor system for simple Bayesian inference. *Behavior Research Methods, Instruments, and Computers*, 29:328–336.

Sedlmeier, P. (1998). The distribution matters: Two types of sample-size tasks. *Journal of Behavioral Decision Making*, 11:281–301.

Sedlmeier, P. (1999). *Improving statistical reasoning: Theoretical models and practical implications*. Mahwah, NJ: Erlbaum.

Sedlmeier, P. & Gigerenzer, G. (2001). Teaching Bayesian reasoning in less than two hours. *Journal of Experimental Psychology: General*, 130:380–400.

Sedlmeier, P., Hertwig, R. & Gigerenzer, G. (1998). Are judgements of the positional frequencies of letters systematically biased due to availability? *Journal of Experimental Psychology: Learning, Memory, and Cognition*, 24:754–770.

Shanks, D. R. (1995). *The psychology of associative learning*. Cambridge, England: Cambridge University Press.

Slamecka, N. J. (1985). Ebbinghaus: Some associations. *Journal of Experimental Psychology: Learning, Memory, and Cognition*, 11:414–435.

Smith, E. E. & Medin, D. L. (1981). *Categories and concepts*. Cambridge, MA: Harvard University Press.

Tversky, A. & Kahneman, D. (1973). Availability: A heuristic for judging frequency and probability. *Cognitive Psychology*, 4:207–232.

Tversky, A. & Kahneman, D. (1983). Extensional versus intuitive reasoning: The conjunction fallacy in probability judgement. *Psychological Review*, 90:293–315.

Watkins, M. J. & LeCompte, D. (1991). Inadequacy of recall as a basis for frequency knowledge. *Journal of Experimental Psychology: Learning, Memory, and Cognition*, 17:1161–1176.

White, P. A. (1991). Availability heuristic and judgements of letter frequency. *Perceptual and Motor Skills*, 72:34.

CHAPTER 10

FREQUENCY, CONTINGENCY AND THE INFORMATION PROCESSING THEORY OF CONDITIONING

C. R. GALLISTEL

Abstract

The framework provided by Claude Shannon's (1948) theory of information leads to a far-reaching, more quantitatively oriented reconceptualization of the processes that mediate what is commonly called associative learning. The focus shifts from processes set in motion by individual events to processes sensitive to the information carried by the flow of events. The conception of what properties of the conditioned and unconditioned stimuli are important shifts from the tangible properties that excite sensory receptors to the abstract and intangible properties of number, duration, frequency and contingency, which are the carriers of the information.

Frequency is an abstraction built on abstractions—one intangible, number, divided by another intangible, time. A contingent frequency raises the pyramid of abstractions still higher. It is the rate at which an event occurs following the onset or offset of a conditioning event. Sensitivity to contingent frequency requires frequency estimation together with the estimation of contingency, which is itself a forbidding abstraction (defined below). Nonetheless, it has begun to appear that brains routinely compute conditional (contingent) frequencies. Their ability to do so may explain much that has been seen as the work of association formation, a process that does not operate on abstractions like number, time and contingency.

In 1968, Robert Rescorla reported a simple experiment that changed in fundamental ways our conception of what has generally been called the associative process, the process that mediates Pavlovian conditioning, and, arguably, much else. The implications were so unsettling that they have not to this day been well digested by the community that studies learning, particularly by those that study it from a neurobiological perspective. What the results suggested was that the simple learning observed in Pavlovian conditioning paradigms arises from an information processing system sensitive to the contingencies between stimuli. If this implication is valid, then it changes our conception of the level of abstraction at which this basic learning process operates.

Rescorla's experiment followed from his reflections on the proper control procedures in Pavlovian conditioning experiments, published the previous year (Rescorla 1967). Excitatory Pavlovian conditioning pairs a motivationally neutral stimulus (CS), like a tone or the illumination of a key, with a motivationally important stimulus (US), like a mildly painful shock to the feet or the delivery of food. The tone comes on and some time later the shock is felt; or, the key is illuminated and some time later food is delivered. After a few or several such experiences, the rat freezes and defecates when the tone comes on, and the pigeon delivers food pecks on the illuminated key, suggesting in both cases that the subject anticipates the imminent appearance of the US. The learning in these paradigms appears to be a simple manifestation of the basic associative conception, which is that new conducting links are created in the nervous system by the temporal pairing of stimulus events. By these links, the CS excites the conditioned response. Therefore, Pavlovian conditioning has long been the paradigm of choice for investigating the properties of association formation.

That temporally pairing the CS and US was a sine qua non for association formation was long assumed to be self-evident (and it still is in some neurobiological circles—cf. Usherwood 1993). To prove that a change in behaviour was a result of association formation, experimenters typically ran an experimental condition in which two stimuli were repeatedly presented together (temporally paired) and a control condition in which they were widely separated in time (not temporally paired). If the change was seen in the experimental condition but not in the control condition, then the underlying process was said to be associative.

Temporal pairing and contingency had been tacitly assumed to be one and the same thing until Rescorla (1967) pointed out that they could be dissociated, but that the commonly used control procedures did not do so. Control conditions in which the CS and US are never paired do not eliminate CS–US contingency, they replace one contingency with another. Rescorla (1967) pointed out that to determine whether it is temporal pairing or contingency that drives the conditioning process, one has to use the truly random control. In this control, the occurrence of the CS does not restrict in any way the time at which the US can occur, so the US must sometime occur together with the CS, assuming that the CS is not a point event. Thus, the truly random control eliminates contingency but not temporal pairing, whereas the usual control eliminates temporal pairing but not contingency.

Rescorla (1968) realized that the most interesting version of the truly random control would be one that did not affect the temporal pairing in the usual experimental condition at all. He achieved this as follows: the experimental condition (really, now, the control) was approximately the usual condition for producing what is called the conditioned emotional reaction (more simply, conditioned fear). Hungry rats were first trained to press a lever in an experimental chamber to obtain food, until they did so readily and steadily throughout two-hour sessions. Then came five sessions during which the lever was blocked. In each of these sessions, twelve 2-minute long tones came on at more or less random intervals during each session. During these sessions, subjects also experienced occasional very short, mildly painful shocks to their feet. What Rescorla manipulated was the distribution of the shocks relative to the tones. For one group, the shocks, 12 of them per session, were completely contingent on the tone: they only occurred when it was on. The rats in a second group, a truly random group, also got 12 shocks each session while the tone was on, and they also got shocks at an equal frequency (0.5/minute) during the intervals when the tone was not on. This protocol did not alter the number or frequency of tone–shock pairings, but it eliminated the tone–shock contingency. It also greatly increased the total number of shocks per session. To check whether that mattered, Rescorla ran a second truly random group: they got 12 shocks (the same total as the first group) but distributed at random without regard to the tone.

Before testing for the extent to which the rats in the different groups had learned to fear the tone, Rescorla first eliminated their fear of the experimental chamber, with two more sessions in which the lever was unblocked and there were no tones and no shocks. By the end of these two sessions, the rats had resumed pressing the lever for food. In several final sessions, the rats' conditioned fear of the tone was then measured by the effect of the tone on their willingness to continue pressing. If they feared the tone, they froze when it came on, and did not resume pressing until it went off. Although the rats in the first two conditions had the same tone–shock pairings, the rats in the contingent condition learned to fear the shock, whereas the rats in the truly random condition did not, nor did the rats in the other truly random condition. Thus, it is contingency and not temporal pairing that drives simple Pavlovian conditioning.

What is commonly called inhibitory conditioning also implies that simple conditioned behaviour is a consequence of contingency, not temporal pairing. It is called inhibitory conditioning because the protocols for producing it are inverses of excitatory conditioning protocols. They are such that when the CS occurs, the US will not occur. The simplest protocol is the explicitly unpaired protocol: the US occurs only when the CS is absent (e.g. Rescorla 1966). The conditioned response—what the animal learns to do or not do—is the inverse of the excitatory response (LoLordo & Fairless 1985; Rescorla 1969). The subject approaches and manipulates an inhibitory CS when it predicts a motivationally negative US. (For example, subjects approach the speaker from which the tone emanates if shocks never occur while the tone is on.) On the other hand, subjects avoid or refuse to manipulate a CS when it predicts that a positive US will not occur. (For example, if food never occurs when the key is illuminated, pigeons learn to distance themselves from the illuminated key.)

Students of basic learning have always recognized that inhibitory conditioning was just as fundamental as excitatory conditioning (Hull 1929; Pavlov 1928), but they have not faced the full implications of the fact that it results from systematically *not* pairing the CS and the US. Between them, inhibitory conditioning experiments and the truly random control experiments demonstrate that temporal pairing is neither necessary nor sufficient to produce associative learning (for reviews of the evidence, see Gallistel 1990; Gallistel & Gibbon 2000). What does appear to be necessary and perhaps sufficient is contingency.

Why has evidence that contingency, not temporal pairing, drives simple conditioning been so unsettling? Rescorla (1972, p. 10) put his finger on the problem, when he wrote:

We provide the animal with individual events, not correlations or information, and an adequate theory must detail how these events individually affect the animal. That is to say that we need a theory based on individual events.

In other words, within the framework in which we have been wont to think about learning, the process had to operate at the level of events. It could not operate at the level of correlations and information. We do in fact provide the animal with correlations and information in conditioning experiments, because when we construct a conditioning protocol we correlate the events in such a way that the CS provides information about the temporal locus of the US. What Rescorla presumably meant was not that we do not provide the animals with information, but rather that we take it for granted that the learning process does not operate at that level of abstraction; we assume that it is sensitive only to the individual (physically definable?) events, not to the information they convey.

In sum, the evidence that conditioning is driven by contingency unsettles us because it challenges what we have taken for granted about the level at which basic learning processes operate. Contingency, like number, arises at a more abstract level than the level of individual events. It can only be apprehended by a process that operates at the requisite level of abstraction—the level of information processing. A US is contingent on a CS to the extent that the CS provides information about the timing of the US.

The information an event provides to a subject is measured by the reduction in the subject's uncertainty (Shannon 1948). This way of defining information leads directly to a rigorous quantification of information. It also accords with our every day intuition that the more information we have, the less uncertain we are. However, a disturbing feature of this definition is its subjectivity. The information conveyed by an event cannot be specified without reference to the subject's prior knowledge. What that means in the present case is that if the CS does not reduce the subject's uncertainty about the timing of the US, then it does not provide information.

The conceptual upheaval in the study of learning in the late 1960s was the result of a series of experiments demonstrating that whether a conditioned response developed or not depended on the information about the US provided by the CS (Kamin 1969a; Rescorla 1968; Wagner, Logan, Haberlandt & Price 1968). These experiments were unsettling because they suggested that the conditioning process manifests the subjectivity inherent in Shannon's definition of information. Learning theorists have been struggling with these implications ever since (Barnet, Grahame & Miller 1993; Mackintosh 1975; Pearce & Hall 1980; Rescorla & Wagner 1972; Wagner 1981). What they have not done, however, is adopt the conceptual framework proposed by Shannon, despite its seeming relevance, its mathematical rigour, and its success in physics and engineering. The reluctance to adopt this framework appears to flow from two sources: dislike of the subjectivity inherent in it and a reluctance to believe that simple learning processes operate at so abstract a level. Here, I analyse some central issues in the study of conditioning from an information theoretic perspective.

Understanding conditioning from an information theoretic perspective

The subject's prior estimate of US frequency is the baseline from which the information provided by a CS may be measured. The information provided by the CS depends on the difference between its uncertainty about the timing of the next US before the CS comes on, and its uncertainty after it has come on. Its uncertainty before the CS comes on depends on the information about US frequency provided by other stimuli experienced up to and including the experience of the CS onset (or offset).

Knowledge of the expected US frequency pure and simple (expected number of USs per unit of time) constitutes one end of an informational continuum about the timing of the US. The other end of the continuum is knowledge of the exact moment at which the US will occur. These two states of knowledge are represented graphically in Fig. 10.1, which plots the cumulative probability that the US will have occurred as a function of the time that will have elapsed. The curve for the random rate process represents the maximum state of uncertainty (given any knowledge at all). At the other of the continuum is perfect certainty, which gives the step function (dashed line) in Fig. 10.1.

When all that a subject knows is the number of USs so far observed and the duration of the interval of observation, its uncertainty about the time of occurrence of the next US is maximal because the probability that the US will occur in the next instant, is—so far as the subject knows (!)—constant, no matter how long it has been since the last US. This is the maximum possible state of uncertainty about the timing of the next US. It may or may not correspond to reality. If a random rate process generates the USs, then greater certainty than this is not attainable. The cumulative probability of the next US under random-rate conditions is given by the random rate curve in Fig. 10.1. The abscissa is time (t), with the origin (time 0) at the present moment. The longer the subject waits (the greater t is), the more certain it is that the next event will have occurred, that is, that t_e, the time of occurrence, is less than t. The probability that the US will have occurred approaches 1 exponentially, at a rate specified by the rate parameter, λ. Random rate processes are specified entirely by this one parameter (in contrast, for example, to Gaussian renewal processes, which require two parameters for their specification, a mean and standard deviation). The reciprocal of λ is T_e, the expected time to the next event.

A defining feature of a random rate process is that the cumulative probability function does not change as the present moment advances into the future. Put another way, the expected interval to the next event, T_e, is the same at every moment, it does not decrease as the interval since the last event lengthens. This property of a random rate

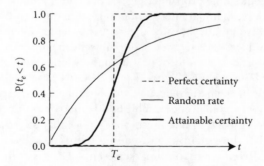

Figure 10.1 These curves represent the limits of certainty regarding the time of occurrence of a future event (t_e) when the expected interval to the next occurrence is T_e. The random rate curve represents the state of affairs when T_e is all that the subject knows. The perfect certainty curve represents the case when the subject knows that the next occurrence is a Gaussian random variable with expectation T_e and zero standard deviation (zero temporal uncertainty). Perfect certainty is not attainable. Attainable certainty is limited by the minimum psychologically attainable value for the standard deviation of the distribution of expected intervals. The attainable certainty curve is the integral of a Gaussian probability density function with expectation T_e and a standard deviation equal to the coefficient of variation (in essence, the Weber fraction) times the expectation.

process is counter-intuitive. Even when we know we are observing an approximation to a random rate process such as the rate of jackpots from a slot machine being played steadily, it is hard to resist the conviction that the longer it has been since the last jackpot, the shorter the time until the next one.

A corollary of this defining property of a random rate process is that the interval to the first observed event gives an unbiased (if often inaccurate) estimate of the expected (average) interval between events, regardless of the moment at which one begins to observe the process. A further corollary is that the subject begins to gain information about a random rate process as soon as it begins to observe it, even before the process generates an observable event. The observer/subject gains information even when nothing is happening, simply from the act of observing!

This startling manifestation of the subjectivity of information arises for the following reason: the subject's uncertainty about the rate it will observe is maximal at the moment it begins to observe, because—absent any prior knowledge about what it is going to observe—any observable rate is as likely as any other. Thus, at the outset of observation, the chances of observing a rate of about 1/s are the same as the chances of observing a rate of about 1/day—so far as the subject knows! (What the chances may in be in point of fact is neither here nor there. When we focus on information, we are concerned not with what is, but with what the subject knows about what is.) After 2 hours of observation without an event, these possibilities are no longer equally likely. The a posteriori probability that the rate is anywhere near 1/s given that one has observed an interval of 2 hours without an event is smaller by many orders of magnitude than the prior probability of such a rate (the probability at the outset of observation). By contrast, this same observation—this same stretch of nothingness—has little effect on the likelihood that the rate the subject is observing is on the order of 1/day, because an interevent interval of two hours or longer is highly likely when the rate being observed is on the order of 1/day (but wildly unlikely when the rate being observed is on the order of 1/s). Thus, the subject gains knowledge of the probable range of base rates even before it observes the first event. The length of an initial interval of observation during which nothing occurs reduces the uncertainty about what value the frequency being observed will turn out to have. Given an interval without any events, the subject can construct a curve representing the lower limit of its current uncertainty about how long it will be to the next event by assuming that the base rate is no higher than 1 divided by the interval of observation.

In sum, an estimate of the base frequency, which estimate can be partly formed even in the absence of any observed events, establishes the maximum uncertainty about the timing of the next event. Further events can only reduce that uncertainty. The limiting case at the other end of the certainty continuum is when the subject knows exactly the expected interval, T_e, from a temporal marker event—CS onset, for example—to the US. If the origin in Fig. 10.1 is at CS onset, then in this case, the cumulative probability that the US has occurred is zero for all $t < T_e$ and 1 for all $t \geq T_e$. In this case, of course, the expected interval to the next event is not constant as the present moment advances into the future; it decreases as t increases, that is, as the moment T_e gets closer.

Perfect certainty about the timing of a future event is unattainable because expected intervals cannot be known with infinite precision. This is a physical truth—or at least a truth of physics—but it takes a form in psychology that helps explain our intuitive sense that temporal pairing is critical to learning. One of the important findings to emerge from the study of human and animal timing is that Weber's law applies to remembered temporal intervals (Gibbon 1977; Killeen & Weiss 1987): the uncertainty regarding the true value of a remembered interval is proportionate to the remembered interval. Expected intervals are always remembered intervals, because expectations depend on information provided by past experience and carried forward in memory. Thus, the shorter the expected interval between a CS and a US, the more the CS reduces our uncertainty about the timing of the US. This information-theoretic consideration may explain why close forward temporal pairing has always seemed to be somehow important in Pavlovian conditioning—despite the extensive experimental evidence showing that it does not drive the conditioning process. If the CS does not precede the US, then it cannot reduce the subject's uncertainty about the timing of the US. That is why forward pairing is important (Barnet, Arnold & Miller 1991; Barnet, Cole & Miller 1997; Matzel, Held & Miller 1988; Miller & Barnet 1993). Secondly, the residual uncertainty about the timing of the US is proportional to the CS–US interval (Weber's law), so the shorter that interval, the smaller the residual uncertainty. More generally, these considerations suggest an explanation for our intuitive sense that close forward temporal pairing creates a strong connection between any two stimuli. This intuition probably accounts for some of the enduring appeal of the associative theory of learning.

Weber's law establishes the limit of attainable certainty, because the Weber fraction establishes the limit to the accuracy with which we can know an unvarying interval (without resort to artificial methods and aids). Because of this irreducible uncertainty, our knowledge of a fixed temporal interval takes the form of a Gaussian probability density function, with a standard deviation proportionate to its mean. The coefficient of variation (in essence, the Weber fraction) is the ratio of the width of the distribution (as measured by its standard deviation) to its mean. The 'attainable certainty' curve in Fig. 10.1 is the integral of this distribution.

These information-theoretic considerations also allow us to understand in a rigorous but intuitively accessible way the experimental results that inspired the conceptual upheaval in learning theory at the end of the 1960s. Rescorla's (1968) experiment, which has already been discussed, was one of them. A second set of results published at the same time came from Kamin's (1967, 1969a, b) experiments establishing the phenomena of blocking and overshadowing.

In a blocking experiment, one CS is paired with the US and when the conditioned response to this CS is well established, a second CS is introduced, which is always presented together with the first. Although it is thereby repeatedly paired with the US, the subject does not learn to respond to it (when it is presented alone on test trials). The conditioning to the first CS is said to block conditioning to the second CS.

As Kamin realized, the essence of the blocking phenomenon is that the second CS does not tell the subject anything it does not already know. More technically, it does not convey any information—in Shannon's (1948) rigorous sense. Recall that the information conveyed by an event—in this case a CS onset—depends on the subject's prior knowledge, because information is defined in terms of its effect on the recipient's uncertainty. The second CS does not reduce the subject's uncertainty about the time of US occurrence, because its onset and offset coincide with those of the first CS.

The blocking phenomenon makes it clear once again that what matters in conditioning is the information conveyed by a CS, not whether it is temporally paired with the US or not. And this is just another way of saying that conditioning is driven by contingency, because a US is contingent on a CS to the extent that the CS conveys information about the timing of the US. Although Kamin often wrote that the CS did not become conditioned to the US unless the subject was 'surprised' by the US, he made no use of the conceptual framework created by Shannon; hence he was not able to give a rigorous, let alone quantitative, formulation of what it meant to say that a US was or was not surprising.

Overshadowing protocols are like blocking protocols except for the omission of the initial phase in which one of the two CSs is alone paired with the US. The two CSs are presented together from the outset of the experiment, but the subject learns to respond to one or the other but not both (Kamin 1969b; Mackintosh 1971; Reynolds 1961). Again, because they have exactly the same temporal relation to the US, both CSs convey exactly the same information. This means that when the information conveyed by one of them alone is taken into account, the other conveys no information. What the conditioning process cares about is the information, not the carriers of that information, so it fastens on one carrier or the other.

Finally, there was a more complex experiment from the laboratory of Allan Wagner (Wagner *et al.* 1968), which showed that when the same information was conveyed both by a single CS and by some combination of CSs, the conditioning process picked out the one CS that by itself conveyed all the information rather than responding to two other CSs that between them conveyed the same information. This result is usually summarized by saying that subjects in conditioning experiments pick out the more valid CS, but what exactly is meant by valid is unclear. A more rigorous formulation is to say that subjects respond to the minimum set of cues that conveys all the available information about US timing. If one cue by itself reduces the uncertainty about the timing of the US to a point where the other cues cannot between them effect any further reduction, then that is the only cue the subject responds to.

What the above amounts to is turning Rescorla's perplexity on its head: what matters in conditioning are not the individual events but rather the information that they convey. We need a theory that deals with the information in the stream of events.

The information theoretic perspective also resolves a perplexity that has confronted thoughtful students of learning without being resolved for so long that they have generally decided to simply ignore it. The perplexity concerns how we are to understand the process of extinction if we take it as given that 'we need a theory based on individual events' rather than on information. When we stop pairing the US with the CS, the subject eventually stops responding to the CS. This is the phenomenon of extinction. Long before it stops responding to the CS, the subject gives unmistakable evidence that it knows that the CS is no longer predicting the US. If the CS predicts a US that the subject likes, for example, food, then the subject gives clear signs of what we empathetically call frustration (Amsel 1962). The perplexity is that the subject in extinction reacts not to events but rather to the failure of events (USs) to occur.

But can we not conceive of this failure as itself an event? This has been suggested as a remedy for our perplexity; it has been suggested that the failure of a US to occur can itself be conceived of as a US—a no-US (Dickinson 1989). On this view, extinction results when we stop pairing a US with the CS and start pairing a no-US. The first objection to this remedy is that it does violence to what has historically been considered the sine qua non of an event, namely, that it generate stimulus fluxes capable of exciting sensory receptors. There can be no sensory receptors sensitive to no-USs, because a no-US has no physical properties. In a materialist neurobiology, only physical stimuli can excite sensory receptors. Part of the motivation for seeking a theory that operates at the level of individual events rather than at the level of information is to have a neurobiologically transparent theory. In the theory as usually understood, the temporal pairing of events that excite different sensory inputs produces changes in synaptic conductances so that signals generated by CSs can excite conditioned responses by way of these new neuronal connections. Introducing events that have no physical properties into such a theory is awkward.

There is a deeper problem with this suggestion. Whatever other properties an event may or may not have, it surely has a locus in time. One can say when it happened. The problem with no-US events is that they may not have a specifiable time of occurrence. Consider an experiment in which we reinforce a CS at a random rate, which is what Rescorla (1968) did in the experiment with which we began this analysis. As long as the CS was on, the US occurred at a random rate. When the CS was not on, there was another random rate of US occurrence (the background rate). The CS provided information about the timing of US occurrences to the extent that these two rates differed.

Under Rescorla's conditions, the CS did not specify where the US would occur within the CS. It was equally likely at any moment. Suppose we put a subject thus conditioned into extinction; that is, when the CS is on, USs no longer occur. If we think that extinction can be explained by the occurrence of no-USs, we must ask when the no-USs occur? It is impossible to say because only moments when the CS is on are relevant and the US was equally likely at all such moments. But if we cannot in principle say when the no-USs occur how can we talk about the processes set in motion by these no USs? When are they set in motion? Must we conclude—falsely—that extinction of the conditioned response to the CS will not be observed under these conditions?

The phenomenon of extinction on its own ought to convince us that the information theoretic perspective is a profitable perspective from which to understand conditioning, because the perplexity it poses vanishes when we consider it from this perspective. Recall that intervals when nothing happens are themselves carriers of information about rate. Rieke *et al.* (1997) apply information theory to the analysis of neural coding in simple preparations and show rather convincingly that the information is carried not by the spikes but rather by the intervals between the spikes, which is why a short neural signal can convey many more bits than there are spikes in the signal. When we put Rescorla's (1968) subjects into the extinction phase of the experiment, they experience a lengthening cumulative interval during which the CS has been on but there has been no US. As already explained, this lengthening interval when nothing happens conveys information about the upper limit on the current rate of US occurrence. There comes a time when the limit implied by this ever-lengthening interval lies well below the rate at which the US used to occur. Thus, the certainty that the rate of US occurrence in the presence of the CS is not what it used to be gets greater and greater the longer the extinction phase of the experiment continues. There is nothing mysterious about extinction when we explain it at the level of information, the mystery only arises when we try to explain it by processes set in motion by individual events. We need to focus instead on processes set in motion by the duration of the intervals in which nothing happens, processes sensitive to the flow of information.

Computing contingent frequency

What is required for a process to be sensitive to the information conveyed by the stream of events? Among other things, the process must time the intervals between events and count the numbers of events, because it is these abstract properties of the event stream that convey the information. Second, it must be capable of the computational operations required to extract from the intervals and counts the information in them. What might these operations look like?

First, let us consider the problem of computing contingency in an environment in which there is more than one CS, more than one stimulus whose presence may affect the expected time to the next US. This is often the case. It was the case, for example, even in Rescorla's (1968) simple experiment. We have so far discussed this experiment as if there was only one CS, the tone. In fact, there was a second CS, namely, the experimental chamber itself. Subjects never experienced foot shocks in their home cages, where they spent most of every day; they only experienced them in the experimental chamber, where they spent only two hours each day. Thus, entry into the experimental chamber changed the rate at which shocks could be expected to occur. One might expect them, therefore, to have learned to fear the chamber in Rescorla's experiment, and, in fact, they did. On the first session following the sessions in which they got tones and shocks, they were so afraid of the chamber that they did little lever pressing. That is why Rescorla gave two sessions with the lever unblocked and no tones or shocks before the sessions in which he tested the effect of the tone; he had to eliminate their fear of the chamber.

Whenever the tone came on, the subjects for which the shocks were contingent on the presence of a tone were also in the experimental chamber, so the shocks they experienced were paired with both the tone and the chamber. By what process is it possible to determine that the shocks were contingent on the tone, not the chamber? Before addressing this question, let us return to an experiment mentioned briefly earlier because it suggests the subtly and power of the process that computes contingency. Wagner *et al.* (1968) ran two groups of rats in an experiment very much like Rescorla's but one that made the problem of multiple possible predictors more apparent and more complex. In their experiment, there were, in addition to the experimental chamber, three intermittent CSs, which, like the tone in Rescorla's experiment, came on and stayed on for 2 minutes at various times during the sessions when the rats got foot shocks. What Wagner *et al.* (1968) manipulated was the correlations among these 3 CSs and the shock. Two of the CSs (designated A & B) were mutually exclusive: each occurred on half of the occasions (trials) on which a CS was presented, but never together with the other. A third CS (designated the X CS) always occurred together with the first two; it came on every time that either of them did, and never otherwise. For one group of rats, the shocks only occurred when the A and X stimuli were on, which means that they occurred on half the occasions on which the X stimulus was present. For a second group, shocks also occurred on half the occasions on which the X stimulus was present, but half the time it was the A stimulus that was also present and half the time the B stimulus. The first group learned to fear only the A stimulus, while the second group learned to fear only the X stimulus. What does it take to extract the 'true' contingencies in the face of this kind of complexity?

Not as much as one might suppose. On the assumption that CSs have independent effects on the observed rates of US occurrence, there is a simple analytic solution to the contingency-computing problem confronted by the rats in Rescorla's (1968) experiment and Wagner et al.'s (1968) experiment and, more generally, by the subjects in any basic Pavlovian conditioning experiment (Gallistel 1990; Gallistel & Gibbon 2000). It is given by the matrix equation $\vec{\lambda}_t = T^{-1} \vec{\lambda}_r$, where $\vec{\lambda}_t$ is the vector (the ordered list) of the rates of US occurrence predicted by each CS acting in isolation, $\vec{\lambda}_r$ is the raw rate vector,

$$\vec{\lambda}_r = \left\langle \begin{array}{c} \frac{N_1}{T_1} \\ \frac{N_2}{T_2} \\ \vdots \\ \frac{N_m}{T_m} \end{array} \right\rangle$$

and T^{-1} is the inverse of the temporal coefficient matrix:

$$T = \left| \begin{array}{cccc} 1 & \frac{T_{1,2}}{T_1} & \cdots & \frac{T_{1,m}}{T_1} \\ \frac{T_{2,1}}{T_2} & 1 & \cdots & \frac{T_{2,m}}{T_2} \\ \vdots & \vdots & \ddots & \vdots \\ \frac{T_{m,1}}{T_m} & \frac{T_{m,2}}{T_m} & \cdots & 1 \end{array} \right|$$

What comes out of this computation is the rate of US occurrence to be ascribed to each CS acting alone. The US is contingent on any CS to which a non-zero rate is attributed, whether negative or positive. When any such CS comes on, it changes the baseline expectation regarding the interval to the next US. If the rate ascribed to the CS is positive, the expected interval to the next US gets shorter; that is, the cumulative probability curve in Fig. 10.1 approaches 1 more rapidly. If the rate ascribed to the CS is negative (as it will be in inhibitory conditioning paradigms), then the expected interval to the next US gets longer; that is, the cumulative probability curve in Fig. 10.1 approaches 1 more slowly.

The quantities that go into the computation are the cumulative numbers of US occurrences $(N_1, N_2 \ldots N_m)$ in the presence of each of the CSs $(CS_1, CS_2, \ldots CS_m)$, the cumulative amounts of time that each CS has been observed to be present $(T_1, T_2, \ldots T_m)$ and the cumulative durations of the pairwise combinations of CSs (the $T_{i,j}$s). These quantities must be obtained by cumulative timing of the intervals when individual CSs and pairwise combinations of them are present and by running counts of how often the US has occurred when each CS was present. These times and counts plus the machinery for doing the requisite computation are all that is needed to solve the contingency problem.

It will often happen that there is more than one solution to the contingency problem, because the animal's experience is inherently ambiguous, that is, it could have arisen from more than one state of the world. For example, in the Wagner *et al.* (1968) condition where the shock occurred on half the AX trials and half the BX trials, the rats concluded that the US was contingent only on X. There are, however, an infinite number of other conclusions consistent both with the data they were given during the experiment and the assumption that the CSs were acting independently. It could have been that A and B were doing the predicting and not X. In that case, the rates of US occurrence ascribed to A and B would together be twice the rate ascribed to X in the solution that the rats favoured. Or, A could have accounted for 1 out of three USs, B for 1 out of 3, and X for 1 out of 3. And so on. There is no way of telling which was in fact the case.

The matrix computation automatically sounds an alarm whenever there is more than one solution, because then the determinant of the temporal coefficient matrix is 0, so the computation cannot be performed. The computation can only be performed when enough alternative predictors (CSs) have been omitted from consideration so as to make the determinant no longer 0. Or, if we prefer to think of the machinery as considering the simplest solutions first and adding to the set of predictors only when more predictors are required, then we can imagine the system doing the computation with successively more inclusive matrices (a successively broader canvass of the potential predictors) until it gets to matrices that have 0 determinant, at which point it has accounted for all the contingency that can be accounted for under the assumption of independent effects. At that point, the system will often have discovered more than one solution. However, there is an information-theoretic consideration that dictates a preferred solution in such cases. Interestingly, this consideration appears to dictate the solution the rats in fact favour in these inherently ambiguous situations.

The consideration is a version of Occam's razor, the principle of explanatory parsimony: that solution is preferred that minimizes the number of predictor variables (CSs to which an information-transmitting capacity is attributed). It is preferred because it is the most powerful solution, where power is measured by the average amount of information conveyed per CS.

The rapidity of conditioning and extinction

Traditional models of classical conditioning have nothing to say about how rapidly conditioning and extinction ought to proceed. In associative models, the rate of conditioning is governed by physically meaningless free parameters, so there are no considerations that suggest even approximately what the values of these parameters should be. The information-theoretic perspective, however, has a natural, meaningful metric, which dictates approximately how rapidly conditioning and extinction should occur. The metric is the amount by which the subject's a priori uncertainty about the current contingency between the CS and the US has been reduced by the flow of events up to a given point in the experiment.

This is the metric that experimentalists use when they test a null hypothesis, the hypothesis that there is no contingency (no reliable relation) between the independent and the dependent variables in their experiment. The assumption is that in the absence of observation (experiment), a zero contingency is as likely as any other. The question is whether the observations made in the course of the experiment make the assumption of zero contingency (no effect) relatively unlikely. If the answer is, Yes, very unlikely, then we proceed to act on the assumption that the independent variable conveys information about the dependent variable. If the answer is, No, not very unlikely, then we proceed on the assumption that the independent variable conveys little information about the dependent variable.

The information gained at any point in a conditioning experiment about the contingency between a CS and a US depends on the durations of the observations in the presence and the absence of the CS and the numbers of USs observed in each condition. Elementary probabilistic reasoning with these quantities gives us a suitable measure, at least in the case where there are only two CSs, one of which is the steady background CS or context (hereafter B), and the other is the usual kind of intermittent CS (a tone or light). (The generalization to more complex cases is not difficult.) Call the intermittent CS A. We want to know how unlikely it is that there is no contingency between A and the US—the odds against the possibility that the rate of US occurrence predicted by A acting alone is 0.

Let T be the total interval of observation, T_a, the interval when A was present, and $T_b = T - T_a$ the interval when only the background was present. Likewise, let N be the total number of USs observed, N_a, the number when A was present, and $N_b = N - N_a$, the number when only the background was present. On the null hypothesis that A had no effect on the rate of US occurrence, the probability that an arbitrarily chosen one of the N USs (ignoring their order) happened to occur when A was present is T_a/T. The probability P_f of observing N_a or fewer events is given by the cumulative binomial probability function, as is the probability P_m of observing N_a or more events. When the number of events in T_a is approximately the expected number, the ratio P_f/P_m is approximates unity and the log of this ratio approximates 0. When the number of USs while A is present becomes improbably high relative to the number observed when A is not present, the ratio between the two cumulative probabilities becomes very large and its log approaches infinity. The growth of this ratio (or its logarithm) measures the extent to which the null hypothesis has become unlikely.

This measure of the information about the CS–US contingency that has so far accumulated works equally well in the case of a negative contingency. In that case, the number of USs observed while A is present becomes improbably low as the experiment progresses, so the ratio of the two cumulative probabilities becomes very small and its log approaches minus infinity. Thus, the absolute value of this logit (log of the odds) measures the growing certainty that there is a contingency; the farther this quantity is from 0, the more certain it is that the CS affects the rate of US occurrence.

Moreover, we know a priori what constitute reasonable values for this measure, because it has an objective probabilistic meaning. Speaking very conservatively, if the conditioned response appears at the point where the subject should be reasonably certain that there is a CS–US contingency, then the response should appear when this measure (the absolute value of the log of the odds) is less than 6, because when the logit equals 6, the odds are already 1,000,000:1 against the null hypothesis.

The empirical problem for the information-theoretic approach is that in the conditioning paradigms from which we have the best data, this measure is unreasonably high at the point in the experiment where the conditioned response is first seen. The odds against the null hypothesis at that point are on order of 10^{65}:1 (Gallistel & Gibbon 2002). Thus, it cannot reasonably be argued that the conditioned response appears when the possibility that the US is not contingent on the CS has become very unlikely. The problem is that it often does not appear until this possibility has become absurdly unlikely. The same is true for extinction. Reasoning very similar to that given above tells us how long it should take for us to observe the cessation of condition responding when we stop delivering USs in the presence of the CS (Gallistel & Gibbon 2002). In fact, however, it often takes much longer than this (Gallistel & Gibbon 2002).

There are, however, paradigms in which subjects demonstrate a much greater sensitivity to the accumulation of information about current contingencies. In the matching paradigm, rewards (USs) become available at random moments at each of two foraging sites. When a reward becomes available at a site, it remains available until the subject comes to that site and harvests it. (Typically, rodent subjects harvest available rewards by pressing a lever, while bird subjects harvest rewards by pecking at keys.) The intervals until the next reward become available are determined by independent random rate (Poisson) processes. This means, as already explained, that the likelihood that a reward will be made available in the next instant is flat. Once made available, a reward remains available until it is harvested; therefore, the longer the subject has gone without visiting a site, the more certain it is that there is a reward waiting for it. The cumulative probability that a reward has become available during the subject's absence is described by the random rate curve in Fig. 10.1, which approaches 1 at a rate determined by the availability rate (usually called the programmed reward rate) for that site. This characteristic means that subjects can increase the total amount of reward harvested per unit of foraging time by moving back and forth between the sites, even when the availability rates at the two sites differ greatly. The longer the subject has stayed at the richer site, the more certain it is that there is a reward waiting to be harvested at the poorer site, and the cost of a quick visit to the poorer site, measured in average time lost in harvesting the next reward at the richer site, is very small.

In a matching paradigm, the experimenter sets the rate parameters of the two sites, usually to unequal values. It is called a matching paradigm because the general finding is that the ratio of the amounts of time the subject spends at the two locations approximately equals the ratio of the rates at which it obtains rewards from them (Davison & McCarthy 1988; Herrnstein 1961). Importantly, the reward rates in this equation are defined by reference to total foraging time (usually session time, time spent in the chamber), not the time the subject spends at a given location. In economists' terms, the rates are incomes (rewards obtained per unit of time), not returns (rewards obtained per unit of time invested in foraging at a site). The relative amounts of time the subject spends at the two sites is determined by the ratio of the expected (average) stay durations at the two sites. These stay durations appear to be themselves determined by random rate processes within the subject, because they exhibit the flat hazard function that is the signature of a random rate process (Gibbon 1995; Heyman 1979, 1982). That is, the likelihood of the subject's leaving a site is independent of how long it has been there.

In information-theoretic terms, the relative amounts of time invested in different foraging sites when the subject is free to move back and forth between them at will is equal to the relative strengths of the contingencies between those sites and food availability. The strength of the contingency between a site and the availability of food is measured by the rate parameter for the cumulative probability curve in Fig. 10.1; the faster this curve approaches 1, the stronger the contingency between that site and the availability of food. We can use this paradigm to ask how rapidly subjects can detect and adjust to changes in the strengths of contingency. When we suddenly decrease the rate of reward at one site and increase it at the other, we produce a step change in the relative strengths of the contingencies. We know that the subject will eventually adjust the expected durations of its stays at the two sites so that the ratio of those expectations approximately equals the ratio of the new rates of reward. How rapidly it does so will depend in part on how sensitive it is to the accumulation of information indicating that the rates of reward have changed.

In more traditional terms, the change in rates of reward will lead to a strengthening of the conditioned response to the site that has become more richly rewarding and a weakening of the conditioned response to the site that has become poorer. These strengthenings and weakenings of the conditioned responses to these sites would appear to be closely analogous to, if not identical with, the strengthening of the conditioned response that occurs during simple acquisition and the weakening that occurs during simple extinction. Thus, this paradigm gives us another approach to the question of the rapidity of conditioning and extinction.

It turns out that when changes in contingency in the matching paradigm are rare, subjects adjust slowly to them (Gallistel, Mark, King & Latham 2001; Mazur 1992, 1995, 1996), but when they are frequent, subjects adjust about as fast as is in principle possible (Gallistel *et al.* 2001). This means that rats can approximate ideal detectors of changes in contingency; they have the maximum possible sensitivity to the flow of information carried by a sequence of events.

These findings suggest that the surprisingly sluggish appearance of conditioned behaviour in some paradigms (e.g. pigeon autoshaping and rabbit eye-blink paradigms) and the equally surprising slow rates of extinction) do not reflect failures to detect the contingency between CS and US and changes in that contingency, but rather some strategy having to do with waiting so see how stable the presently observed contingency will prove to be.

Contingent frequencies are not conditional probabilities

Before concluding, I want to call attention to the important distinction between frequency and probability, or, equivalently, between discrete and continuous frequency. Throughout this essay, frequency and rate mean number of events per unit of time. Because time itself is a continuous quantity (a real-valued variable), this quantity is also continuous (real valued). Frequency is often also used to mean the number of instances per number of observations, in which case it is a discrete (rationally valued) variable. More often than not, when psychologists talk about frequency, they seem to have discrete frequency in mind, although which sense of frequency is intended is often unclear. Also, the driving variable in simple conditioning is usually taken to be the conditional probability of the US, which is a discrete frequency.

It is important to keep the distinction between discrete and continuous frequencies in mind, because the two quantities behave differently. Discrete frequency is equivalent to probability, the conversion involving no more than converting from the rational to the decimal form of the number (for example, from 6/10 to 0.6). Continuous frequency converts to probability only by integrating a probability density function over a finite interval, and the value obtained depends on the interval of integration. When discrete probabilities are independent, they combine multiplicatively to determine joint probabilities. They combine additively only if they refer to mutually exclusive observations, and then their sum cannot be greater than 1. When continuous frequencies are independent, they combine additively, and their sum can be any positive real number.

One reason for recalling this distinction is that the driving variable in simple associative conditioning is often taken to be the conditional probability of the US. Indeed, Rescorla (1967, 1968) defined contingency in terms of conditional probabilities, and many have followed his lead. The contingent frequencies that are the foundation of the analysis given here are not conditional probabilities. Gallistel and Gibbon (2000, p. 333) review the difficulties that arise when one tries to apply an analysis based on conditional probabilities to the many conditioning protocols in which it is difficult or impossible to say what constitutes a trial. The problem is that if trials cannot be defined, then the observed discrete frequencies cannot be defined, because, by the usual definition, the probability of the US is the number of US occurrences in a given number of trials. In an analysis based on contingent frequencies rather than conditional probabilities, these difficulties do not arise.

Conclusions

Information theory provides a powerful, rigorous and quantitative framework within which to analyse the process of conditioning. Contingent frequency is a fundamental notion in such an analysis. A contingent frequency is the rate of US occurrence to be expected in the presence of a given CS (conceived of as operating alone). Working within this framework turns on its head the traditional assumption that conditioning has to be understood in terms of processes set in motion by individual events, more particularly, processes set in motion by the temporal pairing of events. On this analysis, the events are relevant only as carriers of information. Moreover, the information resides in the abstract quantities defined by the events, the intervals between the events, and the numbers of events.

On the usual analysis, the conditioning process does not operate at this level of abstraction; the timing of intervals and the counting of events play no role in traditional models. Although the intervals between CSs and USs and the number of CS–US pairings are assumed to affect the strengths of the resulting associations, the learning mechanism is not imagined to do computations on symbols representing durations and numbers.

If, however, the conditioning process is in fact driven by the information conveyed by the stream of events, then the process must be doing just that. Thus, the validity of the information-theoretic approach to the understanding of conditioning bears strongly on the more general theory of mind and brain that will prevail within psychology. Should we conceive of the brain as a plastic net whose connectivity is molded by its experience so as to adapt its input–ouput function to the exigencies of that experience, without symbolically representing abstract properties of that experience? Or should we conceive of it as an information processing organ doing computations on symbols that represent abstract properties of its experience, like number and duration?

References

Amsel, A. (1962). Frustrative nonreward in partial reinforcement and discrimination learning. *Psychological Review, 69*:306–328.

Barnet, R. C., Arnold, H. M. & Miller, R. R. (1991). Simultaneous conditioning demonstrated in second-order conditioning: Evidence for similar associative structure in forward and simultaneous conditioning. *Learning and Motivation, 22*:253–268.

Barnet, R. C., Cole, R. P. & Miller, R. R. (1997). Temporal integration in second-order conditioning and sensory preconditioning. *Animal Learning and Behavior, 25*(2):221–233.

Barnet, R. C., Grahame, N. J. & Miller, R. R. (1993). Local context and the comparator hypothesis. *Animal Learning and Behavior, 21*:1–13.

Davison, M. & McCarthy, D. (1988). *The matching law: A research review*. Hillsdale, NJ: Erlbaum.

Dickinson, A. (1989). Expectancy theory in animal conditioning. In S. B. Klein & R. R. Mowrer (eds) *Contemporary learning theories: Pavlovian conditioning and the status of traditional learning theory* (pp. 279–308). Hillsdale, NJ: Lawrence Erlbaum Associates.

Gallistel, C. R. (1990). *The organization of learning*. Cambridge, MA: Bradford Books/MIT Press.

Gallistel, C. R. & Gibbon, J. (2000). Time, rate and conditioning. *Psychological Review, 107*:289–344.

Gallistel, C. R. & Gibbon, J. (2002). *The symbolic foundations of conditioned behavior*. Hillsdale, NJ: Lawrence Erlbaum Associates.

Gallistel, C. R., Mark, T. A., King, A. P. & Latham, P. E. (2001). The rat approximates an ideal detector of changes in rates of reward: implications for the Law of Effect. *Journal of Experimental Psychology: Animal Behavior Processes, 27*:354–372.

Gibbon, J. (1977). Scalar expectancy theory and Weber's Law in animal timing. *Psychological Review, 84*:279–335.

Gibbon, J. (1995). Dynamics of time matching: Arousal makes better seem worse. *Psychonomic Bulletin and Review, 2*(2):208–215.

Herrnstein, R. J. (1961). Relative and absolute strength of response as a function of frequency of reinforcement. *Journal of the Experimental Analysis of Behavior, 4*:267–272.

Heyman, G. M. (1979). A Markov model description of changeover probabilities on concurrent variable-interval schedules. *Journal of the Experimental Analysis of Behavior, 31*:41–51.

Heyman, G. M. (1982). Is time allocation unconditioned behavior? In M. Commons, R. Herrnstein & H. Rachlin (eds) *Quantitative analyses of behavior, vol. 2: matching and maximizing accounts*, vol. 2 (pp. 459–490). Cambridge, MA: Ballinger Press.

Hull, C. L. (1929). A functional interpretation of the conditioned reflex. *Psychol. Rev., 36*:498–511.

Kamin, L. J. (1967). 'Attention-like' processes in classical conditioning. In M. R. Jones (ed) *Miami symposium on the prediction of behavior: aversive stimulation* (pp. 9–33). Miami: University of Miami Press.

Kamin, L. J. (1969a). Predictability, surprise, attention, and conditioning. In B. A. Campbell & R. M. Church (eds) *Punishment and aversive behavior* (pp. 276–296). New York: Appleton-Century-Crofts.

Kamin, L. J. (1969b). Selective association and conditioning. In N. J. Mackintosh & W. K. Honig (eds) *Fundamental issues in associative learning* (pp. 42–64). Halifax: Dalhousie University Press.

Killeen, P. R. & Weiss, N. A. (1987). Optimal timing and the Weber function. *Psychological Review,* 94:455–468.

LoLordo, V. M. & Fairless, J. L. (1985). Pavlovian conditioned inhibition: The literature since 1969. In R. R. Miller & N. E. Spear (eds) *Information processing in animals.* Hillsdale, NJ: Lawrence Erlbaum Associates.

Mackintosh, N. J. (1971). An analysis of overshadowing and blocking. *Quarterly Journal of Experimental Psychology,* 23:118–125.

Mackintosh, N. J. (1975). A theory of attention: Variations in the associability of stimuli with reinforcement. *Psychol. Rev.,* 82:276–298.

Matzel, L. D., Held, F. P. & Miller, R. R. (1988). Information and expression of simultaneous and backward associations: Implications for contiguity theory. *Learning and Motivation,* 19:317–344.

Mazur, J. E. (1992). Choice behavior in transition: Development of preference with ratio and interval schedules. *JEP:ABP,* 18:364–378.

Mazur, J. E. (1995). Development of preference and spontaneous recovery in choice behavior with concurrent variable-interval schedules. *Animal Learning and Behavior,* 23(1):93–103.

Mazur, J. E. (1996). Past experience, recency, and spontaneous recovery in choice behavior. *Animal Learning and Behavior,* 24(1):1–10.

Miller, R. R. & Barnet, R. C. (1993). The role of time in elementary associations. *Current Directions in Psychological Science,* 2:106–111.

Pavlov, I. V. (1928). *Lectures on conditioned reflexes: The higher nervous activity of animals* (H. Gantt, Trans.). London: Lawrence & Wishart.

Pearce, J. M. & Hall, G. (1980). A model for Pavlovian learning: Variation in the effectiveness of conditioned but not of unconditioned stimuli. *Psychol. Rev.,* 87:532–552.

Rescorla, R. A. (1966). Predictability and the number of pairings in Pavlovian fear conditioning. *Psychonomic Science,* 4:383–384.

Rescorla, R. A. (1967). Pavlovian conditioning and its proper control procedures. *Psychol. Rev.,* 74:71–80.

Rescorla, R. A. (1968). Probability of shock in the presence and absence of CS in fear conditioning. *Journal of Comparative and Physiological Psychology,* 66(1):1–5.

Rescorla, R. A. (1969). Pavlovian conditioned inhibition. *Psychological Bulletin,* 72:77–94.

Rescorla, R. A. (1972). Informational variables in Pavlovian conditioning. In G. H. Bower (ed.) *The psychology of learning and motivation,* vol. 6 (pp. 1–46). New York: Academic.

Rescorla, R. A. & Wagner, A. R. (1972). A theory of Pavlovian conditioning: Variations in the effectiveness of reinforcement and nonreinforcement. In A. H. Black & W. F. Prokasy (eds) *Classical conditioning II* (pp. 64–99). New York: Appleton-Century-Crofts.

Reynolds, G. S. (1961). Attention in the pigeon. *Journal of the Experimental Analysis of Behavior,* 4:203–208.

Rieke, F., Warland, D., de Ruyter van Steveninck, R. & Bialek, W. (1997). *Spikes: Exploring the neural code.* Cambridge, MA: MIT Press.

Shannon, C. E. (1948). A mathematical theory of communicatioin. *Bell Systems Technical Journal, 27*:379–423:623–656.

Usherwood, P. N. R. (1993). Memories are made of this. *Trends in Neurosciences, 16*(11): 427–429.

Wagner, A. R. (1981). SOP: A model of automatic memory processing in animal behavior. In N. E. Spear & R. R. Miller (eds) *Information processing in animals: memory mechanisms* (pp. 5–47). Hillsdale, NJ: Lawrence Erlbaum.

Wagner, A. R., Logan, F. A., Haberlandt, K. & Price, T. (1968). Stimulus selection in animal discrimination learning. *Journal of Experimental Psychology, 76*(2):171–180.

Part II

ESSENTIAL EMPIRICAL RESULTS

CHAPTER 11

EFFECTS OF PROCESSING FLUENCY ON ESTIMATES OF PROBABILITY AND FREQUENCY

ROLF REBER AND NATASHA ZUPANEK

Abstract

People may apply multiple strategies for estimating frequency or probability of occurrence. One of them is the use of the availability heuristic, which is the ease of retrieving instances from memory. However, this kind of processing fluency often is confounded with amount of recall. We present a new paradigm that helps to resolve this issue. In two experiments, participants were repeatedly exposed to two events; one could be processed more easily than the other. Objective probabilities of occurrence for the fluent event were 25%, 50%, and 75%. Participants had to judge probability (Experiment 1) or frequency (Experiment 2) of occurrence. In both experiments, estimates were biased by manipulated fluency of the event. Moreover, biases in estimates paralleled observed biases in reaction times for the two events. These findings support the notion that processing fluency drives frequency estimates.

Different processes underlying the estimation of event frequencies have been identified (see Brown 1995, 1997; this volume). First, people may simply enumerate episodes they are able to retrieve. Second, they may enumerate some episodes and then extrapolate, yielding a frequency greater than the episodes retrieved. Third, people may directly retrieve estimates that are already stored in memory. Fourth, people may use memory assessment strategies, using availability, similarity, or memory strength as a heuristic basis for frequency estimates. Memory assessment strategies are particularly useful if (1) objective absolute frequencies are high and therefore beyond the ability of enumeration; (2) the rate of events in question is too irregular so that enumeration and extrapolation does not yield sufficiently accurate estimates; (3) the objective frequency is unknown. In the remainder of this chapter, we will not discuss strategies other than memory assessment because we are interested in estimation of high frequencies that are beyond the ability to enumerate. Moreover, we do not discuss the distinctions between availability (or processing fluency), similarity, and strength because higher similarity or higher strength may result in the subjective experience of more fluent processing. Support for this notion has been found by Dougherty *et al.* (1999) who were able to simulate availability biases with their MINERVA DM model that is based on calculation of similarity.

Such models suggest that the contradiction between similarity and strength on the one hand and processing fluency on the other may be more apparent than real.

Processing fluency and frequency estimates

Processing fluency is the ease with which a stimulus is processed. Subjective processing fluency denotes the phenomenal feeling of ease of processing, whereas objective processing fluency denotes the underlying objective speed of processing. Recent research by Reber and Wurtz (2001) has shown that the speeds of processing at different stages contribute jointly to the phenomenal experience of processing fluency, supporting the notion that feelings may provide contextual information in a highly condensed form (e.g. Mangan 1993). The phenomenal experience of processing fluency may provide highly condensed information about simplicity, frequency and familiarity of a stimulus, or about the potential effort connected to stimulus elaboration.

An interesting line of research suggests that frequency of exposure influences perceptual fluency. Solomon and Postman (1952) have shown that identification thresholds for pseudowords decreased with increasing frequency of exposure. Similarly, Feustel, Shiffrin and Salasoo (1983), using a clarification procedure for word identification, were able to show that increasing numbers of presentations increased priming. These studies suggest that frequency of exposure results in higher processing fluency.

If higher objective frequency results in a feeling of higher processing fluency, this feeling could be taken as an indicator of absolute frequency: in order to estimate the frequency of occurrence of a stimulus, people may assess the ease with which this stimulus can be brought to mind. After Tversky and Kahneman (1973) performed studies about the availability heuristic, they concluded that participants estimate the frequency of an event, or the probability of its occurrence, 'by the ease with which instances or associations could be brought to mind' (Tversky & Kahneman 1973, p. 208). Presumably, participants infer that a given class of events is frequent when relevant instances are easy to bring to mind, but rare when instances are difficult to bring to mind. However, most studies on the availability heuristic do not allow strong conclusions about the underlying processes (see Schwarz 1998 for a detailed discussion). For example, Tversky and Kahneman's Experiment 3 (1973) found that participants overestimated the number of words that begin with the letter r, but underestimated the number of words with r as the third letter. This result may be interpreted in two ways: first, participants may base their judgement on the subjective experience of fluent processing. If so, they would estimate a higher frequency if the recall task were experienced as easy rather than difficult. Alternatively, they may base their judgement on the amount of recalled words. If they were able to recall more words that begin with a certain letter, they would again estimate a higher frequency. Schwarz (1998) discussed other studies that had the same kind of ambiguity. In some studies, availability has explicitly been defined by amount of recall (e.g. Bruce et al. 1991; Williams & Durso 1986), leaving the discussed ambiguities unresolved.

In sum, manipulations intended to increase the subjective ease of recall are also likely to affect the amount of recall. If recall fluency—in contrast to amount of recall—serves as a source of information, its influence should vary as a function of perceived diagnosticity of the experience, as has been observed for other types of judgements (see Schwarz

1998). Wänke *et al.* (1995) found support for this hypothesis in a modified replication of Tversky and Kahneman's (1973) letter experiment described above. In the control condition, participants had to estimate the number of words beginning with t compared to the number of words with t in the third position. The findings of Tversky and Kahneman were replicated: participants estimated that words beginning with a t are more frequent than words having a t in the third position. To isolate the role of experienced ease, the diagnosticity of the experience was manipulated in two experimental conditions. Specifically, participants had to write down ten words that begin with a t on a sheet of paper that was imprinted with pale but visible rows of t's. Some participants were told that this background would facilitate the recall of t-words, therefore undermining the diagnosticity of experienced ease; participants were expected to attribute experienced ease to the background and not to the frequency of t's. Other participants were told that this background would interfere with the recall task, therefore enhancing the diagnosticity of experienced ease; participants who experienced ease despite the interfering background were expected to attribute this experience to the frequency of t's to an even higher degree than the control group. As expected, participants who could attribute the experienced ease of recall to the impact of the background assumed that there are fewer t-words than did participants in the control condition. In contrast, participants who expected the background to interfere with recall, but found recall easy nevertheless, estimated that there are more t-words than did participants in the control condition. In combination, these discounting and augmentation effects suggest that participants did indeed base their frequency estimates on the implications of their subjective experience of processing fluency, rather than on the number of words they could bring to mind.

However, the notion that frequency estimates are based on processing fluency has been challenged.[1] In a series of studies, Sedlmeier *et al.* (1998) tested the relation between processing fluency and frequency estimates. Specifically, their participants had to estimate the frequency of words that had a certain letter (e.g. D) at the first or second position in German language. Participants estimated frequency for letters at certain positions with great fidelity, except for the well-known phenomenon of the regression to the mean. In contrast to Wänke *et al.* (1995), the authors did not find any relation between frequency estimates and processing fluency, as measured by both speed and quantity of recall of words that contained the letter in question in the first or second position. These findings contradicted the processing fluency hypothesis and supported the notion that frequency processing is largely unbiased. Another problem pertains to the objective frequencies used: researchers usually used low to moderate frequencies, enabling the use of enumeration strategies (see Conrad *et al.* 1998). In some studies, base frequencies were used that were higher than usual, but difficult to determine

---

[1] Some authors claimed that availability (and thus processing fluency) influences only judgements of set size, but not estimates of frequency of occurrence (Manis *et al.* 1993). However, recent experiments by Betsch *et al.* (1999) have shown that under some conditions, availability may influence frequency of occurrence. In these studies, recall data were used to assess availability. Therefore, these studies can not be taken as direct evidence for effects of processing fluency.

(Sedlmeier *et al.* 1998). These authors used letter frequency counts at the first and second positions of German words; as the authors acknowledge, such frequency counts may be quite unrepresentative for the respective frequencies the participants encountered in real life.

In sum, evidence for the use of the feeling of processing fluency is relatively weak. There are very few studies that addressed the problem of processing fluency versus amount of recall directly. One study found effects of processing fluency (Wänke *et al.* 1995); the other did not (Sedlmeier *et al.* 1998). Moreover, objective frequencies were either low but fully determined, or high but not fully determined. In order to resolve these issues, we developed a new experimental paradigm that is described in the next section.

Manipulating processing fluency: a new paradigm

Participants were presented prime–target pairs on a computer screen. They first saw a prime for 200 ms, followed by a target that appeared on the screen 100 ms after the prime disappeared. Processing fluency was manipulated by congruency of the prime–target pairs: Congruent pairs were presented on the same side of a computer screen, either both on the left or both on the right side. Incongruent pairs were presented on different sides of the screen, either prime on the left and target on the right side or prime on the right and target on the left side (see Fig. 11.1). Sides on the screen were balanced for congruent and incongruent pairs, respectively. Participants had to press a button to indicate the side of the target. We presented 120 prime–target pairs, either 30 (25% group), 60 (50% group) or 90 (75% group) of them were congruent. The number of incongruent pairs was 120 minus the number of congruent pairs.

We expected two mechanisms to have an effect on priming: first, congruent prime–target pairs should be processed faster than incongruent pairs. This is what we mean by processing fluency. Second, we predicted an expectancy effect: if 75% of the prime–target pairs were congruent, participants would begin to expect that congruent pairs are more probable than incongruent pairs. This expectancy of a high probability for congruent pairs facilitates their processing. If, however, only 25% of the prime–target pairs were congruent, participants would—after some trials—expect a high probability of incongruent pairs. Therefore, processing of incongruent pairs is facilitated. Please note that expectancy effects alone would result in a symmetric bias: absolute amount of

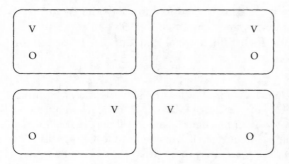

Figure 11.1 Illustration of the four possible prime–target pairs: in the top row are congruent pairs (left–left or right–right), in the bottom row are incongruent pairs (right–left or left–right).

priming would be identical, but in different directions. However, we will discuss that this is the case only if expectancies were not moderated by other factors, and we will present analyses to show whether or not there were such moderating factors.

In combination, the 75% group should show much higher absolute priming than the 25% group. In the 75% group, prime–target congruence adds to the expectancy effect. In the 25% group, however, prime–target congruence attenuates the effect of expectancy.

We also expected two mechanisms to have an effect on frequency estimates: first, high processing fluency was expected to result in estimates of higher frequency or higher probabilities of congruent rather than incongruent pairs. Second, frequency estimates are often biased toward the mean frequency of a specific class (see Sedlmeier *et al.* 1998), a phenomenon known as 'regression to the mean'. Estimates are quite accurate if actual frequency is around the mean; we even expected that participants in the 50% group were accurate and unbiased because they may use 50% as an anchor (see Tversky & Kahneman 1974). Regression to the mean denotes the fact that estimates are too high if actual frequency is substantially lower than the mean and too low if actual frequency is substantially above the mean. Please note that regression to the mean alone would result in a symmetric bias: absolute deviations of estimates from true frequencies are identical. In combination, the estimate of the 25% group is predicted to be closer to 50% than the estimate of the 75% group. The estimation bias due to processing fluency in the 25% group is expected to add to the regression to the mean. If actual probability is at 75%, the two effects are predicted to cancel out each other: The effect due to processing fluency may attenuate the effect due to the regression to the mean.

This asymmetry in estimates parallels the asymmetry in reaction times (RTs) discussed above: the smaller the absolute RT difference, the smaller the absolute difference between estimate and 50% anchor is expected to be. As mentioned above, both expectancy effects in RTs and regression to the mean in estimates result in symmetrical biases. Only effects of fluency result in asymmetrical biases of both RTs and frequency estimates, resulting in both higher priming effects and higher frequency estimates. If frequency estimates are based on subjective experience of processing fluency, absolute estimates are expected to reflect the observed absolute priming.

This paradigm solves three problems associated with existing research. First, fluency, manipulated by ease of processing at encoding, is objectively measurable in terms of priming effects for congruent compared to incongruent prime–target pairs. Therefore, we are able to assess whether frequency estimates parallel the observed priming effects. Second, we used higher frequencies than usual in research on frequency processing, assuming that people need to assess processing fluency if enumeration at retrieval is beyond their ability. Third, frequencies in our experiments were fully determined. Hence, frequency estimates can be compared to actual frequency of occurrence.

In Experiment 1, half of the participants had to estimate the probability of congruent prime–target pairs. As outlined above, we expected that estimates were closer to 50% if actual probability was 25% rather than 75%. In Experiment 1, we expected that estimates for congruent prime–target pairs were closer to 50% if actual probability was 25% rather than 75%. However, such an effect may be due to a bias that regression to the mean tends to be more pronounced if actual probability is low rather than high (see Arkes & Harkness 1983 for a related phenomenon). To address this possibility, we added a condition with different question focus: half of the participants had to estimate the probability that prime and target were on the same side (*congruent focus*), the other half had to indicate the probability that prime and target were on different sides (*incongruent focus*).

Experiment 1

Method

Seventy-two undergraduate students in psychology of the University of Bern participated in the experiment for course credit. Four participants were excluded because they made more than 12 (= 10%) errors, and were replaced by new participants. Thirty-six participants each were in the congruent and incongruent focus condition, respectively. Twelve participants each within both the congruent and the incongruent focus condition were randomly assigned to the three groups, the 25%, 50%, and 75% groups, respectively. The group label always denoted the actual probability of congruent pairs: for the incongruent focus condition, the accurate estimate for participants in the 25% group was 75%; for participants in the 75% group, the accurate estimate was 25%. This yielded a 3 (group) $\times$ 2 (focus)-factorial design, both factors manipulated between participants.

The experiment was run individually on Macintosh computers, using PsyScope, Version 1.0.2b.4 (Cohen *et al.* 1993). In an exposure phase, 120 prime–target trials were presented. The prime was a letter 'V', appearing for 200 ms either on the left or on the right side of the screen (3.5 cm left or right of the centre and 0.8 cm above the centre of the screen). Each prime was preceded by a fixation point presented for 500 ms in the centre of the screen. The interval between fixation point and onset of the prime was 200 ms. The target was a letter 'O', appearing 100 ms after the end of the prime. The 'O' was presented either on the left or on the right side of the screen (3.5 cm left or right of the centre and 0.8 cm below the centre of the screen). For congruent pairs, the target position was 1.6 cm below the prime position. For each of the four prime–target combinations, the sides on the screen were balanced. An illustration of the four resulting prime–target pairs is given in Fig. 11.1.

The participants were instructed to react as fast as possible to the target letter 'O' that was preceded by a prime letter 'V', sometimes on the same side and sometimes on the other side of the screen. Participants should press the left key ('z' on the American keyboard) if the 'O' appeared on the left side, and the right key ('/' on the American keyboard) if the 'O' appeared on the right side. If participants pressed the wrong key, the target stimulus remained on the screen. As soon as participants reacted accurately, the target stimulus disappeared, and the next trial followed.

After 120 trials, participants had to estimate the percentage, either that target ('O') and prime ('V') were on the same side (congruent focus condition) or that target ('O') and prime ('V') were on different sides (incongruent focus condition).

Results and discussion

Estimates

Estimated probabilities of the incongruent focus group were reversed (100 minus estimated probability) in order to get estimates that were comparable to the estimates of the congruent focus group. The results are shown in the first two rows of Table 11.1. We first performed a 3 × 2 factorial analysis of variance, with probability group and question focus both manipulated between subjects. There was a significant effect of probability group, $F(2, 66) = 30.77$, $p < 0.001$. Both the effect of question focus and the probability group x question focus interaction were not significant, $Fs < 1$. As there were no significant effects of question focus, we turn directly to the analyses of the biases.

If processing fluency biased the estimate, the estimates of the 25% group should be closer at 50% than the estimates of the 75% group. For both groups, the absolute differences between estimates and 50% were calculated. The 50% group was not included in this analysis. As can be seen in the first row of Table 11.2, the 25% group was closer

Table 11.1 Means and standard deviations (in parenthesis) for probabilities of Experiment 1, frequencies for congruent and incongruent pairs and resulting probabilities of Experiment 2

	Probability of congruent pairs		
	25%	50%	75%
Experiment 1			
Probability congruent focus	42.5 (12.2)	52.0 (9.8)	74.6 (12.3)
Probability incongruent focus	40.5 (12.7)	52.1 (17.9)	67.3 (12.7)
Experiment 2			
Frequency congruent pairs	38.8 (32.6)	47.5 (17.2)	67.2 (42.4)
Frequency incongruent pairs	47.3 (28.0)	40.8 (11.1)	28.5 (20.0)
Probability	42.5 (9.5)	53.5 (8.2)	69.8 (7.6)

Note: Probability in Experiment 2 was: P(con) = F(con)/(Fcon) + F(incon), where P(con) = probability of congruent pairs, F(con) = frequency of congruent pairs, and F(incon) = frequency of incongruent pairs.

Table 11.2 Estimate difference and RT difference for each group for both experiments

	Probability of congruent pairs			
	25%	75%	df	t
Experiment 1				
Estimate difference	8.5 (12.2)	20.9 (12.8)	46	3.44**
RT difference	21 (36)	92 (43)	46	6.15***
Experiment 2				
Estimate difference	7.5 (9.5)	19.8 (7.6)	10	2.48*
RT difference	19 (39)	89 (61)	10	2.40*

*: $p < 0.05$; **: $p < 0.01$; ***: $p < 0.001$; all probabilities two-sided
Note: Shown are means and standard deviations (in parenthesis) for absolute differences between estimates and the 50% anchor, in % (estimate difference), and absolute differences between incongruent and congruent prime-target pairs in milliseconds (RT difference) for the 25% and 75% groups of both experiments. Analyses of differences between the two groups are shown in the last two columns (df's and t-values). Compared to the data presented in Table 11.2, differences may be different at the last position due to rounding.

to 50% than the 75% group. This finding supports the notion that processing fluency affected probability estimates.

Reaction times

In order to get further support for a processing fluency account, we examined whether RT data paralleled the observed bias in probability estimates. RTs above 1500 ms, below 100 ms, and RTs of errors were removed before analyses. Corrected RTs are presented in the first two rows, average number of errors in the next two rows of Table 11.3.

As discussed above, the bias of expectancy should be the same for the 25% group and for the 75% group. Therefore, any bias in reaction times due to the expectancy effect should be symmetrical; an asymmetry in reaction time differences then can be attributed to a difference in processing fluency. In other words: an unbiased expectancy effect is a prerequisite for the conclusion that biases in estimates are due to biases in processing fluency. We tested this assumption in the following way. We first segmented the 120 prime–target trials into five parts of 24 trials each. In the 25% group, there were 6 congruent and 18 incongruent trials in each part; in the 50% group, there were 12 congruent and 12 incongruent trials per part, and in the 75% group, there were 18 congruent and 6 incongruent trials per part. We calculated priming effects within groups for each part by subtracting mean RTs of congruent trials from mean RTs of incongruent trials. The results are presented in Table 11.4.

Before we analysed whether RT differences paralleled biases in estimates, we checked whether there is an expectancy effect at all before we did further analyses to assess whether this effect is symmetrical or not. If there were an expectancy effect, we would expect a significant interaction between group and the parts of the exposure phase: the 25% group should start with lower RTs for congruent than for incongruent trials and later expect more incongruent trials than congruent trials, reversing the RT pattern. The 75% group, on the other hand, should start with some moderate amount of priming which then increases because participants begin to expect more congruent than incongruent trials. We performed a 3×5-factorial analysis of variance, with the factor probability

Table 11.3 Mean and standard deviations (in parenthesis) for reaction times in milliseconds, and for number of errors for congruent and incongruent prime–target pairs, for both experiments

	Probability of congruent pairs		
	25%	50%	75%
Experiment 1			
Time congruent pairs	411 (70)	363 (46)	347 (59)
Time incongruent pairs	390 (78)	387 (62)	438 (64)
Errors congruent pairs	0.5 (1.2)	0.5 (0.9)	0.4 (0.8)
Errors incongruent pairs	1.4 (1.9)	1.0 (1.3)	1.7 (2.0)
Experiment 2			
Time congruent pairs	400 (61)	370 (48)	327 (40)
Time incongruent pairs	381 (72)	399 (74)	417 (78)
Errors congruent pairs	0.5 (0.8)	0.8 (1.2)	0.2 (0.4)
Errors incongruent pairs	2.0 (1.7)	1.3 (1.5)	0.8 (0.8)

Table 11.4 Priming of each part (24 trials each) for all groups and both experiments

	Probability of congruent pairs				
	1	2	3	4	5
Experiment 1					
Priming 25% group					
Mean	−21	−17	−13	−20	−15
SD	(72)	(47)	(35)	(37)	(46)
Priming 50% group					
Mean	16	37	20	21	12
SD	(52)	(36)	(35)	(37)	(33)
Priming 75% group					
Mean	91	96	94	89	96
SD	(41)	(65)	(68)	(53)	(54)
Experiment 2					
Priming 25% group					
Mean	9	−6	−32	−25	−21
SD	(50)	(63)	(34)	(65)	(68)
Priming 50% group					
Mean	41	33	33	11	34
SD	(28)	(8)	(57)	(45)	(46)
Priming 75% group					
Mean	81	85	77	85	101
SD	(95)	(68)	(53)	(45)	(58)

group manipulated between subjects and the factor part within subjects. There was a significant effect of probability group, $F(2, 68) = 56.66$, $p < 0.001$. The other effects were not significant, $Fs < 1$. There was no expectancy effect at all; therefore, no further tests were necessary to assess whether or not expectancy effects were symmetrical across the probability conditions. In conclusion, any effect in the following analysis can be interpreted as an effect of processing fluency and not as an effect of expectancy.

Next, we calculated absolute RT differences between incongruent and congruent items for the 25% and the 75% group. As for the analysis of absolute differences in estimates reported above, the 50% group was left out for this analysis. If processing fluency influenced estimates, one would expect a lower absolute difference for the 25% group than for the 75% group. This was indeed the case, as can be seen in the second row of Table 11.2, indicating that relative fluency for incongruent pairs in the 25% group was smaller than relative fluency for congruent pairs in the 75% group. These results further supported the processing fluency view: the smaller absolute processing fluency of the 25% group relative to the 75% group paralleled the smaller average difference between the 25% group's estimate and the 50% anchor.

Experiment 2

Using problems of Bayesian reasoning, Gigerenzer (1991) and Sedlmeier (1999) found that the framing of questions in terms of frequency instead of probability increased accuracy of answers. Although our experimental task is different from the problems used by Gigerenzer and by Sedlmeier, their findings show that generalizations from results about probability estimates to frequency estimates are not warranted. Experiment 2 tested whether the bias for probability estimates found in Experiment 1 can be replicated if participants have to estimate frequencies of occurrence.

Method

Eighteen undergraduate students in psychology of the University of Bern participated in the experiment for course credit. Six participants each were randomly assigned to three groups, the 25%, 50%, and 75% groups, respectively.

Materials and procedure were identical to Experiment 1, except the questions after the 120 trials of the exposure phase. Participants were asked how many times prime ('V') and target ('O') were on the same side (first question) and how many times on different sides of the screen (second question).

Results and discussion

Estimates

Frequencies were transformed into probabilities for each participant. Frequencies and resulting probabilities are shown in the last three rows of Table 11.1. As shown in the third row of Table 11.2, estimates of the 25% group were significantly closer to 50% than estimates of the 75% group, replicating the finding of Experiment 1.

Reaction times

As in Experiment 1, RTs above 1500 ms, below 100 ms, and RTs of errors were removed before analyses. Corrected RTs are presented in the fifth and in the sixth row, average number of errors in the last two rows of Table 11.3.

We checked again whether there was an expectancy effect on priming. We used the same kind of segmentation of trials and the same statistical analysis as in Experiment 1. The 3 × 5-factorial analysis of variance with probability group as a between-subjects factor and with part as a within-subjects factor yielded only a significant main effect of probability group, $F(2, 15) = 9.63$, $p = 0.002$. Other effects were not significant, $Fs < 1$.

We again examined whether corrected RT data paralleled the observed bias in estimates. We calculated absolute RT differences as in Experiment 1, and the analysis was the same, again without the 50% group. As can be seen in the last row of Table 11.2, absolute RT differences were lower for the 25% group than for the 75% group, indicating again that relative fluency for incongruent pairs in the 25% group was smaller than relative fluency for congruent pairs in the 75% group. Replicating the findings in Experiment 1, these results further supported the processing fluency view: the smaller absolute processing fluency of the 25% group relative to the 75% group paralleled the smaller average difference between the 25% group's estimate and the 50% anchor.

General discussion

The experiments have shown that manipulations of processing fluency at encoding influenced probability and frequency estimates. In both experiments, biases in the estimates were reflected in biases in RT differences between incongruent and congruent pairs: absolute RT differences as well as absolute differences between estimates and the 50% anchor were smaller for the 25% group than for the 75% group.

One limitation has to be acknowledged: we do not know whether frequency is coded at encoding or constructed at retrieval. One could imagine two processes that both yield the results reported above. First, frequency may be coded at encoding (see Hasher & Zacks 1979; 1984; Jonides & Jones 1992). If so, our experiments would suggest that frequency coding is not simply counting the incoming events but related to phenomenal experiences of fluency. Alternatively, encoding of events may lead to memory traces that are retrieved if a participant is asked to estimate frequency. The total similarity of these memory traces (Dougherty et al. 1999; Hintzman 1988) or total ease of recall of the event (Schwarz 1998; Tversky & Kahneman 1973) may determine the estimate at hand. In the experiments reported above, participants would have stored each single event. When the estimate was asked, people assessed their memory (see Brown 1995, 1997, this volume). As congruent pairs were easier to retrieve than incongruent pairs, participants overestimated the frequency of occurrence of congruent pairs. From the reported data, it is impossible to determine which of the two alternatives hold. This is not an easy task, however, and it may even be impossible in principle to resolve this issue, as Barsalou (1990) has shown for abstractionist versus exemplar-based models.

Studies that may help to disentangle this issue are under way. Specifically, we use the spontaneous guess versus accuracy instruction manipulation used by Haberstroh et al. (this volume). They found that participants used automatically encoded information or the availability of information depending on task requirements: frequency judgements were based on automatically encoded information when made under a spontaneous guess instruction or under time pressure. When participants were instructed to be accurate, frequency judgements were influenced by the availability of information. If processing fluency at retrieval determined estimates in our paradigm, we would expect that accuracy instruction would result in the biases observed in the experiments reported above, whereas a spontaneous guess instruction would remove or at least attenuate these biases. If, however, fluency at encoding determined estimates, we would expect no effect of the instruction.

Our results support the conclusion drawn from the study of Wänke et al. (1995) which demonstrated an impact of processing fluency on frequency estimates. Moreover, our findings supplement existing findings on the impact of processing fluency on several tasks, such as recognition (Johnston et al. 1985; Whittlesea et al. 1990), affective judgements (Bornstein & D'Agostino 1994; Reber et al. 1998; Van den Bergh & Vrana 1998; Whittlesea 1993), judgements of truth (Begg et al. 1992; Reber & Schwarz 1999), self-evaluations (Schwarz et al. 1991), and metacognitive judgements (Begg et al. 1989; Benjamin et al. 1998; Winkielman et al. 1998). Further research is needed to establish whether perceptual fluency has an impact on other tasks, such as contingency judgements (see Shanks 1995).

Objective frequency of occurrence has been shown to play an important role in memory processes and in several kinds of judgements (see Hasher & Zacks 1984). When frequency was manipulated systematically, interesting effects have been observed. For example, mere exposure effects on affect (see Zajonc 1968, 1980) depend on the frequency of exposure. In his meta-analysis, Bornstein (1989) found that liking normally increases up to about 50 to 100 stimulus exposures and then declines (see also Kail & Freeman 1973). These findings suggest that affective judgements were related to objective frequency of exposure. As frequency of exposure increases perceptual fluency (Feustel et al. 1983; Solomon & Postman 1952), it may be that the mere exposure effect on affective judgements is mediated by perceptual fluency (see Jacoby et al. 1989; Seamon et al. 1983). Indeed, recent research has shown that perceptual fluency is an important variable that influences affective judgements (Reber et al 1998; Reber & Schwarz 2001). It will be a challenge for further research to reveal the interconnections between frequency, fluency, and affect.

In conclusion, we were able to bias probability and frequency estimates by manipulation of processing fluency. Does this mean that people are bad at estimating frequencies? We do not think so. Tversky and Kahneman noted that 'availability is an ecologically valid clue for the judgement of frequency because, in general, frequent events are easier to recall or imagine than infrequent ones' (Tversky & Kahneman 1973, p. 209). Only rarely does processing fluency at recall result in errors. Therefore, there is no contradiction between our findings and the conclusion that 'estimates of the relative frequencies of many kinds of events reflect their actual frequencies with great fidelity' (Sedlmeier et al. 1998, p. 768; see also Hasher & Zacks 1979, 1984). It may turn out that the use of processing fluency as a clue for judgements of frequency is ecologically rational (see Gigerenzer 2000; Gigerenzer et al. 1999).

Acknowledgement

This research was supported by the Swiss National Science Foundation (grant no. 1114–50947.97). We thank Tilmann Betsch, Josef Krems and Peter Sedlmeier for comments on an earlier version of this chapter, and Andrea Haerter and Bernhard Sollberger for data collection.

References

Arkes, H. R. & Harkness, A. R. (1983). Estimates of contingency between two dichotomous variables. *Journal of Experimental Psychology: General*, 112:117–135.

Barsalou, L. W. (1990). On the indistinguishability of exemplar memory and abstraction in category representation. In T.K. Srull & R.S. Wyer Jr. (eds) *Content and process specificity in the effects of prior experiences. Advances in social cognition*, vol. 3 (pp. 61–88). Hillsdale, NJ: Lawrence Erlbaum.

Begg, I. M., Anas, A. & Farinacci, S. (1992). Dissociation of processes in belief: Source recollection, statement familiarity, and the illusion of truth. *Journal of Experimental Psychology: General*, 121:446–458.

Begg, I., Duft, S., Lalonde, P., Melnick, R. & Sanvito, J. (1989). Memory predictions are based on ease of processing. *Journal of Memory and Language*, 28:610–632.

Benjamin, A. S., Bjork, R. A. & Schwartz, B. L. (1998). The mismeasure of memory: When retrieval fluency is misleading as a metacognitive index. *Journal of Experimental Psychology: General*, 127:55–68.

Betsch, T., Siebler, F., Marz, P., Hormuth, S. & Dickenberger, D. (1999). The moderating role of category salience and category focus in judgments of set size and frequency of occurrence. *Personality and Social Psychology Bulletin*, 25:463–481.

Bornstein, R. F. (1989). Exposure and affect: Overview and meta-analysis of research 1968–1987. *Psychological Bulletin, 106*:265–289.

Bornstein, R. F. and D'Agostino, P. R. (1994). The attribution and discounting of perceptual fluency: Preliminary tests of a perceptual fluency/attributional model of the mere exposure effect. *Social Cognition, 12*:103–128.

Brown, N. R. (1995). Estimation strategies and the judgment of event frequency. *Journal of Experimental Psychology: Learning, Memory, and Cognition, 21*:1539–1553.

Brown, N. R. (1997). Context memory and the selection of frequency estimation strategies. *Journal of Experimental Psychology: Learning, Memory, and Cognition, 23*:898–914.

Bruce, D., Hockley, W. & Craik, F. I. M. (1991). Availability and category-frequency estimation. *Memory & Cognition, 19*:301–312.

Cohen, J. D., MacWhinney, B., Flatt, M. & Provost, J. (1993). PsyScope: A new graphic interactive environment for designing psychology experiments. *Behavioral Research Methods, Instruments & Computers, 25*:257–271.

Conrad, F. G., Brown, N. R. & Cashman, E. R. (1998). Strategies for estimating behavioural frequency in survey interviews. *Memory, 6*:339–366.

Dougherty, M. R. P., Gettys, C. F. & Ogden, E. E. (1999). MINERVA-DM: A memory process model for judgments of likelihood. *Psychological Review, 106*:180–209.

Feustel, T. C., Shiffrin, R. M. & Salasoo, A. (1983). Episodic and lexical contributions to the repetition effect in word identification. *Journal of Experimental Psychology: General, 112*:309–346.

Gigerenzer, G. (1991). How to make cognitive illusions disappear: Beyond 'heuristics and biases'. *European Review of Social Psychology, 2*:83–115.

Gigerenzer, G. (2000). *Adaptive Thinking*. Oxford: Oxford University Press.

Gigerenzer, G., Todd, P. M. & the ABC Research Group (1999). *Simple Heuristics that make us smart*. Oxford: Oxford University Press.

Hasher, L. & Zacks, R. T. (1979). Automatic and effortful processes in memory. *Journal of Experimental Psychology: General, 108*:356–388.

Hasher, L. & Zacks, R. T. (1984). Automatic processing of fundamental information: The case of frequency of occurrence. *American Psychologist, 39*:1372–1388.

Hintzman, D. L. (1988). Judgments of frequnecy and recognition memory in a multiple trace memory model. *Psychological Review, 95*:528–551.

Jacoby, L. L., Kelley, C. M. & Dywan, J. (1989). Memory attributions. In H.L. Roediger & F.I.M. Craik (eds) *Varieties of memory and consciousness: Essays in honour of Endel Tulving* (pp. 391–422). Hillsdale, NJ: Erlbaum.

Johnston, W. A., Dark, V. & Jacoby, L. L. (1985). Perceptual fluency and recognition judgments. *Journal of Experimental Psychology: Learning, Memory, and Cognition, 11*:3–11.

Jonides, J. & Jones, C. M. (1992). Direct coding for frequency of occurrence. *Journal of Experimental Psychology: Learning, Memory, and Cognition, 18*:368–378.

Kail, R. V. & Freeman, H. R. (1973). Sequence redundancy, rating dimensions and the exposure effect. *Memory and Cognition, 1*:454–458.

Mangan, B. (1993). Taking phenomenology seriously: The 'fringe' and its implications for cognitive research. *Consciousness and Cognition, 2*:89–108.

Manis, M., Shedler, J., Jonides, J. & Nelson, T. E. (1993). Availability heuristic in judgments of set-size and frequency of occurrence. *Journal of Personality and Social Psychology, 65*:448–457.

Reber, R. & Schwarz, N. (1999). Effects of perceptual fluency on judgments of truth. *Consciousness and Cognition, 8*:338–342.

Reber, R. & Schwarz, N. (2001). The hot fringes of consciousness: perceptual fluency and affect. *Consciousness and Emotion, 2*:223–231.

Reber, R., Winkielman, P. & Schwarz, N. (1998). Effects of perceptual fluency on affective judgments. *Psychological Science, 9*:45–48.

Reber, R. and Wurtz, P. (2001). *Exploring 'fringe' consciousness: The case of perceptual fluency.* Manuscript submitted for publication.

Schwarz, N. (1998). Accessible content and accessibility experiences: The interplay of declarative and experiential information in judgment. *Personality and Social Psychology Review, 2*:87–99.

Schwarz, N., Bless, H., Strack, F., Klumpp, G., Rittenauer-Schatka, H. & Simons, A. (1991). Ease of retrieval as information: Another look at the availability heuristic. *Journal of Personality and Social Psychology, 61*:195–202.

Seamon, J. G., Brody, N. & Kauff, D. M. (1983). Affective discrimination of stimuli that are not recognized: Effects of shadowing, masking, and central laterality. *Journal of Experimental Psychology: Learning, Memory and Cognition, 9*:544–555.

Sedlmeier, P. (1999). Improving statistical reasoning: Theoretical models and practical implications. Mahwah, NJ: Erlbaum.

Sedlmeier, P., Hertwig, R. & Gigerenzer, G. (1998). Are judgments of the positional frequencies of letters systematically biased due to availability? *Journal of Experimental Psychology: Learning, Memory, and Cognition, 24*:754–770.

Shanks, D. R. (1995). *The psychology of associative learning.* Cambridge: Cambridge University Press.

Solomon, R. L. & Postman, L. (1952). Frequency of usage as determinants of recognition thresholds for words. *Journal of Experimental Psychology, 43*:195–201.

Tversky, A. & Kahneman, D. (1973). Availability: A heuristic for judging frequency and probability. *Cognitive Psychology, 5*:207–232.

Tversky, A. & Kahneman, D. (1974). Judgment under uncertainty: Heuristics and biases. *Science, 185*:1124–1131.

Van den Bergh, O. & Vrana, S. R. (1998). Repetition and boredom in a perceptual fluency/attributional model of affective judgments. *Cognition and Emotion, 12*:533–553.

Wänke, M., Schwarz, N. & Bless, H. (1995). The availability heuristic revisited: Experienced ease of retrieval in mundane frequency estimates. *Acta Psychologica, 89*:83–90.

Whittlesea, B. W. A. (1993). Illusions of familiarity. *Journal of Experimental Psychology: Learning, Memory, and Cognition, 19*:1235–1253.

Whittlesea, B. W. A., Jacoby, L. L. & Girard, K. (1990). Illusions of immediate memory: Evidence of an attributional basis for feelings of familiarity and perceptual quality. *Journal of Memory and Language, 29*:716–732.

Williams, K. W. & Durso, F. T. (1986). Judging category frequency: Automaticity or availability? *Journal of Experimental Psychology: Learning, Memory, and Cognition, 12*:387–396.

Winkielman, P., Schwarz, N. & Belli, R. F. (1998). The role of ease of retrieval and attribution in memory judgments: Judging your memory as worse despite recalling more events. *Psychological Science, 9*:124–126.

Zajonc, R. B. (1968). Attitudinal effects of mere exposure. *Journal of Personality and Social Psychology Monograph Supplement, 9*:1–27.

Zajonc, R. B. (1980). Feeling and thinking: Preferences need no inferences. *American Psychologist, 35*:151–175.

… CHAPTER 12

FREQUENCY JUDGEMENTS OF EMOTIONS: THE COGNITIVE BASIS OF PERSONALITY ASSESSMENT

ULRICH SCHIMMACK

Abstract

Frequency judgements of emotions are routinely used in personality assessment and clinical diagnoses. However, relatively little is known about the cognitive processes underlying these judgements and their validity. This chapter examines this issue, drawing on cognitive psychological theories of frequency processing. The reviewed evidence suggests that frequency judgements of emotions are made in a fast, intuitive manner that does not rely on the conscious retrieval of individual episodes. These judgements show high discriminant validity for frequency of different emotions but low absolute accuracy. Implications for the validity of personality assessments are discussed.

Frequency judgements of emotions: the cognitive basis of personality assessment

Frequency estimates of emotional experiences are often used to assess individual differences. Personality psychologists use frequency of emotions to assess major personality dimensions such as extraversion and neuroticism (cf. Schimmack et al. 2000). Hedonic psychologists rely on frequency estimates of pleasant versus unpleasant emotions to assess individual differences in subjective well-being (Diener et al. 1991). Clinical psychologists rely on abnormally high or low frequencies of emotions to assess personality disorders (e.g. Andreasen & Black 1991). For these purposes, frequency judgements of emotions need to reflect actual individual differences in emotional experiences fairly accurately. This chapter examines the accuracy of frequency judgements of emotions and the cognitive processes underlying these judgements.

What is an emotion?

I define emotions as intentional affective states that are directed at something (Ortony et al. 1988; Reisenzein & Schönpflug 1992; Schimmack & Diener 1997). This definition has several implications. First, the definition excludes other affective experiences such as a moods (e.g. cheerful, downhearted, grouchy; Schimmack 1997a) and global dimensions of core affect (pleasant–unpleasant, awake–tired, tense–calm; Schimmack & Grob 2000). Most people feel some degree of pleasure and wakefulness most of the time. Hence, it would be odd to assess the *frequency* of these states ('I feel awake three times a day'). Second, the definition implies that emotions, in contrast to other affective states, have a cognitive component (Ortony *et al.* 1988). For example, pride entails a certain set of cognitions that partially defines an experience as pride. Typically feeling pride entails thinking about some positive attribute of oneself or a close other. It is this cognitive component that makes it possible to *count* the number of times an emotion is experienced. Other components of emotions do not have a categorical structure. For example, pride may increase one's heart rate but one's heart was already beating before the onset of pride. For most emotions, the categorical difference between feeling an emotion and not feeling an emotion is rooted in the cognitive component of emotions (Ortony *et al.* 1988).

Finally, the definition of emotions implies that frequency judgements of emotions are likely to be based on memories of the cognitive aspects of emotional experiences. The reason is that cognitions are the most salient difference between specific emotions (Reisenzein 1994). For example, jealousy and anger are both arousing and unpleasant emotions but people discriminate between the frequencies of the two emotions in direct frequency judgement tasks and in conditional probability judgements (Schimmack & Reisenzein 1997). As the two emotions are similar in many respects except their cognitive elements, sensitivity to the different frequencies of the two emotions is likely to be based on the different cognitive aspects of the two emotions. Hence, I propose that frequency judgements of emotions draw on representations of the 'cold' cognitive aspects of emotional experiences rather than on representations of the 'hot' affective aspects of emotional experiences. Based on this assumption—and I am well aware of the fact that it is an assumption—I present two lines of research on frequency judgements of emotions, namely frequency judgements of naturally occurring emotional experiences in everyday life (Schimmack & Reisenzein 1997; Schimmack *et al.* 2000) and frequency judgements of emotions in hypothetical scenarios (Schimmack & Hartmann 1997).

Absolute accuracy of frequency judgements of emotions

Schimmack (1997b) distinguished three types of accuracy, namely (a) absolute accuracy, (b) discriminative accuracy across emotions, and (c) discriminative accuracy across participants. Absolute accuracy compares the absolute level of a frequency judgement to an observed frequency (e.g. the actual frequency was 3, whereas the estimated frequency was 2). Schimmack (1997b) examined the absolute accuracy of frequency estimates of emotions using daily diary data as a standard of comparison. Participants estimated frequencies of emotions once a day for three weeks. Before and after the daily diary study, participants estimated how often they experienced the same emotions during the past three weeks (e.g., 'I experienced ANGER—times a week during the past three weeks'). Daily estimates were transformed into weekly rates. Extended frequency judgements underestimate the frequencies of emotions derived from daily diary data (Fig. 12.1). Participation in a daily diary study reduced this bias, but did not completely

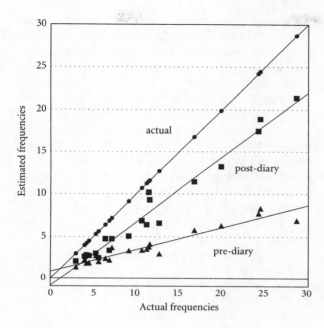

Figure 12.1 Plot of estimated frequencies as a function of daily diary frequencies.

eliminate it. Finally, frequent emotions were underestimated more severely than infrequent emotions. For example, the least frequent emotion was romantic jealousy ($M = 2.98$ times a week), which was only slightly underestimated before ($M = 1.39$) and after the daily diary study ($M = 1.71$). In contrast, contentment was the most frequent emotion ($M = 28.65$), which was severely underestimated before ($M = 6.94$) and after the daily diary study ($M = 21.41$). In sum, the results indicate that the absolute accuracy of frequency judgements of emotions is low. This finding is consistent with previous studies in cognitive psychology, in which frequent events were also increasingly underestimated (e.g. Alba *et al.* 1980).

The effect of time frames and response formats

The following study provides further evidence regarding absolute accuracy. I manipulated the response format of absolute frequency judgements of emotions. If response formats have a strong influence on frequency judgements, then these judgements have low absolute accuracy. Participants were asked to provide absolute frequency estimates of happiness *about something* and sadness *about something*. 'About something' was added to communicate that the researcher was interested in emotions and not in happy and sad moods. Five different response formats were used and 50 participants responded to each format. One format asked about absolute frequencies in the past month. The second format also asked about the past month but included special instructions that the researcher was 'interested in all experiences of happiness and sadness about something, no matter how mild or intense the experience was'. These instructions were used because Winkielman *et al.* (1998) found that participants limit frequency judgements over longer time periods to intense experiences (see also Schwarz & Wänke, this volume). The next format asked about the frequency of emotions in the past month in terms of weekly rates. This format should also lead participants to include milder experiences because inquiries about weekly rates suggest that the researcher is interested in experiences that occur on a weekly basis. The next format asked about experiences in the past week, which should produce the same results as monthly judgements in weekly rates. Finally, one response format asked for seven separate judgements for each day in the past week. This category-split format was expected to produce especially high frequency estimates (see Fiedler, this volume). If necessary, judgements were transformed into weekly rates. As expected, all response formats yielded significantly higher frequencies than the plain monthly format (Fig. 12.2). The special instructions, weekly rates, and past week formats did not differ significantly from each other, but all three formats yielded significantly lower frequencies than the daily-split format. Without any validation data it is not possible to determine which estimates provided the most accurate information. However, the frequencies in the daily diary study (happiness 24, sadness 11) suggest that the daily-split format may provide the highest absolute accuracy.

The effects of different response formats provide further evidence that people have very little sense of the absolute frequencies of emotions, in that monthly frequency judgements grossly underestimated absolute frequencies. This effect is partly due to different inclusion criteria for shorter and longer time frames (Winkielman *et al.* 1998). However, even when instructions informed participants that they should consider all experiences, the absolute estimates were considerably lower than aggregated category-split frequencies. Future research needs to examine whether category-split judgements can increase absolute accuracy (see Fiedler, this volume). Another open question is whether different types of category splits have different effects on frequency judgements of emotions. For example, splitting emotion categories by objects (e.g. gratitude towards partner, co-worker, etc.) may produce even higher frequency estimates because objects are better retrieval cues than days of the week.

Discriminative accuracy across emotions

Discriminative accuracy across emotions refers to the correspondence in the rank ordering of emotions between daily and extended frequency judgements. Discriminant accuracy for

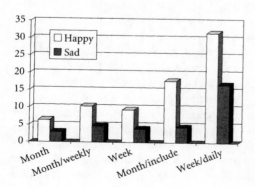

Figure 12.2 Absolute estimates of happiness and sadness with different response formats.

aggregated judgements, which can be seen in Figure 12.1, is high. However, aggregation may mask poor accuracy at the level of individuals. Hence, a more appropriate measure of accuracy is the average correlation between daily and extended frequency estimates for each individual, which is $r = 0.79$ for estimates before the daily diary study and $r = 0.85$ for estimates after the daily diary study. This finding indicates that people have a good sense of how often they experience one emotion compared to another (see also Schimmack & Reisenzein 1997). This finding is again consistent with the cognitive literature on frequency judgements (Alba et al. 1980; Sedlmeier, this volume; Zacks & Hasher, this volume).

Salience effects in everyday life

Although frequency judgements of emotions show high discriminant accuracy across emotions, they are also likely to be influenced by some systematic biases. Several studies in the cognitive literature have demonstrated that salience during encoding influenced frequency judgements (e.g. Bruce et al. 1991; Greene 1989). Schimmack and Hartmann (1997) obtained preliminary evidence of salience effects on frequency judgements of emotions. The following two studies examined salience effects on frequency judgements of emotions more thoroughly.

People can have the same emotional experience but they may label them differently or may not apply an emotion label at all (Schimmack & Hartman 1997). For example, somebody can think 'I am grateful for your help', 'I am thankful for your help', or 'I am happy that the task is done'. How do the different ways of labelling an experience influence frequency estimates of gratitude? This question was addressed within the daily diary study that I described earlier (Schimmack 1997b). The pre-diary and post-diary judgement task included 10 additional emotions that were not included in the daily study. Furthermore, the additional emotions were matched to emotions that were included in the diary study (daily–not daily: joy–happiness, affection–love, anxiety–fear, anger–rage, contempt–dislike, guilt–regret, embarrassment–shame, sadness–depression, hopelessness–helplessness). As demonstrated earlier, participation in the diary study boosted frequency judgements of those emotions that were on the daily diary form. If this effect was due to the greater salience of these emotions during the daily diary period, then emotions that were not included on the form should not receive the same boost. An analysis of variance with salience (salient vs. non-salient) and time (pre vs. post) as within-subject variables revealed a significant interaction, $F(1,145) = 16.56$, $p < 0.01$. As expected, the interaction was due to a stronger increase in frequency estimates of salient emotions compared to non-salient emotions (Fig. 12.3).

Salience effects in the laboratory

Daily diary studies provide insights into frequency judgements of real emotional experiences. However, it is not possible to manipulate the frequencies of emotions experimentally in these studies. Furthermore, diary data are already frequency data that can share biases with the extended frequency estimates. Shared biases would lead to an overestimation of accuracy. To overcome these limitations, I use a scenario-rating task (SRT) (Schimmack & Diener 1997; Schimmack & Hartmann 1997). The SRT is based on word-list studies in the cognitive literature (Alba et al. 1980), in which participants read

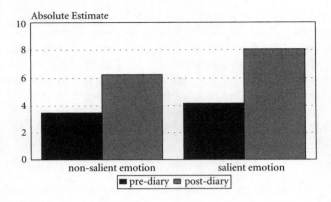

Figure 12.3 Absolute estimates of salient and non-salient emotions before and after a daily diary study.

a list of words (e.g. birch, oak, and maple) and then estimate how many instances of a category (i.e. trees) occurred in the word list. In the SRT, emotional scenarios are the instances and emotion concepts are the categories. Scenarios are stories such as 'I always encouraged a friend to pick up jogging. Recently, I heard that she actually started jogging'. Following each scenario, participants rate the intensity of several emotions (pride, anger, sadness, etc.) to indicate how they would feel in this situation. Ratings are made on a scale ranging from 0 to 6. Zero-ratings reflect the absence of an emotion, whereas ratings from 1 to 6 indicate the presence of an emotion (cf. Schimmack & Diener 1997, for more detail). In the above example, most people respond with a rating greater than 0 to pride, and zero ratings to anger and sadness. The number of non-zero ratings across all scenarios is used as a measure of the actual frequency of emotions in the scenarios. For example, if an individual made 5 pride-ratings greater than zero, the frequency of pride would be 5.

The SRT differs in several ways from traditional word-list studies. First, in existing word-list studies instances belonged to only one category (e.g. birch—tree). In contrast, emotional scenarios often describe situations that elicit multiple emotions because the occurrence of a single emotion is quite rare (Schimmack & Hartmann 1997; Schimmack & Reisenzein 1997). For example, in the above example, participants would report pride and joy, and sometimes gratitude. Hence, instances (scenarios) belonged to multiple categories (emotion concepts).

Word-list studies vary the frequencies of categories to eliminate any effects of prior experience with the categories. For example, some participants see three different trees and five different fruits, whereas others see five different trees and three different fruits. It is much harder to change the frequencies of emotions in the SRT. For example, replacing an anger scenario with a fear scenario also affects frequencies of disappointment. Hence, I used a different approach to control for prior experiences of emotions, using one set of scenarios in which the frequencies of emotions were orthogonal to the frequencies of emotions in everyday life.

Another problem with the SRT paradigm is the subjectivity of the rating task. In cognitive experiments, the relation between instances and categories (e.g. birch—tree) is universal and researchers can count instances to obtain an objective frequency measure.

However, individuals can differ in the appraisal of emotional situations. For example, some participants may feel relief when a friend picks up jogging after their recommendation, whereas others may not feel relief. Hence, studies of frequency judgements of emotions do not have an objective measure of actual frequencies. In the present study, I relied on scenario ratings by a different group of participants to derive the criterion frequencies. This approach ignores valid individual differences in emotion reactions to the same scenario. Fortunately, these individual differences are relatively small for unambiguous scenarios (cf. Reisenzein & Spielhofer 1994). Nevertheless, readers should keep in mind that valid individual differences in the appraisal of the scenarios attenuate the accuracy measures reported below.

The following experiment used the design of a word-list study by Bruce et al. (1991) that examined the influence of category-salience on frequency judgements. During the encoding stage, the authors presented instances with (salient) or without (non-salient) category labels. During the retrieval period, the authors varied the order of two retrieval tasks, namely (a) cued recall of instances with category labels as cues and (b) frequency estimates of category members. Bruce et al. (1991) found that frequency judgements in the salient condition had a steeper slope than frequency judgements in the non-salient condition and that task order had no effect on frequency estimates.

The study was run on computers to measure the judgement times of the frequency judgements because judgement times provide insights into participants' judgement strategies (Brown, this volume). If the judgement times increase with the magnitude of the judgements and if they are relatively slow (up to 10 s), then people typically rely on the recall of instances. However, if the judgement times are fast (about 5 s or less) and are not related to the magnitude of the frequency judgements, then participants seem to rely on a fast, intuitive, recall-free strategy. For example, participants may experience a mild or intense feeling of familiarity that serves as the basis of the frequency estimate (Dougherty & Franco-Watkins; this volume; Hintzman 1988; Schimmack & Reisenzein 1997).

Eighty students at the University of Berlin participated in this study for course credit. Participants were randomly assigned to the four conditions. In the salient condition, participants rated each scenario for the presence and intensity of 12 emotions (Table 12.1). In the non-salient condition, participants rated the presence and intensity of 13 moods taken from the Everyday Language Mood Questionnaire (Schimmack 1997a). These 13 moods were pleasant [German: angenehm], unpleasant [unangenehm], elated [ausgelassen], relaxed [entspannt], aroused [erregt], annoyed [gereizt], indifferent [gleichgültig], cheerful [heiter], melancholic [melancholisch], grumpy [mürrisch], nervous [nervös], downhearted [niedergeschlagen], and calm [ruhig]. These moods do not have the rich cognitive structure of the emotion concepts (pride, gratitude, etc.) and therefore it is not possible to infer the frequency of specific emotions from mood ratings. In the frequency judgement task, participants entered an absolute number. The computer recorded the time between onset of the emotion word and the first keystroke. In the cued-recall task, participants had 10 s to recall as many scenarios as possible. For each recalled scenario they pressed the space bar and the computer recorded the time between onset of the cue and pressing of the space bar. For the present article, the number of keystrokes is used as a measure of cued recall (see Schimmack et al. 2000, for additional measures that can be derived from this task).

Table 12.1 Aggregated data for the first scenario rating task study

Emotions	Everyday frequency	Scenario frequency	Frequency judgement		Recalled scenarios	
			Salient	Non-salient	Salient	Non-salient
Anger	2.72	10.75	9.15	8.08	3.43	3.03
Anxiety	1.94	10.55	9.15	5.85	3.03	2.30
Embarrassment	1.90	9.60	6.35	4.48	2.48	2.10
Sadness	2.60	9.45	8.18	6.65	2.95	2.28
Disappointment	2.12	9.05	7.83	6.05	2.70	2.43
Contempt	0.85	7.90	5.60	3.90	2.23	2.03
Love	3.90	7.45	6.00	5.10	3.05	2.63
Joy	4.04	7.30	7.48	5.45	2.78	1.88
Pride	2.80	6.85	5.13	3.33	2.35	1.35
Disgust	1.04	6.20	4.28	3.08	2.35	2.38
Gratitude	2.76	5.85	5.80	3.65	2.48	2.03
Envy	1.33	2.70	3.60	3.38	1.45	1.13
Average	—	—	6.54	4.91	2.60	2.13

Note: everyday frequencies (range 0 = never to 6 = nearly always), scenario frequencies (range 0 to 20), frequency judgements (range 0 to 20), number of recalled scenarios (0 to 20), frequency judgement times (seconds).

Aggregated results

Table 12.1 shows the means in the salient and non-salient condition for the 12 emotions. Salience produced significantly higher frequency estimates, $t(11) = 7.42$, $p < 0.01$, and shorter judgement times, $t(11) = 7.14$, $p < 0.01$. Salience also produced significantly more recalled scenarios in the cued-recall task, $t(11) = 5.54$, $p < 0.01$. This pattern of results indicates that salience renders instances more accessible, which produces faster and higher frequency judgements. Frequency judgements in the salient and non-salient condition were significantly correlated with scenario frequencies ($rs = 0.87$, 0.76, respectively), but not with everyday frequencies ($rs = 0.36$, 0.38). The cued-recall task also reflected scenario frequencies quite accurately ($rs = 0.83$, 0.71). Judgement times were fast ($< 4.2s$) and did not increase with the magnitude of the frequency judgements ($rs = 0.08$, 0.26).

Individual-level results

The aggregated analyses are informative but they do not reveal the full picture. For example, aggregated data provide inflated estimates of accuracy that may mask poor accuracy at the level of individuals. Hence, it is important to examine the data at the level of individuals, which is the standard approach in word-list studies (e.g. Bruce et al. 1991). Typically researchers regressed actual frequencies on frequency estimates and then averaged the regression coefficients. This traditional approach is increasingly replaced by more sophisticated hierarchical models (Kreft & Leeuw 1998). I used Hierarchical Linear Modeling 5 (HLM5; Raudenbush et al. 2000) to analyse the data. HLM 5 differentiates *levels* of data. In the present application, HLM models the relation between variables within individuals at *Level 1*. To illustrate, the first model used frequency judgements as a criterion variable and scenario frequencies and everyday frequencies as predictors, plus an intercept that accounts for different absolute levels in the frequency judgements. HLM estimates the population regression coefficients B0, B1, B2 that reflect the average relation between criterion and predictor variables. At *Level 2*, HLM decomposes the Level 1 parameters into Level 2 components. For example, B0 is decomposed into G00, an intercept, G01, a regression coefficient that captures differences in the Level 1 intercept due to Salience (0 = non-salient, 1 = salient), G02, a regression coefficient that captures differences in the Level 1 intercept due to Task Order (0 = judgement first, 1 = cued recall first), and a Random Effect that captures random variability between participants.

Level 1 Model
Frequency judgements = B0 + B1 * Scenario frequencies + B2 * Everyday frequencies
Level 2 Model
B0 = G00 + G01 * Salience + G02 * Task order + Random
B1 = G10 + G11 * Salience + G12 * Task order + Random
B2 = G20 + G21 * Salience + G22 * Task order + Random

The model revealed a significant Intercept (G00 = 1.09, Standard Error [SE] = 0.48), which was not moderated by salience or task order. It also revealed a significant slope for scenario frequencies (G10 = 0.72, SE = 0.06), which was moderated by salience (G11 = 0.18, SE = 0.08). Everyday frequencies revealed no significant effect. The positive intercept and the slope smaller than one imply that low frequencies are overestimated, whereas higher frequencies are underestimated (see Table 12.1). More important, the significant effect of salience on the slope reveals that participants in the salient condition were more accurate. The absence of task-order effects indicated that frequency judgements were not influenced by a preceding cued-recall task. This pattern of results replicates Bruce *et al.*'s (1991) findings.

The second model focused on the judgement times and their relations to the magnitude of the frequency estimates. The relation was allowed to be moderated by the two experimental variables salience and task order.

Level 1
Judgement times = B0 + B1 * Frequency judgements
Level 2
B0 = G00 + G01 * Salience + G02 * Task order + Random
B1 = G10 + G11 * Salience + G12 * Task order + Random

The model revealed a significant intercept (G00 = 2961 ms, SE = 137), and a moderating effect of salience on the intercept (G01 = −542 ms, SE = 204). Judgement times were not significantly related to the magnitude of the frequency judgements and the relation was not moderated by salience. Hence, the data do not indicate that participants switched to a recall-based strategy in the non-salient condition. However, they needed about half a second more time to make judgements in this condition. At present, none of the existing memory models predicts a constant increase in judgement times in the non-salient condition. However, several models can incorporate this finding if one assumes that a non-salient probe needs more time to produce an output that can serve as the basis for a frequency estimate (see Dougherty & Franco-Watkins, this volume; Sedlmeier, this volume).

In sum, the results are consistent with findings in the frequency judgement literature (Alba *et al.* 1980; Bruce *et al.* 1991). Judgements reflect the different frequencies of emotions quite accurately, but they are also influenced by a manipulation of the salience of category labels and they increasingly underestimate higher frequencies. The fast and flat judgement times suggest that people relied on a fast, intuitive strategy and did not enumerate scenarios (see Brown, this volume). The results are also inconsistent with the idea that frequency judgements of emotions are based on ease of retrieval, which would be reflected in negative correlations between frequency estimates and judgement times. Once more, this conclusion is consistent with the evidence in the cognitive literature (Watkins & LeCompte 1991).

Meta-memory: a shortcut to recall-based judgement strategies

Tversky and Kahneman's (1973) famous article on the availability heuristic has stimulated most of the research on frequency judgements in social psychology. As noted by Schwarz and Wänke (this volume), the availability heuristic is not a unitary process but a family of strategies that provide shortcuts to a cumbersome enumeration of all available instances. Schwarz and Wänke (this volume) discuss two versions of the availability heuristic, namely a strategy that relies on the number of recalled instances and another strategy that relies on the ease of recall. The previous study suggested that participants used neither of these two strategies. However, this finding is not as inconsistent with Tversky and Kahneman's (1973) availability heuristic(s) as it might seem. In fact the authors also proposed a recall-free strategy.

To assess availability it is not necessary to perform the actual operations of retrieval. It suffices to assess the ease with which these operations *could* [italics added] be performed, much as the difficulty of a puzzle or mathematical problem can be assessed without considering specific solutions. (Tversky and Kahneman, 1973, p. 208)

Tversky and Kahneman (1973) also presented the first evidence that people can assess frequency information without recalling instances from memory. They asked people to predict, for various categories (mammals, Russian writers), how many instances they could recall within 2 minutes. The predictions had to be made within 7 seconds. Predictions were surprisingly good predictors of performance in the actual recall task ($r = 0.93$). The study did not reveal how participants assessed availability under time pressure. They may have relied on the retrieval of a few instances. They may also have relied on knowledge of their expertise in the various domains ('I never read Russian novels' or 'I study Russian literature'). Or they may rely on meta-memory, that is, knowledge about memory that is not based on retrieval (Hart 1967; Metcalfe 1993). The present study examined this question using the SRT paradigm.

Twenty-four students at the Free University Berlin, Germany, took part in this study for course credit. The scenarios and emotions were the same as in the previous study, and the SRT was identical to the salient condition in the previous study. However, due to a programming error the item 'Anxiety' was not displayed during the scenario-rating task. For this reason, Anxiety was excluded from all analyses. After the scenario-rating task, participants worked on a meta-memory task in which they estimated for 20 diverse categories (Indian languages, US Presidents, Bundesliga soccer clubs) how many instances they could recall within 10 s. Participants were instructed to respond as fast as possible and judgements had to be entered within a 5 s response window. Otherwise the computer automatically displayed the next item. This task served as a distracter task and it familiarized participants with the meta-memory task. Afterward, participants worked on the critical meta-memory task for emotions. In this task, participants estimated for each emotion how many scenarios they could recall within 10 s. These estimates had to be made within 5 s, and participants were instructed to respond as fast as possible without recalling instances. The next task was a practice task of the recall task, which asked for the recall of instances of the same 20 categories that were used in the meta-memory practice task. Afterwards, participants worked on the actual cued-recall task in which they recalled for each emotion as many scenarios as possible in which the emotion occurred. The task was limited to 10 s for each emotion.

Aggregated-level results

Table 12.2 shows the averaged scenario frequencies, meta-memory judgements, number of recalled scenarios and judgement times. Consistent with Tversky and Kahneman (1973), aggregated meta-memory judgements were good predictors of the aggregated number of recalled scenarios ($r = 0.90$). Furthermore, both meta-memory judgements and number of recalled scenarios were highly related to scenario frequencies (0.87, 0.86, respectively). This finding indicates that often explicit and implicit measures of accessibility reflect the availability of instances with relatively high accuracy. Hence, in many instances, the availability heuristics will provide reasonably accurate frequency estimates (Tversky & Kahneman 1973). An inspection of the meta-memory judgement times revealed that responses were made in less than 3 s for each emotion, suggesting that participants followed instructions to provide intuitive estimates.

Table 12.2 Aggregated data for the meta-memory study

Emotion	Scenario frequencies	Meta-memory judgements	Recalled scenarios	Judgement times
Anger	12.71	3.71	3.25	2.20
Joy	7.33	3.33	2.75	2.42
Disappointment	10.79	3.26	3.08	2.51
Sadness	11.29	3.13	3.29	2.40
Contempt	8.38	2.70	2.58	2.45
Embarrassment	9.46	2.67	2.63	2.44
Love	7.50	2.58	3.00	2.62
Gratitude	5.79	2.57	2.67	2.41
Pride	7.04	2.43	2.42	2.24
Disgust	5.67	2.33	2.29	2.16
Envy	2.71	1.33	1.13	2.49

Note: Everyday frequencies see Table 12.1; scenario frequencies (range 0–20), meta-memory judgements (range 0–20), recalled scenarios (range 0–20), judgement times (seconds).

Individual-level results

Nine data points (out of 264, i.e. 24 participants × 11 emotions) were missing because participants failed to respond within the 5 s response window. This is not a problem for HLM analyses, because HLM can estimate population parameters from data sets with missing values. In the first model, meta-memory judgements were regressed onto the number of recalled scenarios to test the accuracy of meta-memory judgements.

Level 1
Meta-memory judgements = B0 + B1 * Recalled scenarios + B2 * Everyday frequencies
Level 2
B0 = G00 + Random
B1 = G10 + Random
B2 = G20 + Random

The model revealed a significant intercept (G00 = 1.06, SE = 0.18) and a significant slope for recalled scenarios (G10 = 0.52, SE = 0.11), whereas everyday frequencies did not influence meta-memory judgements. Finally, the relation between meta-memory judgements and their judgement times was examined. Meta-memory judgement times were regressed onto meta-memory judgements. No significant relation emerged, providing further support for the use of a recall-free basis of the meta-memory judgements.

Frequency judgements of emotions: findings and implications

The present chapter examined frequency judgements of emotions in real life and in hypothetical scenarios. The results can be summarized as follows.

1. Frequency judgements of emotions have poor absolute accuracy. In particular, they underestimate the frequencies of frequently occurring emotions.
2. Frequency judgements of emotions show fairly good discriminant accuracy across emotions.
3. Frequency judgements of emotions show context-sensitivity. That is, people do not confuse the frequency of emotions that occurred in a set of scenarios with the frequency of emotions in their own life.
4. Frequency judgements of emotions are influenced by the salience of emotion concepts during the encoding stage.
5. Frequency judgements of emotions are made quickly, and speed is unrelated to the magnitude of the judgement.
6. Participants can anticipate the number of emotion memories that are accessible without recalling individual instances.

The judgement times provide the most diagnostic information about the cognitive processes underlying frequency judgements. Fast and flat judgement times suggest that people did not rely on the recall of instances to estimate frequencies of emotions (Brown, this volume). Several other findings also support a recall-free judgement process. For example, Schimmack and Reisenzein (1997) observed that participants judged conditional probabilities of emotions ('How often do you feel angry when you feel jealous?') within 5 s. For a recall-based strategy, participants would first need to recall jealousy experiences and then check whether anger co-occurred in these situations; a process that would take more time than 5 s. On the other hand, recall-free models can account for the accuracy and speed of conditional probability judgements (Dougherty & Franco-Watkins, this volume; Greene 1990; Sedlmeier, this volume). Moreover, memories of emotional experiences in everyday life are not very accessible. Schimmack et al. (2000) report retrieval latencies for the first memory that participants could recall. The average retrieval latencies were 7.76 s for pleasant emotions and 11.25 s for unpleasant emotions. However, the judgement times of frequency judgements (not reported in Schimmack et al. 2000) were 4.65 s for pleasant and 4.36 s for unpleasant emotions. Hence, frequency judgements were much faster than the recall of a single emotional event. This was particularly true for infrequent emotions like disgust, which yielded an average retrieval latency of 17.71 s and a judgement time of 3.85 s! Given the difficulty of recalling even a single instance of infrequent emotions, it is not surprising that participants revert to an intuitive recall-free estimation strategy.

Implications for personality assessment

Three aspects of frequency judgements of emotions are most relevant for personality assessment, namely (1) the low absolute but relatively high discriminant accuracy across emotions, (2) the influences of salience on frequency judgements, and (3) the context-sensitivity of frequency judgements.

Accuracy of personality assessments

The low absolute accuracy implies that people have to guess the reasonable range of emotion frequencies. For example, they may infer this information from the response format (Schwarz and Wänke, this volume). This process will necessarily bias frequency judgements because people are bound to use different ranges for their frequency estimates. Fortunately, this bias can be removed from the data by computing difference scores. For example, well-being researchers often subtract frequencies of unpleasant emotions from frequencies of pleasant emotions to derive a general index of hedonic balance (Schimmack et al. in press). Hedonic balance scores can be expected to be more accurate than direct measures of emotion frequencies because they reduce the bias in people's choices of absolute frequencies. However, when researchers are interested in frequencies of specific emotions, they should use a multi-method approach. For example, Schimmack et al. (2000) asked participants to estimate absolute frequencies of emotions and to recall emotional experiences. Absolute estimates and number of recalled experiences are influenced by different biases, but they also share some common variance due to true individual differences in the frequency of emotional experiences. This shared variance can be separated from the biases in a multi-trait–multi-method study.

Salience and labelling effects on frequency judgements

The second important finding was the effect of salience on frequency judgements of emotions. In the present study, salience was experimentally manipulated. However, it is reasonable to assume that individuals differ in their use of emotion categories

(Schimmack & Hartmann 1997). These individual differences influence frequency judgements of emotions and the accessibility of emotion memories (cf. Schimmack & Hartmann 1997). A study by Schimmack and Lechuga (1999) suggests that the exact emotion label is less critical than the activation of the emotion concept at the time of encoding. To separate concepts from labels, the authors examined frequency judgements of emotions with Spanish/English bilinguals. Participants completed an SRT in English or in Spanish (scenario descriptions and labels were translated into Spanish) and then made frequency judgements of the same 12 emotions in English or in Spanish. Frequency judgements were as high in the condition when the language was changed as in the condition when the same language was used for both tasks. This finding shows that it is not the semantic label but the activation of emotion concepts that produces salience effect.

Context sensitivity of frequency judgements of emotions

Another important finding was the context sensitivity of frequency judgements. Personality researchers typically ask for frequencies of emotions in general. However, in some circumstances it may be interesting to assess frequencies of emotions in specific circumstances. For example, organizational psychologists may wish to assess frequencies of emotions at the workplace. The present results suggest that respondents should be able to distinguish how often they experience emotions at work in contrast to other situations (see also Dougherty & Franco-Watkins; this volume). This conclusion is also supported by conditional probability judgements of emotions (Schimmack & Reisenzein 1997). These judgements show that participants are sensitive to the fact that some emotions (e.g. anger) are more likely to occur *in the context* of some emotions (e.g. jealous) than in the context of others (e.g. guilt). Future research should examine the practical implications of the context sensitivity of frequency judgements more thoroughly.

Conclusion

In conclusion, I presented evidence that many findings in the extensive cognitive literature on frequency judgements generalize to frequency judgements of emotions in scenario rating tasks and everyday life. Frequency judgements of emotions are often based on a fast, intuitive assessment of associations in episodic memories (which may be modelled in a multiple-trace or a neural network model). These judgements are neither infallible nor hopelessly biased. A better understanding of the cognitive basis of frequency judgements can lead to the development of more precise measures of individual differences in the frequency of emotional experiences.

References

Alba, J. W., Chromiak, W., Hasher, L. & Attig, M. S. (1980). Automatic encoding of category size information. *Journal of Experimental Psychology: Learning, Memory, and Cognition*, 6:370–378.

Andreasen, N. C. & Black, D. W. (1991). *Lehrbuch Psychiatrie* [Psychiatry Textbook]. Weinheim: Beltz.

Bruce, D., Hockley, W. E. & Craik, F. I. M. (1991). Availability and category-frequency estimation. *Memory and Cognition, 19*:301–312.

Diener, E., Sandvik, E. & Pavot, W. (1991). Happiness is the frequency, not the intensity, of positive versus negative affect. In F. Strack, M. Argyle & N. Schwarz (eds) *Subjective well-being* (pp. 119–139). Pergamon Press, Oxford.

Greene, R. L. (1989). On the relationship between categorical frequency estimation and cued recall. *Memory and Cognition, 17*:235–239.

Greene, R. L. (1990). Memory for pair frequency. *Journal of Experimental Psychology: Learning, Memory, and Cognition, 16*:110–116.

Hart, J. T. (1967). Memory and memory-monitoring process. *Journal of Verbal Learning and Verbal Behavior, 6*:685–691.

Hintzman, D. L. (1988). Judgments of frequency and recognition memory in a multiple trace memory model. *Psychological Review, 95*:528–551.

Kreft, I. & Leeuw, J. (1998). *Introducing multilevel modeling.* Sage, London.

Metcalfe, J. (1993). Novelty monitoring, metacognition and control in a composite holographic associative recall model: Implications for Korsakoff amnesia. *Psychological Review, 100*:3–22.

Ortony, A., Clore, G. L. & Collins, A. (1988). *The cognitive structure of emotions.* Cambridge University Press, Cambridge.

Raudenbush, S., Byrk, A., Cheong, Y. F. & Congdon, R. (2000). *HLM5.* Lincolnwood, Illinois: Scientific Software International.

Reisenzein, R. (1994). Pleasure-arousal theory and the intensity of emotions. *Journal of Personality and Social Psychology, 67*:525–539.

Reisenzein, R. & Schönpflug, W. (1992). Stumpf's cognitive-evaluative theory of emotion. *American Psychologist, 47*: 34–45.

Reisenzein, R. & Spielhofer, C. (1994). Subjective salient dimensions of emotional appraisal. *Motivation and Emotion, 18*:31–75.

Schimmack, U. (1997a). Das Berliner-Alltagssprachliche-Stimmungsinventar (BASTI): Ein Vorschlag zur kontentvaliden Erfassung von Stimmungen [The Every-Day Language Mood Inventory (ELMI): Toward a content valid assessment of moods]. *Diagnostica, 43*:150–173.

Schimmack, U. (1997b). *Frequency judgments of emotions: How accurate are they and how are they made?* Unpublished dissertation, Free University Berlin, Germany.

Schimmack, U. & Diener, E. (1997). Affect Intensity: Separating intensity and frequency in repeatedly measured affect. *Journal of Personality and Social Psychology, 73*:1313–1329.

Schimmack, U., Diener, E. & Oishi, S. (in press). Life-satisfaction is a momentary judgment and a stable personality characteristic: The use of chronically accessible and stable sources. *Journal of Personality.*

Schimmack, U. & Grob, A. (2000). Dimensional models of core affect: A quantitative comparison by means of structural equation modelling. *European Journal of Personality, 14*:325–345.

Schimmack, U. & Hartmann, K. (1997). Interindividual differences in the memory representation of emotional episodes: Exploring the cognitive processes in repression. *Journal of Personality and Social Psychology, 73*:1064–1079.

Schimmack, U. & Lechuga, J. (1999, February). *Bilinguals' representation of emotions: Disgust versus asco.* Paper presented at the Annual Meeting of the Society for Cross-Cultural Research (SCCR), Santa Fe.

Schimmack, U., Oishi, S., Diener, E. & Suh, E. (2000). Facets of affective experiences: A new look at the relation between pleasant and unpleasant affect. *Personality and Social Psychology Bulletin, 26*:655–668.

Schimmack, U. & Reisenzein, R. (1997). Cognitive processes involved in similarity judgments of emotion concepts. *Journal of Personality and Social Psychology, 73*:645–661.

Tversky, A. & Kahneman, D. (1973). Availability: A heuristic for judging frequency and probability. *Cognitive Psychology, 5*:207–232.

Watkins, M. J. & LeCompte, D. C. (1991). Inadequacy of recall as a basis for frequency knowledge. *Journal of Experimental Psychology: Learning, Memory, and Cognition, 17*:1161–1176.

Winkielman, P., Knäuper, B. & Schwarz, N. (1998). Looking back at anger: Reference periods change the interpretation of (emotion) frequency questions. *Journal of Personality and Social Psychology, 75*:719–728.

CHAPTER 13

ONLINE STRATEGIES VERSUS MEMORY-BASED STRATEGIES IN FREQUENCY ESTIMATION

SUSANNE HABERSTROH AND TILMANN BETSCH

Abstract

We distinguish two kinds of strategies by which people can arrive at a frequency estimate: online strategies and memory-based strategies. Online strategies are based on a frequency record, which is formed during encoding. Subsequent frequency judgements can be directly based on this record. In contrast, when people use memory-based strategies, they base their estimate on a memory sample drawn at the time of judgement. In line with a multiple bases approach to frequency estimation, we assume that people can capitalize on both mechanisms. The Strategy Application Model predicts under which conditions online strategies and memory-based strategies are used. We describe an experiment in which we tested one of the key assumptions of the Strategy Application Model regarding the role of cognitive effort on the selection of judgement strategies. We provide evidence indicating that people tend to use an online strategy when the cognitive effort to be invested in the estimation task is low (guessing condition). Conversely, individuals tend to rely on a memory-based strategy, i.e. on recall content, when they thoroughly deliberate on the estimation problem (thinking condition).

Introduction

Imagine you recently attended a party, where you met many people. Today a friend asks you how many women joined the party and how many men. How do you proceed in answering such a question? Psychological research has proposed many different strategies which people may use in estimating frequencies. One possibility would be to recall instances of the categories to be judged for frequency. For example, you might remember that you met Bill, John, Tom, Bob, and Mary at the party, hence, four males and one female. You might either directly use this count as an estimate (simple enumeration; Brown 1995, this volume) or you might extrapolate from the enumeration of the instances in the sample (enumeration and extrapolation; Brown 1995, this volume). In this case, perhaps your resulting estimation is that about 20 men and 5 women joined the party.

While attending the party, however, you might have noticed that the majority of guests were male and that there were only four or five females. Later, you can simply retrieve the judgement you explicitly formed before, i.e. at the time of encoding (for a similar strategy regarding social judgements see Hastie & Park 1986). Other models suggest that the online formation of frequency records does not require explicit counting but rather can happen automatically (Hasher & Zacks 1984). Accordingly, frequency of occurrence is assumed to be automatically incremented into a memory structure representing frequency information (Underwood 1969). The magnitude of this memory structure can later be used in frequency judgement.

These few examples suggest that people can employ different strategies in frequency estimation, which rely on distinct sources of knowledge (Betsch *et al.* 1999; Brown 1995, this volume). However, it does not suffice to identify different strategies and sources of knowledge on which judgements can be based. In order to predict frequency estimation, we need to discover moderating conditions, which provoke individuals to employ a particular knowledge base.

In this chapter we cluster strategies for frequency estimation into two broad categories. These two types of strategies differ regarding the central mechanism used to estimate frequencies. The strategies we subsume under the label *online strategies* are based on the assumption that the integration of frequency information is an *online process*, which happens at the time of encoding. In contrast, *memory-based strategies* are based on the retrieval of exemplars at the time of judgement. We will then present the Strategy Application Model (SAM), which allows predictions about the utilization of different strategies depending on the judgement context. The experiment we describe tests one of the key assumptions: frequency judgements are based on an online strategy under low cognitive effort and they are based on a memory-based strategy under high cognitive effort.

Online strategies and memory-based strategies

A considerable amount of frequency estimation strategies can be classified into two broad categories with respect to the time when frequency information is assumed to be integrated into a quantitative judgement: online strategies and memory-based strategies. A similar distinction was suggested earlier to classify judgement tasks. Hastie and Park (1986) demonstrated that a great deal of variance in social judgement can be accounted for if one considers whether the judgement was formed online or memory-based. Although these authors did not primarily focus on quantitative estimation, their fundamental distinction on the process level helps to cluster most of the approaches to frequency judgement.

Hastie and Park describe online judgements as intentionally and explicitly formed during encoding of relevant stimuli. We extend this definition of online processes to the integration of information during encoding, which is unintentional and implicit.

A similar distinction of types of estimation models is drawn by Schimmack (this volume). He distinguishes between models, which assume either direct or indirect encoding of frequency information. Direct encoding models relate to what we call online strategies. His second distinction differentiates indirect encoding models regarding the recall at the time of judgement (recall versus no recall). Indirect encoding models, assuming recall at the time of judgement, correspond to our memory-based strategies.

Online strategies

When using online strategies, people base their frequency estimate on a memory structure, which was—explicitly or implicitly—formed during encoding of the stimuli (see Fig. 13.1, left column). There are several models on frequency estimation which propose online strategies and, therefore, the online formation of such a memory structure. These models assume that organisms are highly sensitive to event and category frequency. They posit that memory records of frequency are established automatically during encoding of the stimuli. According to this view, the memory system does not only store concrete event information but also aggregates of information, such as implicitly or explicitly formed records of event repetition and category size. Names of party guests, images of faces, figures, parts of conversation and so forth represent concrete information to be stored in memory. In addition, such models assume that memory structures are established which represent quantitative aspects of experience, such as the number of male and female guests at the party. Online strategies are then based on these aggregated memory structures.

There are different notions of how such a memory structure is formed. First of all, it can be formed in an explicit manner, for instance, if the individual is interested in the quantitative aspect of an episode. Accordingly, one can count the number of male and female guests at a party and keep the numbers in mind. The count can be retrieved later and used for judgement. Recall of concrete instances, such as male or female names, is not necessary in order to arrive at a judgement (e.g. simple enumeration, Brown 1995, this volume). Thus, online strategies can utilize such an explicitly formed memory structure.

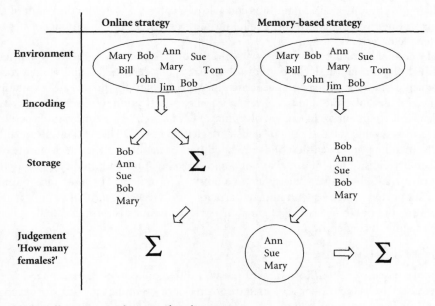

Figure 13.1 Online strategy and memory-based strategy.

However, the explicit counting of frequencies is no necessary prerequisite for the use of online strategies. The automatic encoding models, which are prominent especially in the experimental literature, assume that frequency records can be established implicitly. They propose that frequency of occurrence is a fundamental aspect of experience, which is continuously and automatically encoded and incremented into a unitary memory structure (e.g. Alba *et al.* 1980; Hasher & Zacks 1979, 1984; Watkins & LeCompte 1991; Whalen *et al.* 1999; Zacks & Hasher, this volume). In a subsequent judgement, the individual can simply read out the magnitude of this structure. Encoding of information is assumed to be a sufficient condition for this process. According to the automatic encoding models, recall of instances or exemplars is not necessary for arriving at frequency judgements. Accordingly, frequency judgements need not be constructed on the spot, but can rather be directly based on such pre-established memory structures (Church 1984; Gibbon 1992). Thus, online strategies can also be based on the utilization of this implicitly and online formed memory structure.

Automatic encoding models differ regarding their assumption about the properties of online formed frequency representations. Some use the metaphor of a 'cognitive counter' to illustrate the accumulative nature of the underlying memory process (e.g. Alba *et al.* 1980; Underwood 1969). Counters are assumed to be attached to superordinate categories in memory and to tag instances of occurrence of category exemplars. At any time, the individual may read the current setting of the counter to form a frequency judgement. However, these models do not propose that instances of every category encountered in the environment are automatically aggregated. Only those categories that are activated during encoding are counted (Betsch *et al.* 1999). For example, an ambiguous word such as 'Violet' can only be retrieved and can only be counted as an instance of the category, which was salient during encoding, colours or female names (Tulving & Thompson 1973). If it was encoded as a colour, it increases the count of this category, but it will not contribute to a frequency judgement of the non-salient category, i.e. female names.

The animal literature provides us with a variant of the automatic encoding approach in which the formation of the memory structure is not described as a discrete process in which each occurrence of an event increments equally to the mental representation. Rather, each event is thought to deliver an impulse of activation to the encoding system which varies in intensity dependent on context factors, such as the time elapsed between consecutive events (Church 1984; Gallistel & Gelman 1992). This variant of the automatic encoding approach describes the hypothetical memory structure in terms of a probability density function. This notion allows explanation and prediction of a great deal of variance in frequency learning in a noisy environment. The model has recently been supported in studies with human participants (Whalen *et al.* 1999).

In sum, online strategies utilize a memory structure, which is formed either explicitly or implicitly during encoding.

Memory-based strategies

Memory-based strategies, however, utilize a very different knowledge basis in arriving at a frequency judgement: they are based on the recall of exemplars at the time of judgement (see Fig. 13.1, right column). Authors proposing this kind of strategy assume that if a person is confronted with a number of stimuli in the environment, part of this information is encoded and stored in memory. However, they do not propose that frequency records are formed online. Rather, the integration of frequency relevant information into a quantitative judgement is assumed to happen at the time of recall. For example, if a person has to estimate the number of female guests at a recent party, he or she can recall as many women as possible and base the judgement on this memory sample. There are several ways the result of a recall process can be translated into a frequency estimate. The count can be transformed directly into a judgement, a constant can be added to or multiplied with the count, or the ordinal relation among counts can be used to extrapolate a frequency judgement (Brown 1995; Watkins & LeCompte 1991). Apart from the content of recall like the actual number counted, other features of the recall process itself can be used such as the ease with which instances came to mind (availability heuristic; Tversky & Kahneman 1973; see also Schwarz, this volume, and Betsch & Pohl, this volume, for a critical discussion). Hence, if a person can easily recall a number of women he met at the party, he can use this experience to infer that there have been many female guests. In sum, these recall integration models share the assumption that frequency estimates are based on the recall of exemplars at the time of judgement.

There is empirical evidence for both online strategies (e.g. Alba *et al.* 1980; Barsalou & Ross 1986; Hasher & Chromiak 1977; Williams & Durso 1986) and memory-based strategies (e.g. Tversky & Kahneman 1973; Manis *et al.* 1993; Schwarz 1990; Stapel *et al.* 1995). Consequently, the notion that people can use multiple bases in estimating frequencies has been raised by various researchers (e.g. Betsch *et al.* 1999; Betsch & Pohl, this volume; Brown 1995, this volume). However, the critical question remains: under which conditions are frequency estimates more likely to be based on a memory structure, which has been established during encoding or based on memory samples drawn at the time of judgement?

In this chapter we do not want to argue that all estimation strategies can be exhaustively captured by the classification in online and memory-based strategies. There are strategies which do not fit into these categories, because they describe frequency judgements as based on both online and memory-based processes (e.g. PASS, Sedlmeier 1999; MINERVA2, Hintzman 1988). However, we think that the share of estimation strategies which is comprised by our distinction is sufficiently large to justify the broad classification into these two kinds of strategies.

Diagnosing online strategies and memory-based strategies

In order to determine which judgement strategy is used in a particular situation, we have to derive contradictory implications from the two classes of strategies which can then be tested experimentally. In fact, the implications of the two strategies yield at least two different predictions.

Judgements using memory-based strategies rely on a memory sample, e.g. the recall of exemplars. Therefore, if a frequency judgement is based on recall, the estimate and the number of exemplars recalled should be positively correlated (e.g. Watkins & LeCompte 1991). Furthermore, because recall mediates between true frequency and estimation, the correlation between the recall and the estimate should be higher than the correlation between the estimate and the true frequency. In contrast, if judgements reflect frequency records that were established online, then recall content and frequency estimates should not covary systematically, because the recall of exemplars does not influence frequency judgements.

Second, and most importantly, the two types of strategies have divergent implications regarding the relative accuracy of frequency judgements. Online strategies utilize the fact that encoding of information is sufficient for the automatic aggregation of frequency information. This implies that, given complete encoding of information, frequency judgements should be high in relative accuracy, i.e. that at least the rank order of frequencies in the environment will be mirrored by the estimations. It is important to note that online strategies do not necessarily lead to absolute accuracy in frequency judgement, because the impulse that each event contributes to the formation of the memory structure can vary in intensity (Church 1984; Gallistel & Gelman 1992). However, since each encoded event delivers an impulse, estimates are predicted to be high in relative accuracy. Memory-based strategies, however, are based on a memory sample drawn at the time of judgement. Hence, relative accuracy can only occur if the memory sample is representative for the encoded stimuli. From research on memory, we know that recall can be biased by many factors, such as primacy or recency of encoding (Glanzer & Cunitz 1966), the availability of exemplars (Tversky & Kahneman 1973) or depth of processing (Craik & Lockhart 1972). Since frequency judgements are based on recall, variables that influence the recall process should also affect frequency judgement. This means that frequency estimates are open to be influenced by the same biasing variables that influence the recall process. Therefore, memory-based strategies do not imply relative or even absolute accuracy in frequency judgements.

To summarize, online strategies imply that frequency judgements are generally high in relative accuracy, given that the relevant category has been focussed during encoding. Memory-based strategies imply that frequency judgements are distorted by the same factors, which bias recall. These different implications of the strategies allow us to examine factors which determine utilization of different knowledge bases in frequency judgement.

Some variables moderating the reliance on different sources of knowledge have been considered in the literature on frequency estimation so far. Betsch and Pohl (this volume) review evidence for factors that determine the use of the availability heuristic (Tversky & Kahneman 1973) as a basis for frequency judgements. The availability heuristic is a prominent variant of memory-based strategies. However, there are only few studies that test the utilization of online strategies and memory-based strategies competitively (for more details see Betsch & Pohl, this volume). Betsch and his colleagues (1999), for example, found that category salience during recall enhances the likelihood that people rely on recall content. We will now describe the central assumptions of the Strategy Application Model (Haberstroh 2001), which predicts the use of judgement strategies depending on the judgement context.

The Strategy Application Model (SAM)

The Strategy Application Model allows predictions about the utilization of online or memory-based strategies depending on the judgement context (Haberstroh 2001). In this model we assume that, whenever a relevant category is salient, a memory structure ('implicit tally,'

IT) is formed online, in which the frequency information is implicitly and automatically aggregated. Regarding the process of aggregation, it shares the assumptions of Whalen and his colleagues (1999). Thus during encoding frequency information is aggregated in the IT and stored in memory, in addition to the exemplar information about the stimuli. SAM differentiates between two modes of judgement: spontaneous and deliberate. Figure 13.2 shows the most important factors leading to spontaneous and deliberate judgements, respectively. In psychological experiments, spontaneous judgements are usually evoked by inducing time pressure or by a guessing instruction (see e.g. Haberstroh 2001; Betsch et al. 2001). Deliberate judgements, on the other hand, are usually triggered by an accuracy instruction or by financially rewarding correct answers (e.g. Hertwig & Ortmann 2001; for a critical discussion see Betsch & Haberstroh 2001). In this model, the two modes of judgement are postulated to lead to the use of different estimation strategies.

Spontaneous judgements are assumed to be based on highly accessible pieces of information, because the judgement situation does not allow for an extensive information search. The implicit tally contains frequency information in an aggregated format, so that an estimate can be directly based on the count of the IT without the necessity of further elaboration or editing of the information. Furthermore, the IT is assumed to be associated with the respective stimuli and should therefore come to mind first when a frequency estimation is required. Therefore, spontaneous judgements are assumed to be based on this online formed memory structure.

On the other hand, when forming a deliberate judgement people are highly motivated to arrive at valid and correct estimates. However, people do not have a metacognitive

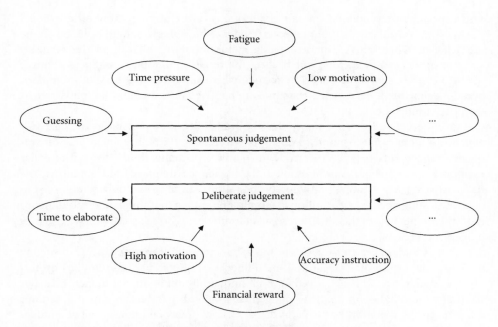

Figure 13.2 Factors leading to spontaneous and deliberate judgements.

insight into the online aggregation of frequency information, as a great deal of research has shown (e.g. Hasher & Zacks 1984). They do not know that they are equipped with this very good basis for frequency estimations and, consequently, doubt the 'first guess' that comes to their mind, i.e. the highly accessible count of the implicit tally. On the contrary, people think that the best way to arrive at an accurate estimate is to consider as many pieces of information as possible. Especially under high motivation, people should try to activate all knowledge about the judgement object and then base their frequency judgement on the IT plus these additional pieces of information. Therefore, in the deliberate judgement context, other features of the objects, which are activated during the information search, can influence frequency judgements. These features might be the valence of the stimuli, knowledge about the base rates of the stimuli or the ease by which instances of a category come to mind (Haberstroh 2001). However, these additional features are not necessarily correlated with the presentation frequencies. If this additional information is not correlated with the frequencies, but is integrated in the estimates, this judgement should deviate more from the objective frequency than a judgement based on the IT. Taken together, deliberate judgements should be based on a memory sample drawn at the time of judgement. Moreover, the higher the motivation, the higher the likelihood that people engage in a more thorough memory search. Therefore, SAM makes the counter-intuitive prediction that accuracy of frequency judgements can decrease when people think carefully.

Experiment: strategy selection in spontaneous and deliberate frequency judgements

This experiment tests one of the key assumptions of the Strategy Application Model (SAM). SAM postulates that spontaneous frequency judgements should be based on the implicit tally. Therefore, spontaneous judgements should be high in relative accuracy and not influenced by factors, which affect the recall of exemplars. Deliberate judgements, however, should be biased by factors, which influence the recall of exemplars, because deliberate judgements are assumed to be based on a memory sample drawn at the time of judgement.

A great deal of research has shown that the availability of exemplars, i.e. the ease with which instances come to mind (Tversky & Kahneman 1973), has an impact on frequency judgements (e.g. Tversky & Kahneman 1973; Manis et al. 1993). The easier exemplars of a category can be recalled, the higher the frequency judgement for the respective category. However, SAM predicts that this effect should only be observed in deliberate judgements. Availability of exemplars should not influence spontaneous judgements, because in this situation recall of exemplars is not necessary for a frequency judgement.

Is the distinction between deliberate and spontaneous judgements capable of solving the contradictory findings regarding the influence of availability of exemplars on frequency judgements, which emerged in some studies but not in others? In fact, an analysis of previous studies on frequency judgements showed that the experimental settings differ regarding the effort participants are demanded to expend in arriving at a judgement. The experimental procedures in those studies that support online strategies often invite participants to guess rather than to thoroughly think about the estimation problem. Some studies employed a time pressure manipulation or asked participants to give a spontaneous judgement (e.g. Alba *et al.* 1980; Barsalou & Ross 1986; Whalen *et al.* 1999). Most of the studies that support memory-based strategies encouraged participants to think before making their judgements. Accordingly, there were no time limits and often participants additionally were told to be accurate in their judgements (e.g. Bruce *et al.* 1991; Brown 1995).

The research paradigm

The following research paradigm was set up to test this hypothesis: spontaneous frequency judgements are not influenced by the availability of exemplars, but availability biases deliberate judgements. More specifically, we presented participants with two categories consisting of a number of exemplars and we manipulated the availability of these exemplars. Both categories were presented the same number of times. We expected that participants would notice that both categories were presented equally often when they had to judge the frequencies spontaneously. However, when they were instructed to think about the estimate, judgements were expected to be influenced by the availability of exemplars. Then participants should judge the category with a high availability of exemplars to have been presented more often than the category with a low availability of exemplars.

In order to keep participants from explicitly counting exemplars, they were told that the experimenters were interested in reaction times when people work with a computer mouse. Therefore, the task required them to click on several boxes with the computer mouse and they were instructed to do so as quickly as possible. In order to avoid influences of participants' prior knowledge on their judgements, the scenario was situated in a completely unknown location: in the future and on an unknown planet. The use of an unknown environment had proven to be successful in another tool we developed for research on decision making (Betsch *et al.* 2000). Participants were told that when exploring a new planet biologists had discovered animals on this planet. These animals belonged to two fictitious genera, 'Amanepes' and 'Oropholus'. The genera differed with respect to their feeding habits. Amanepes only eat meat and Oropholus only eat plants. The participants' task was to 'feed' the animals.

One feeding trial consisted of the following steps (see Fig. 13.3): a box appeared on the computer screen. It had a red background and showed the name of an 'animal', e.g. Stong ('animal box'). After two seconds the background colour changed to green. When this happened, the participant had to click on this box as fast as possible by using the computer mouse. Then on a random position a second box appeared, again with a red background. This second box indicated whether the animal belonged to the Oropholus or the Amanepes ('genera box'). Again after two seconds the background colour changed to green and the box had to be clicked on as fast as possible. Then two new boxes appeared in the two lower corners of the monitor, which were labelled 'meat' and 'plants'. Now participants had to 'feed' the animal by clicking on the appropriate box. When the correct box had been clicked on, the next trial began by automatically presenting the next animal box. When the wrong box had been clicked on, a message box appeared which said, 'This was the wrong box' and participants had to click on the correct

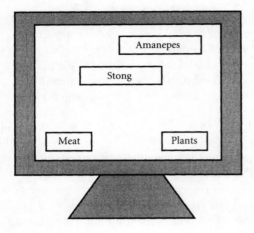

Figure 13.3 One trial in the feeding task.

box in order to continue to the next trial. Thus, the feeding task per se was very easy and we could instruct participants to perform the task as quickly as possible in order to maintain the premise of our cover story.

When the feeding trials were completed, participants were given a questionnaire. First, they were asked how often they performed the two behaviours, feeding plants and feeding meat. In the instruction to these questions the cognitive effort participants were requested to invest in the task was manipulated by either asking them to guess or to think about their judgement. Second, participants were asked to recall as many animals as possible and to classify them as belonging to Amanepes or Oropholus.

The animal names used in the four experiments were selected by these criteria: sixty unfamiliar names for 'animals from a foreign planet' were pre-tested regarding four criteria: valence, familiarity, carnivore vs. herbivore ('Is this animal a carnivore or a herbivore?') and free associations to each word. Names were selected for which the rating on the three scales ranged between 3 and 5 (on 7-point scales) and for which not more than 4 out of the 21 participants produced the same association. These names were randomly assigned to the two categories, Amanepes and Oropholus.

Method

Participants and Design

Thirty-four psychology undergraduates of the University of Heidelberg participated in a study on 'reaction time measurement' to fulfil a course requirement. They were randomly assigned to one of the two experimental conditions resulting from a 2 (cognitive effort: low/spontaneous judgement versus high/deliberate judgement) × 2 (availability of exemplars: high versus low) factorial design with the second factor varied within participants.

Procedure

Participants were seated individually in front of a personal computer, and worked on the animal feeding task described in the paradigm section. In this encoding phase the availability of exemplars of the two categories was manipulated within subjects by confounding the number of exemplars (= animal names) within each category with the repetition frequency of each exemplar. The category with the low availability of exemplars (Amanepes) consisted of many exemplars (40) which were presented only once during the feeding phase. Due to the unfamiliarity of the exemplars, this manipulation was expected to lead to a low recall rate. The category with the high availability of exemplars (Oropholus) consisted of only few names (4), which were presented more often (10 times) during the encoding phase. We expected this manipulation to lead to a higher recall rate of exemplars in the second condition, in which the exemplars were repeated more frequently. Therefore, availability of exemplars was manipulated by the number of exemplars we expected participants to be able to recall (Curt & Zechmeister 1984; Lewandowsky & Smith 1983; Tversky & Kahneman 1973). Both behaviours were performed the same number of times (42 times including two practice trials, respectively).

After the encoding phase participants estimated the frequencies of the two behaviours ('How often did you feed plants/meat?'). The order of these two questions was balanced. In the instruction to these questions the cognitive effort participants were expected to expend was varied. Participants were either asked 'please guess spontaneously, how often did you feed plants/meat?' or 'please think carefully, how often did you feed plants/meat?'. Next participants were asked to recall the animal names. This measure served as a check for the availability manipulation. We expected a higher recall in the category in which few names were repeated more often than in the other category. After filling out the questionnaire, participants were thanked, debriefed and dismissed.

Results

The recall measure shows that the manipulation of availability was successful: participants recalled more names from the category in which there were only a few names that were repeated often ($M = 1.2$) than from the category in which there were many names which were only presented once ($M = 0.6$, $t(33) = 3.6$, $p < 0.001$).

Regarding the frequency judgements, we expected that spontaneous judgements would not be influenced by availability, but in the deliberate judgement condition a higher availability of exemplars was expected to lead to higher frequency judgements. This should result in an interaction of the two factors cognitive effort and availability. To test this hypothesis, we subjected the frequency judgements to a 2 (cognitive effort) × 2 (availability) analysis of variance with repeated measures on the second factor. We found a significant main effect for the availability of exemplars (low availability: $M = 36.3$, high availability: $M = 40.4$, $F(1, 32) = 5.77$, $p < 0.03$). This main effect was qualified by the expected interaction of task characteristics and availability (see Fig. 13.4). There was no influence of availability on spontaneous judgements, but deliberate judgements were biased by availability ($F(1, 32) = 4.9$, $p < 0.04$). This difference for the deliberate judgements was significant ($t(15) = 3.3$; $p < 0.01$, one-tailed).

Discussion

This experiment supports the hypothesis that cognitive effort moderates the utilization of estimation strategies in frequency judgements. Spontaneous judgements were assumed to be based on an online strategy, i.e. the implicit tally, and should therefore

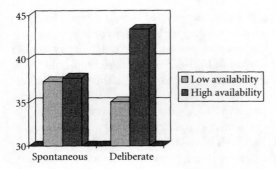

Figure 13.4 Spontaneous and deliberate frequency judgements for categories with either a high or a low availability of exemplars.

not be affected by factors influencing the recall of exemplars. As expected, spontaneous judgements were not influenced by the availability of exemplars. Thinking about the judgement, however, should lead to the use of a memory-based strategy. Consequently, these judgements were influenced by the availability of exemplars. The category with the high availability of exemplars was judged to have been presented more frequently than the category with the low availability of exemplars, although in fact both were presented the same number of times. When participants thought about their judgements, they perceived a difference in the frequencies which did not exist in the objective frequencies.

Thus, regarding the relative accuracy, i.e. the ordinal frequency information, these results provide evidence for the rather counter-intuitive assumption that the accuracy of frequency judgements can decrease when people think carefully about their judgements.

However, these data also show that the estimates of participants in the thinking condition were closest to the true frequency of 42, when exemplars were highly available. This could be interpreted as a high *absolute* accuracy when participants think about the judgements compared to high *relative* accuracy when participants rely on the implicit tally. We do not think that this is an adequate interpretation of these data. Rather, this pattern is due to the mutual compensation of two conflicting tendencies. On the one hand, there is the general tendency to underestimate high frequencies and on the other hand, there is the tendency for a higher frequency estimate in this condition due to the high availability of exemplars. In this specific experimental setting these two tendencies seem to be of a similar strength, so that the estimate accidentally comes closest to the true frequency. In another experiment we showed that if one of these two tendencies is reduced, this frequency is no longer high in absolute accuracy (Haberstroh *et al.* 1999). In this experiment we reduced the first tendency to underestimate high frequency by presenting a much lower total number of animal names. In this case, the tendency to give a high estimate for the category with the more available exemplars outweighs the underestimation of high frequencies, so that this frequency (thinking, high availability) is even overestimated.

Still, there is an alternative explanation for the results of this experiment. It is possible that when instructed to guess, participants did not rely on the implicit tally in estimating frequencies at all. Equal estimates might not indicate a high level of relative accuracy, but rather ignorance of any frequency information. Thus, what we interpret as sensitivity to the ordinal frequency information could only be due to the fact that both behaviours were performed the same number of times in this specific experimental setting. In another study we ruled out this alternative explanation by manipulating the objective frequencies of the two behaviours experimentally (Haberstroh et al. 2002). If participants in the spontaneous guess condition base their judgement on the implicit tally and are, thus, sensitive to relative frequencies, they should be able to notice the difference in the frequencies of the two behaviours. However, if they only expressed their lack of knowledge by estimating the same frequencies in this experiment, they should also do so when the frequencies are unequal. Therefore, we varied the set size of the categories. One genus was more frequent than the other, so that one choice (meat) was made more often than the other (plants). Either the more or the less frequent behaviour had the more available exemplars. Spontaneous frequency judgements mirrored the actual difference between the number of meat and plant choices. Again, judgements were unaffected by the availability of category exemplars, indicating that people relied on online formed frequency records. If participants were instructed to think before making their judgements, estimates were strongly biased by availability. Since, in this study, the exemplars of the infrequent category were made easily available, judgements no longer reflected the actual difference between choice frequencies. Rather, judgements of choice frequencies were identical.

Taken together, these studies show that spontaneous judgements were unbiased by factors which influence the recall process such as the availability of exemplars. Therefore, we can conclude that these judgements are likely to be based on memory structures built at the time of encoding, i.e. that these judgements are arrived at using online strategies. However, thinking before making a judgement led to availability biases in estimates, which reflected the transient state of memory at the time of recall. This shows that they were based on a memory sample drawn at the time of judgement and thus on a memory-based strategy. In this research the very subtle manipulation of the instruction to guess or to think carefully moderated the utilization of judgement strategies, as evident from differential relative accuracy in frequency judgements.

General discussion

We distinguished two broad categories of estimation strategies: online strategies and memory-based strategies. The two classes of strategies are based on different sources of knowledge. Online strategies capitalize frequency records, which were established at the time of encoding. Memory-based strategies are based on memory samples available at the time of judgement. These assumptions yield differential predictions, for example, regarding relative accuracy in frequency judgements. In turn, this allows for identification of strategy employment on an operational level. Under certain conditions, accuracy in judgements indicates that the individual relied on online established frequency records. In contrast, judgements that systematically reflect recall biases indicate that the individual has formed the estimate on the basis of recall.

The Strategy Application Model (SAM) (Haberstroh 2001) predicts conditions under which the strategies should be used. Spontaneous judgements were expected to be based on online strategies, whereas deliberate judgements should evoke memory search and, therefore, lead to a memory-based strategy.

We presented research that was based on this rationale and investigated the moderating role of cognitive effort (low: guessing vs. high: thinking) in frequency judgement. In line with our hypothesis, spontaneous judgements were highly accurate, whereas deliberate judgements were strongly biased by the availability of exemplars at the time of recall. These results support the assumptions of the SAM. In particular, the results indicate that individuals are likely to rely on online strategies when cognitive effort is low. Conversely, if the cognitive effort invested in the estimation task is high, people are prone to rely on concrete information, which is available at the time of recall.

The use of the availability of exemplars as a basis for frequency judgements is usually described as a heuristic approach, which does not require much cognitive effort (Kahneman, Slovic & Tversky 1982; Tversky & Kahneman 1973). The results presented here contradict this common assumption that heuristics are 'rules of thumb', which are employed when people lack motivation or ability to deliberate. When judgements and decisions are made intuitively by relying on heuristics, biases are likely to occur (Kahneman, Slovic & Tversky 1982). This reasoning implies that thorough thinking before making a judgement may result in higher accuracy. However, our experiments suggest that this is not always the case (Haberstroh et al. 1999; Haberstroh et al. 2002). In contrast, there might be situations in which thinking can decrease accuracy and quality of judgement and decision-making (e.g. Betsch et al. 2001; Wilson & Schooler 1991).

How can this contradiction be solved? We want to suggest that it first depends on the characteristics of the encoding situation. If the encoding situation allows for complete encoding of a representative sample, an IT can be formed which can later be used for a frequency estimation. However, if the encoding is incomplete or the sample is not representative, such an online formed memory structure will probably not lead to accurate judgements. Second, it depends on the retrievability of a memory sample, which is representative for the stimulus distribution during the judgement phase.

In the animal-feeding paradigm we used for our experiments, participants could not profit from more effort, because they could not access a representative sample of prior actions, which they had to judge in terms of frequency. In many other tasks employed in the heuristics-and-biases research, such as the permutation task (How many lines can be drawn through two different structures of Xs and Os?), the combination task (How many groups can be formed from a population of 10 persons?), or variations of arithmetical problems (What is the product of $8 \times 7 \times 6 \cdots \times 1$ versus $1 \times 2 \times 3 \cdots \times 8$?; Tversky & Kahneman 1973) participants lack a pre-established solution to the problem. In this case, thorough reasoning and systematic application of formal rules may indeed lead to better judgements than relying on the first guess. However, increasing the motivation by accuracy instruction or monetary payoffs does not guarantee that people think in a more rational fashion (Kahneman & Tversky 1972; Tversky & Kahneman 1973).

We demonstrated that when a memory structure has been developed which reflects the distribution of variables in the environment, judgements are more accurate when arrived at in a spontaneous manner, because guessing maximizes the probability that this source of knowledge is used. If people lack such memory structures and the task requires application of formal rules, judgement accuracy might increase with the degree of elaboration.

References

Alba, J. W., Chromiak, W., Hasher, L. & Attig, M. S. (1980). Automatic encoding of category size information. *Journal of Experimental Psychology: Human Learning and Memory, 6*:370–378.

Barsalou, L. W. & Ross, B. H. (1986). The roles of automatic and strategic processing in sensitivity to superordinate and property frequency. *Journal of Experimental Psychology: Learning, Memory and Cognition, 12*:116–134.

Betsch, T., Glöckner, A. & Haberstroh, S. (2000). COMMERCE—A micro-world simulation to study routine maintenance and deviation in repeated decision making. *Methods of Psychological Research, 5*.

Betsch, T. & Haberstroh, S. (2001). Financial incentives do not pave the road to good experimentation. *Behavioral and Brain Sciences, 24*:404.

Betsch, T., Plessner, H., Schwieren, C. & Gütig, R. (2001). I like it, but I don't know why: A value-account model to implicit attitude formation. *Personality and Social Psychology Bulletin, 27*:242–253.

Betsch, T., Siebler, F., Marz, P., Hormuth, S. & Dickenberger, D. (1999). The moderating role of category salience and category focus in judgements of set size and frequency of occurrence. *Personality and Social Psychology Bulletin, 25*:463–481.

Brown, N. R. (1995). Estimation strategies and the judgement of event frequency. *Journal of Experimental Psychology: Learning, Memory and Cognition, 21*:1539–1553.

Bruce, D., Hockley, W. E. & Craik, F. I. M. (1991). Availability and category frequency estimation. *Memory & Cognition, 19*:301–312.

Church, R. M. (1984). Properties of the internal clock. *Annals of the New York Academy of Sciences, 423*:566–582.

Craik, F. I.M. & Lockhart, R. S. (1972). Levels of processing: a framework for memory research. *Journal of Verbal Learning and Verbal Behavior, 11*:671–684.

Curt, C. L. & Zechmeister, E. B. (1984). Primacy, recency, and the availability heuristic. *Bulletin of the Psychonomic Society, 22*:177–179.

Gallistel, C. R. & Gelman, R. (1992). Preverbal and verbal counting and computation. *Cognition, 44*:43–74.

Gibbon, J. (1992). Ubiquity of scalar timing with a Poisson clock. *Journal of Mathematical Psychology, 36*:283–293.

Glanzer, M. & Cunitz, A. R. (1966). Two storage mechanisms in free recall. *Journal of Verbal Learning and Verbal Behavior, 5*:351–360.

Haberstroh, S. (2001). Die Abhängigkeit der Erwartung vom Wert in Entscheidungen: Das Frequenz-Valenz-Modell. [The dependency of expectation on utility in decisions: The frequency-valence-model]. Unpublished dissertation, University of Heidelberg.

Haberstroh, S., Betsch, T. & Aarts, H. (1999). The influence of the judgement context on judgements for low frequencies. Unpublished raw data.

Haberstroh, S., Betsch, T., Pohl, D. & Aarts, H. (2002). *When guessing is better than thinking: The Strategy Application Model for frequency judgement*. Manuscript submitted for publication.

Hasher, L. & Chromiak, W. (1977). The processing of frequency information: An automatic mechanism? *Journal of Verbal Learning and Verbal Behavior, 16*:173–184.

Hasher, L. & Zacks, R. T. (1979). Automatic and effortful processes in memory. *Journal of Experimental Psychology: General, 108*:356–388.

Hasher, L. & Zacks, R. T. (1984). Automatic processing of fundamental information: The case of frequency of occurrence. *American Psychologist, 12*:1372–1388.

Hastie, R. & Park, B. (1986). The relationship between memory and judgement depends on whether the judgement task is memory-based or on-line. *Psychological Review, 93*:258–268.

Hertwig, R. & Ortmann, A. (in press). Experimental practices in economics: A challenge for psychologists. *Behavioral and Brain Sciences, 24*:383–451.

Hintzman, D. L. (1988). Judgements of frequency and recognition memory in a multiple-trace memory model. *Psychological Review, 95*:528–551.

Kahneman, D., Slovic, P. & Tversky, A. (1982). *Judgements under uncertainty: Heuristics and biases.* New York: Cambridge University Press.

Kahneman, D. & Tversky, A. (1972). Subjective probability: A judgement of representativeness. *Cognitive Psychology, 3*:430–454.

Lewandowsky, S. & Smith, P. W. (1983). The effect of increasing the memorability of category instances on estimates of category size. *Memory & Cognition, 11*:347–350.

Manis, M., Shedler, J., Jonides, J. & Nelson, T. E. (1993). Availability heuristic in judgements of set-size and frequency of occurrence. *Journal of Personality and Social Psychology, 65*:448–457.

Schwarz, N. (1990). Assessing frequency reports of mundane behavior: Contributions of cognitive psychology to questionnaire construction. In C. Hendrick & M. S. Clark (eds) *Research methods in personality and social psychology* (pp. 98–119). Newbury Park, CA: Sage.

Sedlmeier, P. (1999). *Improving statistical reasoning: Theoretical models and practical implications.* Mahwah, NJ: Erlbaum.

Stapel, D. A., Reicher, S. D. & Spears, R. (1995). Contextual determinants of strategic choice: Some moderators of the availability bias. *European Journal of Social Psychology, 25*:141–158.

Tulving, E. & Thompson, D. M. (1973). Encoding specificity and retrieval processes in episodic memory. *Psychological Review, 80*:352–373.

Tversky, A. & Kahneman, D. (1973). Availability: A heuristic for judging frequency and probability. *Cognitive Psychology, 5*:207–232.

Underwood, B. J. (1969). Attributes of memory. *Psychological Review, 76*:559–573.

Watkins, M. J. & LeCompte, D. C. (1991). Inadequacy of recall as a basis for frequency knowledge. *Journal of Experimental Psychology: Learning, Memory and Cognition, 17*:1161–1176.

Whalen, J., Gallistel, C. R. & Gelman, R. (1999). Nonverbal counting in humans: The psychophysics of number representation. *Psychological Science, 10*:130–137.

Williams, K. W. & Durso, F. T. (1986). Judging category frequency: Automaticity or availability. *Journal of Experimental Psychology: Learning, Memory, and Cognition, 12*:387–396.

Wilson, T. D. & Schooler, J. W. (1991). Thinking too much: Introspection can reduce the quality of preferences and decisions. *Journal of Personality and Social Psychology, 60*:181–192.

CHAPTER 14

FREQUENCY LEARNING AND ORDER EFFECTS IN BELIEF UPDATING

MARTIN BAUMANN AND JOSEF F. KREMS

Abstract

Belief updating is the process of revising one's beliefs in the light of new information with the goal of maintaining a consistent and up-to-date belief system in a dynamically changing environment. This process plays a major role in many decision-making situations, like medical diagnosis, where uncertain information has to be taken into account. This process is highly connected to frequency learning, as the knowledge of how to interpret different pieces of information is often acquired by learning the frequency of occurrence. Nevertheless, belief updating and frequency learning are usually studied separately. We conducted a series of experiments bringing together these two processes in order to investigate order effects. Order effects describe the phenomenon that not only the content of information but also their order affects the final belief, possibly leading to distorted beliefs and a deterioration in decision-making performance. We asked two questions regarding these order effects: first, does own experience with the frequencies of relevant events help reducing order effects in belief updating? Second, do the predictions of a well established model of belief updating (Hogarth and Einhorn 1992) hold true both if belief updating is measured by rating scales, and by behavioural decisions in a classification task? Our results demonstrate order effects in belief updating tasks both for probability ratings and classification decisions as dependent variables. We found no order effects for ratings of belief strength and absolute frequencies. Perhaps more importantly, the results show that even directly experiencing the frequencies of relevant events does not prevent order effects in probability judgements.

Introduction

Belief updating is the process of revising one's own beliefs in the light of new information. The goal is to maintain a belief system that is reasonably up-to-date and consistent. Belief revision is necessary as beliefs can be or can become wrong in a dynamically changing environment. As time passes new pieces of information become available and may suggest that one's own beliefs are incomplete, incorrect, inconsistent or outdated and therefore have to be revised. Suppose, for example, you are interested in whether your favourite German soccer team will win the German championship. Every week you will receive new pieces of information, namely the results of the different matches. These pieces of information are relevant to your belief that your favourite team will win the championship and you will have to update your belief after every match of your team.

Research on belief updating has shown that the result of the believe updating process often depends on the *order* of how pieces of evidence are presented. However, from a normative (for example, Bayesian) point of view the order of information should not matter, i.e. the diagnosis of a physician should not depend on whether the patient first mentions symptom A and then symptom B instead of vice versa. The same holds true for the soccer example mentioned above. How strongly you believe that your favourite team will win the German championship should not depend on whether you hear first that your team lost the match versus Team A, but won the match versus Team B or vice versa. But people tend to adjust their beliefs according to the order of data presentation. It is in this sense that order effects can be considered as a special kind of cognitive bias.

In many belief updating tasks the meaning of a piece of information with regard to a belief depends on how often the respective two events—the piece of information and the event stated in the hypothesis—occur together. If it happens very often that the soccer team that beats last year's champion wins this year's championship, the fact that your team beat last year's champion increases your belief that your team will win this year's championship. Therefore information about the frequencies of co-occurrences of events are necessarily used in belief updating tasks. This information is acquired through frequency processing. This is the important connection between the processes of belief updating and frequency processing: therefore both processes have to be studied together.

Studies on frequency processing suggest that people can process frequency information automatically (Hasher & Zacks 1984) and can use this information in judgement task quite accurately (Christensen-Szalanski & Beach 1982). Therefore it seems plausible to assume that belief-updating performance is more accurately if it is based on information acquired through frequency processing.

The major objective of this paper is to give an overview of our studies. We examined whether order effects in belief updating tasks will disappear if the meaning of the respective pieces of information relevant for the belief updating task is acquired through the processing of frequencies of co-occurrences of events in real event sequences.

First we will give a short overview on studies addressing order effects in belief updating. Then the belief–adjustment model of Hogarth and Einhorn (1992) will be described. We use this model to identify situations where order effects are to be expected. We will then describe some results of studies on frequency processing that are relevant for to own work. Finally results from a series of experiments will be summarized, testing the impact of learning of frequencies on order effects in belief updating tasks.

Order effects

Order effects with regard to belief updating have been the focus in many areas including impression formation (Asch 1946), attitude modification (Haugtvedt & Wegener 1994), deductive reasoning (Johnson-Laird & Steedman 1978), causal inference

(Hogarth & Einhorn 1992), and abductive reasoning (Johnson & Krems 2001) etc. In this study we address the following situation. Two pieces of evidence, A and B, are given to participants and both are related to a certain hypothesis; one group receives data in the order A–B, the other one in the order B–A. If the two groups believe in the hypothesis with a different strength depending on the sequence of the pieces of information, an order effect occurred. This definition of order effects follows the definition given by Hogarth and Einhorn (1992).

If one compares results from different studies on order effects, contradictions and inconsistencies are found. It is unclear what kind of order effect is most important in belief updating in different domains. Nisbett and Ross (1980, p. 172) conclude 'that primacy effects in information processing are the rule'. Davis (1984), however, found an important recency effect in a review on juridical decisions. Prior studies of belief updating in audit judgements also showed a recency effect (Johnson 1995). Anderson (1981) found both recency and primacy effects.

The belief–adjustment model of Hogarth and Einhorn (1992)

Hogarth and Einhorn (1992, p. 8) suggested a mathematical model for the process of belief updating. They regard belief updating as '... a sequential anchoring-and-adjustment process in which current opinion, or the anchor, is adjusted by the impact of succeeding pieces of evidence'. In algebraic terms the model can be written in the following form (the description follows Hogarth and Einhorn 1992).

$$S_k = S_{k-1} + w_k[s(x_k) - R]$$

S_k : degree of belief in some hypothesis after kth piece of evidence
S_{k-1} : anchor or prior opinion before kth piece of evidence
$S(x_k)$: subjective evaluation of the kth piece of evidence
R : the reference point or background
w_k : the adjustment weight for the kth piece of evidence

The model was developed in order to explain the seemingly inconsistent results from different domains and tasks regarding order effects during belief updating by the interaction of certain task features and subprocesses involved in belief updating. The major task features were identified by Hogarth and Einhorn (1992) as the complexity of individual pieces of evidence to be processed, the amount of different items of information that has to be processed (length of series), and the response mode. There are two response modes addressed by the model: (a) Step-by-Step (SbS) where participants have to express their beliefs after every new piece of evidence, and (b) End-of-Sequence (EoS), where opinions are reported after all pieces of evidence are given. According to the model there are three important subprocesses involved in belief updating: encoding of information, processing of information and belief adjustment. Resulting from the combination of task features and subprocesses involved the model makes predictions about order effects in belief updating for many situations. We will restrict a detailed description of the model to the situation realized in our experiments.

This situation can be characterized as follows: the participants had to evaluate each piece of evidence in a sequence of two pieces as supporting or refuting a hypothesis regarding the category membership of an object. The experimental procedure ensured that the participant had to update his belief about the category membership of the object after each piece of evidence.

This belief updating task is described by the *evaluative* form of the belief adjustment model. 'In evaluation tasks, people encode evidence as positive or negative relative to the hypothesis under consideration' (Hogarth and Einhorn 1992, p. 9). Independent of the current level of belief, a person believes more in a hypotheses if it is supported by new evidence, and she believes less if new evidence contradicts it. In this case the reference point is R=0, and evidence is evaluated on a bipolar scale, such that $-1 \leq s(x_k) \leq +1$. The algebraic form of the model in this case is

$$S_k = S_{k-1} + w_k s(x_k).$$

Regarding the belief adjustment Hogarth and Einhorn (1992) assume that the adjustment weight, w_k, should depend on the sign of the impact of the evidence, $[s(x_k)-R]$, and on the level of the anchor, S_{k-1}. For negative evidence with regard to the current hypothesis the adjustment is proportional to the current anchor; i.e. the more a person already beliefs in a hypothesis the higher the adjustment weight. Consequently, strong anchors are weakened more by means of the same evidence than weak anchors, which implies a contrast effect. Therefore:

$$w_k = \alpha S_k, \quad \text{when} \, s(x_k) \leq R$$

yielding

$$S_k = S_{k-1} + \alpha S_{k-1}[s(x_k) - R].$$

The analogous holds true for pieces of positive evidence. In this case the adjustment weight is inversely proportional to the anchor. Thus the current belief is strengthened more if the already existing level is low. Therefore:

$$w_k = \beta(1 - S_{k-1}), \quad \text{when} \, s(x_k) > R$$

yielding

$$S_k = S_{k-1} + \beta(1 - S_{k-1})[s(x_k) - R].$$

where α and β are constants representing the sensitivity towards pieces of negative or positive evidence.

In this situation Hogarth and Einhorn's model predicts different order effects depending on the relationship between the pieces of information presented in a sequence. If the sequence of evidence is inconsistent, i.e. if it contains a positive and a negative piece of evidence, the model predicts a recency effect. This is caused by the contrast effect described above. The following example will show how the prediction follows from the model. The formal proof is given in Hogarth and Einhorn (1992).

Let us consider the situation of a participant in one of our experiments presented with an inconsistent sequence of information. This sequence consisted of a positive and a negative piece of information. Assume that her opinion about the relevant hypothesis is neutral at the beginning, that is $S_0 = 0.5$. Let us assume additionally that the first information of the sequence favors the hypothesis and the second contradicts it. The first piece of information evaluates to $s(+) = 0.5$, the second to $s(-) = -0.5$. For the sake of simplicity the sensitivity parameters α and β are set to 1.

In evaluation mode the belief adjustment model is described by

$$S_k = S_{k-1} + w_k s(x_k).$$

As the first piece of information is positive, w_k for the first piece of information is defined as

$$w_1 = 1 - S_0,$$

therefore

$$S_1 = S_0 + (1-S_0)s(+) = 0.5 + 0.5 \cdot 0.5 = 0.75.$$

The second piece of information is negative, therefore the adjustment weight for the second piece of evidence is defined by $w_2 = S_1$, and the final belief in the hypothesis after the second piece of evidence equals

$$S_2 = S_1 + S_1 s(-) = 0.75 + 0.75 \cdot (-0.5) = 0.375.$$

If the order of information is vice versa, with the negative information first, followed by the positive information, the adjustment process is described as follows. The adjustment weight for the first, now negative piece of information is $w_1 = S_0$, therefore

$$S_1 = S_0 + S_0 s(-) = 0.5 + 0.5 \cdot (-0.5) = 0.25.$$

When the positive piece of information is presented now, the respective adjustment weight is larger than in the first sequence as the current level of belief, S, is smaller than S_0, specifically $w_2 = 1 - S_1$. Therefore with this sequence the final level of belief is

$$S_2 = S_1 + (1 - S_1)s(-) = 0.25 + (1-0.25) \cdot 0.5 = 0.625.$$

As the predicted level of belief is larger if the information order is negative–positive than if it is positive–negative, the predicted order effect is a recency effect. As one can see from this example it is the definition of the adjustment weights that is responsible for the prediction of the recency effect. This definition leads to a stronger weighting of the positive piece of information, if it is preceded by the negative piece of information, than if the positive information is the first piece of information in the sequence. After the negative piece of information the level of belief is smaller than before its presentation. Therefore a succeeding piece of positive information receives a higher weighting and, consequently, a piece of positive information is emphasized more if presented as the second than as the first one. The analogue holds true for the negative piece of information, which is also emphasized more if displayed after than before the positive piece of information as its weight is proportional to the current level of belief. This definition establishes a contrast effect which in turn leads to the recency effect.

If the sequence of evidence is consistent, i.e. only positive or only negative pieces of evidence are presented, the model predicts no order effect. Again the formal proof is shown in Hogarth and Einhorn (1992). The calculation of an example with $S_0 = 0.5$, s (strong positive) = 0.6, s (weak positive) = 0.3 is left up to the reader.

Major predictions of the Hogarth and Einhorn model were evaluated in a couple of studies. Ashton and Ashton (1990) and Johnson (1995) found a recency effect in the domain of auditing. Similar results were obtained by Adelman et al. (1993). Similarly, Tubbs et al. (1993) found a recency effect, but not only—as predicted by the model— for inconsistent but also for consistent cases. A more unclear picture was obtained by Adelman and Bresnick (1992). They found a recency effect as well as a primacy effect and no order effect at all, depending on task characteristics. In a detailed replication of this study, Adelman et al. (1996) found a primacy effect in early processing steps as well as in later ones. Plach (1998) did not find an order effect at all. So the general picture with regard to the predictions of the Hogarth and Einhorn model is quite inconsistent.

Belief updating and frequency learning

There is at least one problem with the studies mentioned above that could be responsible for the contradictions and inconsistencies found. Normally tasks and scenarios were used where participants could not base their judgement on personal experience with the task domain. For example, Hogarth and Einhorn (1992) forced participants to rate the probability that a new coaching programme was responsible for the improvement of a baseball player's hitting rate. It is improbable that participants could base their decision on personal experience about how often a change in a coaching programme would cause an improvement in the hitting rate. Also in most of the studies conducted to evaluate predictions of Hogarth and Einhorn's model participants did not have personal experience on the task domain used in the experiments (e.g. Adelman et al. 1996; Tubbs et al. 1993).

Many studies on frequency processing have demonstrated that well known biases in judgement tasks, like base rate fallacy, conjunction fallacy, or overconfidence, disappear or are at least reduced if the relevant information in those tasks is not presented as probabilities but as absolute frequencies or as real event sequences (Carroll & Siegler 1977; Christensen-Szalanski & Beach 1982; Christensen-Szalanski & Bushyhead 1981; Cosmides and Tooby 1996; Gigerenzer & Hoffrage 1995; Manis et al. 1980). Christensen-Szalanski and Beach (1982) for example could show that the base rate fallacy disappeared when people experienced the probabilistic relationship between symptom and disease in form of real event sequences. Their participants were presented with a series of learning trials, each representing a patient suffering from disease A and showing symptom S or not. After these learning trials the participants were asked to rate the probability of suffering from disease A given symptom S. The participants estimated this probability quite accurately.

But most of the studies on frequency learning and the use of frequency information do not deal with the sequentiality of single pieces of evidence during belief updating (e.g. Christensen-Szalanski & Beach 1982; Christensen-Szalanski & Bushyhead 1981; Gigerenzer & Hoffrage 1995). In most cases either only one piece of information is presented in a task or if more are presented they are presented all at once. Therefore the sequential updating of belief as information is presented sequentially plays no role in these studies and order effects cannot be studied. Recently Zhang et al. (1998) jointly considered frequency learning and order effects in belief updating. Their participants had to complete a training consisting of 50 classification trials. On each trial the participants had to decide whether an approaching aircraft was hostile or commercial. This decision had to be made on the basis of two pieces of information about the approaching plane, the route of the plane and the response to the request for identification, These two pieces of information were presented sequentially, yielding some probabilistic information about the category membership of the plane. So, for example, if a plane flew on a commercial route the probability that the plane was commercial was 0.8. After the second piece of information the participants had to assign the plane to one of the two possible categories. Thereafter they received feedback about the true identity of the plane. The aim of this training was to let the participants directly experience the probabilistic relationship between pieces of information and category membership through frequency learning. Zhang *et al.* (1998) report a recency effect based on the order of evidence that deviates from the normative values. The recency effect was found in the classification task during frequency learning as well as in a belief evaluation questionnaire after training. Participants had to give a rating as to how strongly they believed that the plane belonged to one of the two categories given the sequentially presented pieces of information about route and identification. However, their study shows some methodological problems: for example, the different sequences of route and identification information were connected with a different timing of information presentation. Whereas information about the route was displayed immediately after the respective button had been pressed, information about the identification was delayed for 30 sec after the request for identification was 'issued'. An effect of this different timing on participants' belief evaluation cannot be excluded.

Experimental studies

Predictions

The belief adjustment model predicts a recency effect for inconsistent sequences of information and no order effect for consistent sequences. On the other hand, studies on frequency learning suggest that people can use probabilistic information much more accurately with regard to a normative model if they are able to acquire knowledge about the relevant probabilities (base rates, conditional probabilities) by experiencing a set of examples of the relevant events. This learned knowledge about the probabilistic relationship between evidence and the truth of a given hypothesis could be used in the belief updating task in order to assess the impact of a certain piece of evidence to the current belief level. No participants had such kind of knowledge in the mentioned studies on order effects (Adelman & Bresnick 1992; Adelman *et al.* 1993; Adelman *et al.* 1996; Hogarth & Einhorn 1992; Tubbs *et al.* 1983). In these studies participants had to rely on common and very general knowledge about the task domain in order to assess how a presented piece of information should alter their current level of belief in the hypothesis under consideration. Recently Wang 1999 showed that confidence might play an important role in belief updating, where confidence is mainly determined by the experience one has with the relevant hypothesis. His results suggest that experience with the task domain might reduce order effects. From these results and from the findings on the use of frequency information in decision tasks (Christensen-Szalanski & Beach 1982; Christensen-Szalanski & Bushyhead 1981; Gigerenzer & Hoffrage 1995) we formulate the following hypothesis contrary to the Hogarth and Einhorn belief adjustment model: judgements in a belief updating task in a situation where pieces of evidence are encoded in evaluation mode and where belief is adjusted sequentially show no order effect if participants learn the relationship between evidence and hypothesis from real event sequences.

The experimental paradigm

In order to give you an idea about the general procedure, the following section will first describe the experimental paradigm, then one of the experiments will be reported in more detail. We will then give a summary of recent findings from a set of experiments described elsewhere (Baumann & Krems 2001).

The following cover story was used for all of the experiments: the participants should imagine they are a captain on a navy ship located in an area with both commercial planes and hostile military planes. If a plane approached the ship they had to decide to which of the two categories the plane belonged. The decision had to be made on the basis of two criteria, information about the route of the plane (R) and information about the pilot's response to the request for identification (ID). With regard to the route information there were two possibilities. The plane could fly on a commercial route (R+) or not (R−). There were also two possible responses to the request for identification. The pilot could identify the plane as commercial (ID+) or give no response at all (ID−). All four possible pieces of information were associated with each category with a certain probability. See Table 14.1 for an exemplary probability distribution. The table shows the conditional probabilities of each category given the respective pieces of information. This probability distribution was used in the experiment described below. The table shows that two pieces of information were in favour of the 'commercial hypothesis', namely flying on a commercial route and identifying itself as commercial, whereas the other two possible pieces of information made the 'military hypothesis' more probable.

In all our experiments the participants had first to complete a training phase. The training consisted of a series of trials. On each trial two pieces of information were presented sequentially, one about the route of the plane, the other about the ID. After the

Table 14.1 Conditional probabilities for the two types of planes given the different pieces of information

Plane	Route	ID	Probability
Commercial	Commercial	Positive	0.94
Commercial	Commercial	Negative	0.50
Commercial	Not commercial	Positive	0.50
Commercial	Not commercial	Negative	0.06
Hostile	Commercial	Positive	0.06
Hostile	Commercial	Negative	0.50
Hostile	Not commercial	positive	0.50
Hostile	Not commercial	Negative	0.94

second piece of information the participants had to assign the plane either to the category 'commercial' or to the category 'hostile'. After their decision the participants immediately received feedback about the plane's category. The goal of this training phase was that participants acquire knowledge about the probabilistic relationship between each piece of information and the category membership of the object through real event sequences.

Afterwards, participants had to perform a rating task within the same belief-updating scenario used in the training phase. They were presented again with the information that a plane was approaching the ship and with two additional pieces of information, one about the route and one about the ID. After each piece of information the participants were asked to give a rating, e.g. how strongly they believed that the plane was hostile given the pieces of information displayed until then.

In all experiments the main factor we manipulated was the order of information in the inconsistent sequences of information. There were two possible orders: positive information first, followed by negative information, or vice versa. The dependent variables measured in all experiments were the relative frequencies of category choices in the training phase and the ratings in the questionnaire.

This paradigm allowed us first to train the participants on an arbitrary probabilistic relationship between pieces of information about an object and its category membership. Second, we could examine whether the participants acquired the relevant frequencies correctly. Third, we could use the behavioural decisions in the training phase to examine whether these classification decisions show a recency effect. Such an effect was predicted by the belief adjustment model for the belief-updating situation realized in the experiment—evaluative encoding, adjustment after each piece of information, and inconsistent sequences, as shown above. Former studies examined the predictions of the belief adjustment model mainly with rating scales as dependent measures and not with behavioural decisions. And fourth, we could use the ratings in the questionnaire to verify our prediction that order effects are reduced or disappear if people acquired knowledge about the probability distribution underlying the task through real event sequences.

The goal of the experiment described now in more detail was to examine the two questions mentioned above: first, is there an order effect for behavioural decisions? Second, does the training on the probabilistic relationship between pieces of information about an object and its identity in the form of real event sequences reduce or even prevent order effects in ratings of belief strength? By avoiding the methodological problems of the Zhang *et al.* (1998) study clearer answers to these two questions should be possible.

Method

Participants

Eighty-seven undergraduate students from the University of Regensburg and the Chemnitz University of Technology participated in the experiment. Six participants had to be excluded from the data analysis because of an error in the data sampling procedure.

Procedure, material, and design

The experiment was run on a PC. At the beginning of the experiment the participants received an instruction which introduced the cover story and the task. Then participants started the experiment, which consisted of three parts: training phase with categorical decisions, questionnaire on belief strength and manipulation check.

The training phase consisted of 100 trials. On each trial the participants received two pieces of information, one about the route and one about the ID. As each information type could adopt two different values, flying on a commercial route or not and answering to the request for identification or not, four possible sequences of information were possible, two consistent and two inconsistent. Each piece of information and consequently each of the four sequences was associated with the two plane categories with a predefined probability, shown in Table 14.1.

The two pieces of information were presented sequentially as text strings, like 'The plane is flying on a commercial route', separated with a blank screen in order to make sure that the participants update their belief after each piece of information. After the second piece of information the participants had to assign the plane to one category, commercial or hostile, and received feedback thereafter. The order of the information was varied across participants.

After the training phase one sequence of three pieces of information was sequentially presented to the participants in the questionnaire:

Base information: An unknown plane is approaching your ship.
ID+: The plane identifies itself as commercial.
R−: The plane does not fly on a commercial route.

After each piece of information the participants had to rate from 0 to 100 how strongly they believe that the plane was a hostile military plane. The order of the last two pieces of information (ID+, R−) was varied across the participants.

At the end of the experiment the participants had to rate how strongly each single piece of information supported the hypothesis that the approaching plane was hostile. Here they had to give a rating between −100 to 100. The purpose of this test was to verify whether the participants interpreted R− as favouring the hostile hypothesis and ID+ as contradicting the hostile hypothesis. The order of information both in the training phase and in the post-test was a between-subjects factor. Therefore we got four different experimental groups.

Data analysis

In order to analyse the categorical decision data of the training phase, a loglinear analysis was computed. Trials with consistent and inconsistent sequences of information were analysed separately. The loglinear analysis of the classification responses in trials with inconsistent sequences included three factors: the response of the participants, the content of the sequence (R+ and ID− or R− and ID+) and the order of information in a sequence. There were two possible orders of information: the first piece of information favoured the commercial hypothesis and the second piece of information contradicts the commercial hypothesis (positive–negative) and vice versa (negative–positive). For trials with consistent sequences of information the analysis also included the classification response, the content of the sequence and the order of information in a sequence as factors. But in this case order of information was defined as the sequence of route and ID information in a trial as these trials contained only positive or negative pieces of information with regard to the commercial hypothesis. The significance of the independent factors (content of sequence and order of information) and their interaction was determined by computing the difference between the likelihood ratios of two loglinear models: Model A that did not contain the factor and Model B that was identical to Model A but additionally involving the respective factor. This statistic is called DeltaLQ.

For evaluating the data of the questionnaire, the difference between the ratings after the first piece of information (base) and after the third piece was computed. This difference was then used in an ANOVA as the dependent variable. The order of information in the training phase (Route–ID vs. ID–Route) and in the questionnaire (Route–ID vs. ID–Route) were used as the independent factors. The data of the manipulation check were also analysed by an ANOVA.

Results and discussion

The results of the manipulation check show that the participants interpreted the data roughly in the way we expected. The average rating of the base information, 'A plane is approaching the ship', was 0.90, the rating for ID+, 'The plane identifies itself as commercial', was −9.22, and for R−, 'The plane does not fly on a commercial route', was 25.01. That is the base information was interpreted as neutral, ID+ as speaking against the hostile hypothesis and R− as speaking for the hostile hypothesis. These differences were significant, as an ANOVA with the factor information content revealed, $F(2, 160) = 27.354$, $p < 0.001$.

The results of the training phase are shown in Fig. 14.1. The column labelled 'Bayes' shows the relative frequency of trials with a commercial plane for the four possible sequences used in the experiment. The remaining two columns depict the relative frequency of the answer 'commercial' for the four sequences depending on the order of route and ID information. The data show a significant recency effect for inconsistent sequences of information. For example, the group of columns labeled 'R + ID−' in Fig. 14.1 represent the relative frequency of the decision 'commercial plane' for sequences where the

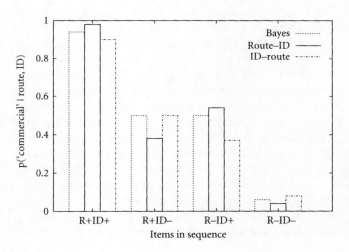

Figure 14.1 Results of the training phase, showing the relative frequency of response 'commercial' given the different pieces of information and different orders.

plane flew on a commercial route and did not answer to the request for identification. As can be seen, participants chose the category 'commercial' more often, if the sequence of information was ID–Route than if it was Route–ID, that is if the last piece of information in the sequence favoured the hypothesis 'commercial plane'. The analogue holds true for the second possible inconsistent sequence represented in Fig. 14.1 by the group of columns labeled 'R−ID+'. That is, for both inconsistent sequences the relative frequency of the response 'commercial' is higher if the sequence is negative–positive than if it is positive–negative. The loglinear analysis confirmed this. The factor order of information was highly significant, DeltaLQ = 55.56, df = 1, p < 0.0001. The content of the sequence had no significant effect on the classification response, DeltaLQ < 1. The interaction of both factors was also not significant, DeltaLQ = 1.30, df = 1, p = 0.25.

For consistent sequences the three-way interaction of classification response, content of sequence and order of information (ID–Route vs. Route–ID) was significant, DeltaLQ = 56.93, df = 1, p < 0.0001. As Fig. 14.1 shows, this can be attributed to a higher proportion of commercial responses in consistently positive sequences (R+ID+) if the order of information is Route–ID than if it is ID–Route. For consistently negative sequences Fig. 14.1 shows that the proportion of the answer 'commercial' is smaller if the sequence of information is Route–ID than if it is ID–Route. The results of the manipulation check suggest that the information about the route of the plane was considered as much more important for the classification of the plane than the information about the identification of the plane. Therefore each consistent sequence contained a strong piece of information (route) and a weak piece of information (identification). The results show that the strong piece of information had more influence on the final decision if it was presented as the first piece of information. Therefore the results for consistent sequences demonstrate a primacy effect. The difference between consistently positive and negative sequences was also highly significant as was to be expected, DeltaLQ = 5286.05, df = 1, p < 0.0001.

These results follow partly the predictions made by the model of Hogarth and Einhorn (1992). A significant recency effect for those trials containing inconsistent sequences of information could be verified. The final decision was significantly determined by the last piece of information presented to the person. This replicates the results of Zhang *et al.* (1998) who also found a recency effect in categorical decisions. But for consistent sequences there was a primacy effect, contrary to the predictions of the model and not found by Zhang *et al.* (1998).

The results of the belief–strength rating are shown in Fig. 14.2. After the presentation of the base information participants assessed the possibility of the plane being commercial or being hostile equally. After the first piece of information the belief was adjusted according to information displayed. Participants receiving R− believed more in the hostile hypothesis (M = 69.7), participants receiving ID+ less (M = 42.84). After the second piece of information participants of both groups believed equally strong in the hostile hypothesis (R+ID−: M = 62.07, ID−R+: M = 57.08). The analysis of the difference in ratings after the base and second information item by an ANOVA showed no significant effects at all. Neither the order of information in the training phase nor the order in the questionnaire influenced the difference in belief strength between the ratings after the base information and the second information item. The participants adjust their belief

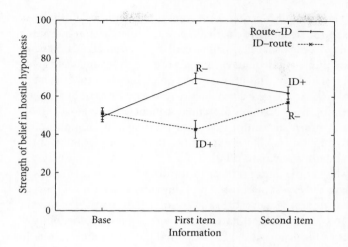

Figure 14.2 Results of questionnaire, showing the ratings after each piece of information for each order of information in the questionnaire.

that the approaching plane is hostile independently of the order of information in the questionnaire. These results follow our hypothesis that knowledge about the underlying probability distribution of the task acquired through own experience leads to the disappearance of order effects. But this interpretation has to be made cautiously. Only 53% of the participants answer with a reduction in the belief strength rating after the presentation of ID+ information and with an increase after R− presentation. This suggests that the participants encountered difficulties when interpreting the questions in the questionnaire. On the other hand the data of the training phase show clearly that the participants are sensitive to the impact of the different pieces of information as the frequencies of the responses are clearly different given the different sequences of information.

Further experimental results

The experiment described above suggests that behavioural decisions show order effects depending on the consistency of the pieces of information in a sequence. It suggests furthermore that experience in a task domain acquired by learning of frequency of occurrences of relevant events prevents order effects in ratings of belief strength. As one would expect that if behavioural decisions show order effects then ratings pertaining to the same task domain should show these effects, too, this pattern of results seemed to be somewhat inconsistent. The aim of further experiments was therefore to exclude methodological characteristics of the experiment as possible causes for the results. First, we wanted to test whether the disappearance of the order effect in the ratings was just a matter of the question format in the questionnaire. Second, we wanted to replicate the recency effect found in the classification data of the experiment described above under various experimental designs.

In four experiments we used the paradigm described in detail above. We focused on the situation in which two pieces of information were encoded as positive or negative with regard to a current hypothesis (evaluation task) and where these pieces of information are processed sequentially. Sequences with consistent and contradictory pieces of information were used. In each experiment the participants had to complete a training phase consisting of a series of trials. In each trial they had to classify an approaching plane as commercial or hostile given information about the plane's route and ID and were given feedback about the plane's real identity. After this training phase they had to complete a questionnaire where they were asked to give a probability rating or a rating of absolute frequency after each piece of information in the presented sequence. The main factor we manipulated was the order of information presented in the training trials or in the questionnaire. We give a short summary of the results of the four experiments here; the details are presented in Baumann and Krems (2001).

In order to investigate whether the disappearance of the order effect in the ratings of the questionnaire was due to the experience in the task domain and not to the special kind of question asked in the questionnaire, the kind of information asked for in the questionnaire was varied. In two experiments participants had to rate the probability of the plane belonging to a certain category given the displayed pieces of information, in the other two experiments participants had to rate the absolute frequency of planes shown in the training phase that matched the ID and route information shown in the questionnaire.

The rating in the questionnaire showed only a significant recency effect for inconsistent sequences if the participants were asked to sequentially rate the probability of the plane belonging to one of two categories. If they had to rate the absolute frequency of planes they had seen in the training phase with the same characteristics, e.g. flying on a commercial route and not answering to the request for identification, as shown in the current question, no order effect was found. But the participants overestimated the real frequencies at least by factor two. This difference between probability and absolute frequency ratings might be attributed to different underlying cognitive processes. Whereas the probability ratings might rely on the integration of different pieces of information as assumed in the belief adjustment model, ratings of absolute frequency might depend more on recall processes. To estimate the number of planes flying on a commercial route and not answering to the request for identification the number of instances with these characteristics shown in the training phase must be recalled. No combination of route and ID information is necessary.

With regard to the origin of the order effect found in the behavioural decisions in the training phase, the experiments showed that this effect cannot be attributed to a special experimental design. The effect could be found both if order of information in the trials of the training phase was varied as between- or within-subjects factor. The results also suggest that it is not lack of knowledge about the underlying probability distribution that causes the order effect. The recency effect for inconsistent sequences was also found if only the second half of the training trials was analyzed, so that participants should have had enough opportunity to acquire knowledge about the underlying probability distribution. In another experiment the diagnostic value of the two kinds of information, route and ID, was different. For one group of participants the information about the route of the aircraft was highly diagnostic for identifying it as a hostile or friendly plane, whereas knowing the ID of the plane gave little information about its 'real' identity. In the second group the role of route and ID was interchanged. In this experiment we could not find any order effect for consistent and inconsistent sequences in the classification data of the training phase. A possible reason could be a reduction in the size of the order effect because of the manipulation of the diagnostic value.

Summarizing these results, in accordance with the predictions of the belief adjustment model the decision data of the training phase repeatedly showed a recency effect for inconsistent sequences of information. We were able to rule out that this effect is caused by some methodological characteristics of the experiments. Contrary to the first experiment, described in detail in the previous section, we found a recency effect for inconsistent sequences in probability ratings. But we found none in ratings of absolute frequency. Therefore the kind of information asked for in the questionnaire seems to affect whether order effects will arise or not in ratings in belief-updating tasks.

Conclusions

The major objective of the research reported in this chapter was to investigate the relationship between frequency learning and order effects in belief updating. Contrary to most of previously published studies (with the exception of Zhang *et al.* 1998) participants were presented with a series of examples from which they could learn the frequency of occurrences of evidences combined with alternative categories. By this, first, we wanted to test if order effects disappear, like other cognitive biases in decision tasks, when people can learn conditional probabilities from a set of events. A second goal was to investigate whether behavioural decisions also show order effects comparable to rating responses which are usually used as the dependent measures in belief updating tasks (Hogarth & Einhorn 1992). The belief adjustment model of Hogarth and Einhorn (1992) was used to make predictions about the occurrence of order effects under specific conditions.

The results partly confirm the predictions of the Hogarth and Einhorn (1992) model. The experiments demonstrated that both behavioural decisions and ratings show order effects, although the participants had experienced the relationship between evidence and hypothesis in the form of real event sequences in a training phase. This is in agreement with the results of Zhang *et al.* (1998) who also found order effects in judgements, although their participants accurately acquired the frequency distribution defining the probabilistic relationship between pieces of evidence and relevant hypotheses. This suggests that order effects do not arise from deficits in knowledge about the task domain, but from the integration of different pieces of information during the belief revision process. This could explain why order effects appeared in the behavioural decisions in the training phase and in probability ratings in the questionnaire. In both cases an integration of different pieces of information is necessary to generate an appropriate response. If the response to the task does not rely on such an integration process, as it is the case for ratings of absolute frequencies, there is no order effect. Here only the recall of information is necessary to answer the question. However, this explanation leaves the results on the ratings of belief strength unexplained. One would expect the same order effects as with probability ratings as the integration of different pieces of information is necessary. But in this experiment it could not be ruled out that the disappearance of the order effect in the rating response was not due to difficulties participants had in interpreting the questions.

In general, our results suggest that order effects in belief updating tasks do not disappear if people acquire relevant knowledge about the task domain by a set of examples. In this respect they are not comparable to other cognitive biases like the base rate fallacy. The results further suggest that the predictions of the belief adjustment model of Hogarth and Einhorn (1992) cannot only be applied to rating responses as dependent measures but also to behavioural decisions. But the inconsistencies in the results, e.g. no order effect for ratings of belief strength, show that further experiments are necessary to obtain more information about the relationship between frequency learning, order effects, rating responses and behavioural decision responses.

References

Adelman, L. & Bresnick, T. A. (1992). Examining the effect of information sequence on Patriot air defense officers' judgements. *Organizational Behavior and Human Decision Processes*, 53:204–228.

Adelman, L., Tolcott, M. A. & Bresnick, T. A. (1993). Examining the effect of information order on expert judgement. *Organizational Behavior and Human Decision Processes*, 56:348–369.

Adelman, L., Bresnick, T. A., Black, P. K., Marvin, F. & Sak, S. G. (1996). Research with Patriot air defense officers: Examining information order effects. *Human Factors*, 38:250–261.

Anderson, N. H. (1981). *Foundations of information integration theory*. New York: Academic Press.

Asch, S. E. (1946). Forming impressions of personality. *Journal of Abnormal and Social Psychology*, 41:258–290.

Ashton, R. H. & Ashton, A. H. (1990). Evidence–responsiveness in professional judgement: Effects of positive versus negative evidence and presentation mode. *Organizational Behavior and Human Decision Processes*, 46:1–19.

Baumann, M. & Krems, J. F. (2001). *The effect of frequency learning on order effects in belief updating*. Manuscript in preparation, Chemnitz University of Technology.

Carroll, J. S. & Siegler, R. S. (1977). Strategies for the use of base rate information. *Organizational Behavior and Human Performance*, 19:392–402.

Christensen-Szalanski, J. J. J. & Beach, L. R. (1982). Experience and the base rate fallacy. *Organizational Behavior and Human Performance*, 29:270–278.

Christensen-Szalanski, J. J. J. & Bushyhead, J. B. (1981). Physicians' use of probabilistic information in a real clinical setting. *Journal of Experimental Psychology: Human Perception and Performance*, 7:928–935.

Cosmides, L. & Tooby, J. (1996). Are human good intuitive statisticians after all? Rethinking some conclusions from the literature on judgement under uncertainty. *Cognition*, 58:1–73.

Davis, J. H. (1984). Order in the courtroom. In D. J. Miller, D. G. Blackman & A. J. Chapman (eds) *Perspectives in psychology and law*. New York: Wiley.

Gigerenzer, G. & Hoffrage, U. (1995). How to improve Bayesian reasoning without instruction: Frequency formats. *Psychological Review*, 102:684–704.

Hasher, L. & Zacks R. T. (1984). Automatic processing of fundamental information: The case of frequency of occurence. *American Psychologist*, 39:1372–1388.

Haugtvedt, C. P. & Wegener, D. T. (1994). Message order effects in persuasion: An attitude strength perspective. *Journal of Consumer Research*, 21:205–218.

Hogarth, R. M. & Einhorn, H. J. (1992). Order effects in belief updating: The belief–adjustment model. *Cognitive Psychology*, 24:1–55.

Johnson, E. N. (1995). Effects of information order, group assistance, and experience on auditors' sequential belief revision. *Journal of Economic Psychology, 16*:137–160.

Johnson, T. & Krems, J. (2001). Use of current explanations in multicausal abductive reasoning. *Cognitive Science, 25*:903–939.

Johnson-Laird, P. N. & Steedman, M. (1978). The psychology of syllogisms. *Cognitive Psychology, 10*:64–99.

Lyon, D. & Slovic, P. (1976). Dominance of accuracy information and neglect of base rates in probability estimation. *Acta Psychologica, 40*:287–298.

Manis, M., Dovalina, I., Avis, N. E. & Cardoze, S. (1980). Base rates can affect individual predictions. *Journal of Personality and Social Psychology, 38*:287–298.

Nisbett, R. & Ross, L. (1980). *Human inference: Strategies and shortcomings of human judgement.* Englewood Cliffs, NJ: Prentice-Hall.

Plach, M. (1998). *Processes of judgement revision. Cognitive modeling of the processing of uncertain knowledge.* Wiesbaden: Deutscher Universitäts–Verlag.

Tubbs, R. M., Gaeth, G. J., Levin, I. P. & Van Osdol, L. A. (1993). Order effects in belief updating with consistent and inconsistent evidence. *Journal of Behavioral Decision Making, 6*: 257–269.

Wang, H. (1999). Order effects in human belief revision (Doctoral dissertation, Ohio State University, 1998). *Dissertation Abstracts International, 59(8–B)*:4504.

Zhang, J., Johnson, T. R. & Wang, H. (1998). The relation between order effects and frequency learning in tactical decision making. *Thinking and Reasoning, 4*:123–145.

CHAPTER 15

THE PSYCHOPHYSICS METAPHOR IN CALIBRATION RESEARCH

GERNOT D. KLEITER, MICHAEL E. DOHERTY, AND
GREGORY L. BRAKE

Abstract

According to the standard paradigm of behavioural decision theory, calibration refers to how well subjective probability assessments match objective frequencies. The present paper argues that that conception results from a misleading metaphor that relates subjective probabilities to objective frequencies in the same way as a psychophysical function relates subjective sensations to physical dimensions. The results of four calibration studies on experts' predictions of the outcomes of baseball games are reported. The results show good calibration and underconfidence. A Brunswikian Lens model was employed to analyse the relationships between the predictability of the baseball games and the subjects' cue utilization on one hand, and calibration on the other. We finally demonstrate that the hard–easy effect can be explained by the proportion of counter-intuitive items contained in the item pool. While with the easy and intuitive items all subjects showed underconfidence, with the difficult and counter-intuitive items all subjects showed overconfidence. The hard–easy effect is a statistical consequence of how the response data in calibration studies are analysed.

The psychophysics metaphor

Overconfidence in probability judgement is usually explained as follows.

When you go to the doctor and the doctor tells you she is 99% sure you are suffering from a certain disease, but it turns out that she is wrong, she should not be criticized too much because you may just have been an exception. When the doctor tells one hundred patients she is 99% sure they are suffering from a certain disease, and if it turns out that she is right in only 40% of the cases, the doctor is said to be overconfident. To be overconfident means to give probability judgements that are systematically higher than the corresponding hit rates. To be well-calibrated means that the probability judgements closely match the hit rates. Underconfidence means that the probability assessments are too low as compared with the corresponding relative frequencies. Overconfidence is the effect that has been found most often.

According to the standard paradigm of behavioural decision theory, the calibration index measures how well subjective and objective probabilities agree. We will call this view the 'psychophysics metaphor'. In psychophysics, each of a set of stimuli is measured by a physical instrument. The values obtained represent a feature of the external world, and in this sense are *objective* (sound energy, light energy, length, duration etc.). They are measured on a physical scale. When presented to a subject, each stimulus evokes a response (reporting a sensation such as loudness, brightness, apparent length, apparent duration etc.) that then is measured on a *subjective scale*. The relationship between the objective and the subjective continuum is described by a function fitted to the data and called a *psychophysical function* or psychophysical law (Stevens' law, for example).

In the psychophysics metaphor for calibration, relative frequencies take on the role of the objective continuum and uncertainty judgements take on the role for the subjective continuum. The relationship between the two is described by a calibration curve. It is drawn in a Cartesian coordinate system much as it is done in psychophysics, with the only exception that the subjective and the objective scales are interchanged. Systematic deviations from the idendity line are interpreted as *biases*. *Overconfidence* is the bias most often reported and investigated. The metaphor takes it for granted that there is one correct reference scale against which the accuracy of the subjective judgements is evaluated. Probability judgements are regarded as appropriate if they match relative frequencies in the long run. Systematic deviations are interpreted as over- or underconfidence. For about thirty years the heuristics and biases approach has reported evidence that predominantly shows overconfidence (Lichtenstein *et al.* 1982; Keren 1991).

Is calibration research really analogous to psychophysics? Lakoff and Núñez (1997) (Núñez & Lakoff 1998) have shown how metaphors in mathematical reasoning may lead to false intuitions about the concept of continuity or the concept of limits. Similarly, the psychophysical model of the relationship between subjective and objective dimensions may arise from a metaphor in our reasoning that leads to inappropriate conclusions.

The basic philosophy of calibration research has been attacked by Lad (1996). He argues that good or bad calibration is a question of *coherence* (de Finetti 1974).

The Lad argument

According to Lad (1984, 1996), the probability judgments of a coherent probability assessor must—by logical necessity—be perfectly calibrated. In other words, poor calibration is a consequence of *incoherence*, and not of a poor match between 'subjective' and 'objective' probabilities. Lad is a mathematical statistician, and as we have not seen his argument being discussed in the psychological literature we will explain his argument in some detail.

Equal probabilities for binary events

Calibration is usually defined as follows: Probability assertions are said to be well calibrated at the level of probability p if the observed proportion of the propositions that are assessed with probability p, equals p. This definition, though, is inappropriate. To illustrate its inappropriateness we first consider the most simple case, namely a binary event, for which we are 50% sure it is true.

A drawing pin (thumbtack), after it has been thrown on a plain surface, may land flat-side up or flat-side down. We denote the two possibilities by U and D. A probability assessor—let him or her be called You—gives equal probabilities to U and to D, $P(U) = P(D) = 0.5$. The pin is thrown 100 times. It lands on U 70 times and on D 30 times. What about Your calibration at $p = 0.5$? Was Your assessment over- or underconfident? Was Your subjective probability too high or too low? We may argue that the assessed probability of the event U, $P(U) = 0.5$, was too low because it is clearly less than the relative frequency $F(U) = 0.70$. But with the same justification we may argue, that the assessed probability of the event D is too high because $P(D) = 0.5$ is clearly larger than the relative frequency $F(D) = 0.30$. When You assessed $p = 0.5$, You *simultaneously* assessed the probability of *two* events, of U and of D. We cannot select one of the two events and attribute good or bad calibration to this event and ignore the second event which is the complement of the first one. If, with respect to U Your probability is too low, it must be too high with respect to D. At $p = 0.5$ You cannot be said to be over- or underconfident, presuming that You are coherent in the sense that Your probabilities for U and D add up to one.

Each time we throw the drawing pin, we have *two* possible events and one of the two must be true, that is, one out of two, or 50%, must turn out to be true and one out of two, or 50%, must turn out to be false. Throwing the thumbtack 100 times and taking all possible events into account leads to 200 possible events, half of which must be true and half of which must be false. Your probability assessment $p = 0.5$ is perfectly calibrated. You assessed the probabilities of two events. The assessed probabilities were 0.5 for each of them. One of the two events turned out to be true. One out of two is just 0.5. Thus, Your probability assessment and the relative frequency agree perfectly.

We see that for binary events at $p = 0.5$ over- and underconfidence cannot be defined in the way the definition given above suggests. The reason is that we cannot assess the probability of a binary event without simultaneously 'co-assessing' the probability of its complement. A probability function is a set function that assigns a value to *all* subsets (events) of a possibility space. In the present case the set of possibilities is $\Omega = \{U,D\}$ and the family of all subsets is $\mathfrak{E} = \{\emptyset, U, D, \Omega\}$. The probability of the impossible event, $P(\emptyset) = 0$ and that of the sure event $P(\Omega) = 1$ is not touched by the assessment, but $P(D)$ is of course $1 - P(U)$. Assigning a probability to one event in $\mathfrak{E}$ has always implications for the probability of the remaining events. In the strict sense, it is not possible to assign a probability to an *isolated* event, but only to assign a probability *distribution*, in the most simple case consisting of $0, p, 1 - p$, and 1. In the psychophysics metaphor there is no room for such a structure. More generally, we may be dealing with a random variable X that (in the finite case) may take on one out of n values, $\Omega = \{x_1, x_2, \ldots, x_n\}$. We may think about the outcome of rolling a die. Assigning a probability to any subset of Ω introduces *linear constraints* for the probability of the other subsets. In the binary case the probability of the second event is completely determined, in the more general case upper and lower probabilities may result.

Is there an explanation as to why we tend to compare probability judgments with the frequency of just one event and why we do not think about the other possibilities? Thinking of only one instead of two or more hypothesis can be explained by the *positivity bias* (Doherty et al. 1979). It has been shown that human subjects tend to represent only one out of two alternative hypotheses (or possibilities). In the binary case neither of the two events is in the 'foreground' or in the 'background'. Mathematically none of the events is more or less important than the other one. Not so psychologically.

We have designated 'marker events' that put one possibility in the foreground and the other one in the background. We suspect that the positivity bias strongly supports the psychophysics metaphor in calibration research.

Frank Lad argues that when we are 50% sure that a binary event will happen, we are, by logical necessity, in 50% of the cases right and in 50% of the cases wrong. Moreover, when this is true for one trial it must remain true for a sequence of n trials. The observed relative frequencies are irrelevant. If, however, Your probabilities $P(U)$ and $P(D)$ do not add up to one, then You obviously may be over- or underconfident. We have to modify the definition by adding the decisive word 'all' so that it excludes the positivity bias that considers only one proposition. It now reads (Lad 1996): Probability assertions are said to be well calibrated at the level of probability p if the observed proportion of **all** the propositions that are assessed with probability p, equals p.

Equal probabilities for the multi-outcome case

You may accept Lad's argument for binary events and $p = 0.5$. What about more general cases? Let us consider an uncertain quantity next. Let x be the proportion of blue sky (not covered by clouds) in Salzburg on August 11 1999 at 12.47 o'clock pm, the time of a complete eclipse of the sun in this area. Seven days before the eclipse Charlotte, the wife of one of the authors, assigned the probability of 0.2 (quantiles) to each of the following intervals: $A = [0,40]$, $B = [40,60]$, $C = [60,70]$, $D = [70,80]$, $E = [80,100]$. The actual value turned out to be about 25%, that is, event A happened. How well was Charlotte calibrated at $p = 0.2$?

In the present multi-outcome case it is even more obvious that we must think in a distributional way. There is not just one possibility, but five, and we must consider all of them. We would fall into the trap of the positivity bias if we would only consider the event that turned out to be true. There are 5 events, 1 of them must turn out to be true, 1 out of 5 is 0.2 so that at $p = 0.2$ the calibration is perfect.

Coherence in the whole event space

To be coherent requires a probability assessor to follow the syntactical rules of probability theory. Up to now we were only concerned with one rule: probabilities add up to one. Let us now take, as an example, the addition rule: the probability of an event that is the union of a set of disjoint events is equal to the sum of the probabilities of these events. If Charlotte is coherent (what I am pretty sure she is not) in the example given above she assesses the probability that x falls into any *two* of the five intervals to be 2/5. What is the relative frequency of such pairs of intervals? There are $\binom{5}{2} = 10$ pairs of events. Four of these must be true when one of the atoms is true. As 4 out of 10 is just 2/5, my wife is perfectly calibrated at 0.4. Similarly, for the union of three events there are $\binom{5}{3} = 10$ triples of events. Six of these must be true, if one of the atoms A, B, C, D or E is true. Six out of 10 is just 3/5. Likewise we obtain for the quadruples the probability 4/5, and for the impossible event that is never true whatever the sky looks like, and the sure event that is always true, the probabilities are 0 and 1, respectively.

In the example, coherence not only requires perfect calibration at the probability explicitly assessed, that is at $p = 0.2$, but also at multiples of this probability, that is, at

$p = 0, 0.2, 0.4, 0.6, 0.8$ and 1.0. In the next paragraph we will employ the property of equally spaced values to demonstrate that coherence implies perfect calibration in the non-symmetric case.

Augmented embedding

To demonstrate that also in the non-symmetric cases coherence implies perfect calibration Frank Lad introduces an *augmented embedding*. Consider a binary event A for which a judge assessed $p(A) = 0.8$. In addition to A we introduce two symmetric coins for which the judge assumes that their probabilities of landing on heads is 0.5 and that their outcomes are independent. We thus have three binary events and

$$P(AHH) = P(AHT) = P(ATH) = P(ATT) = P(\neg A) = 0.2.$$

The two coins play the role of a 'unit meter'. They allow one to concatenate the smallest probability in the domain so that all the probabilities are multiples of this unit. This is an idea well known in the foundations of measurement theory. Here a *standard sequence* is a basic concept (Krantz et al. 1971). It is fundamental, e.g. to the measurement of the length of objects. Suppose $x_1, x_2, \ldots, x_n$ are perfect copies of a rod x, then $x, 2x = x_1 \circ x_2, 3x = (2x) \circ x_3, \ldots nx = ((n-1)x) \circ x_n$ is called a standard sequence based on x. Here $\circ$ denotes the concatenation operator of laying sticks end to end next to each other. 'A meter stick graded in millimetres provides, in convenient form, the first 1000 members of a standard sequence constructed from a one-millimeter rod'. (Krantz et al. 1971)

Embedding an event into a set of events such that each constituent obtains equal probability corresponds to the construction of a standard sequence. We measure the length of an object by concatenating the copies of x until their joint length fits the length of the object, and then we count the number of copies. Similarly, in the example, the probabilities of sets of events may be conceived as multiples of the probability of the augmented embedding. This is a condition for a proper scaling and implies coherence for all the events under consideration.

If the subjective probability scale does not correspond to the multiples of a standard sequence, when, for example, it is biased near one end of the scale, then coherence must be violated and calibration cannot be perfect. Is not this exactly what we mean by the term 'calibration', that the units of a measurement device are equally spaced over the whole domain of the scale?

Scoring rules and the Lad argument

It is important to note that Frank Lad argues against the logic underlying the over/underconfidence concept, not against the logic of scoring rules. We cite from one of the first investigations on the evaluation of probability assessments, from the thesis of a pioneer of empirical investigations of probability assessments, Carl-Axel Staël von Holstein:

We shall assume that an individual's judgment concerning an uncertain quantity X can be represented by a probability distribution F defined on the outcome space $\mathcal{S}$. Let $E_1, E_2, \ldots, E_n$ constitute an n-fold partition of $\mathcal{S}$, i.e., there are a set of mutually exclusive and colletively exhaustive events. The probability mass in E_j will be denoted by p_j, where $p_j = P(X \in E_j) = \int_{E_j} dF(x)$. We shall write $\underline{p}$ for the vector $(p_1, \ldots, p_n)$. $\underline{p}$ thus represents the assessor's true beliefs. We shall assume that he answers by stating the distribution $\underline{r} = (r_1, \ldots, r_n)$, and $\underline{r}$ need not necessarily be equal to $\underline{p}$. An *honest* assessment is one for which $\underline{r} = \underline{p}$. A *scoring rule* is a function of the assessed distribution $\underline{r}$ and of the event which eventually turns out to be true. The assessor receives a score $S_k(\underline{r})$ if the kth event occurs. (Staël von Hostein 1970, p. 25)

Let us emphasize the points that are important in the present context: (1) The judgements are concerned with a random variable X, not with just one point in the outcome space. (2) The assessment is a probability distribution, i.e. the probabilities sum up to one. The definition of scoring rules, thus, *presupposes* a coherent probability assessor (see also, for example, Bernardo and Smith 1994). Moreover, optimizing a proper scoring rule implies that the usual index of calibration in the decomposition of a score is *not* zero, that is, minimizing the penalty function of a proper scoring rule implies over- or underconfidence (Blattenberg and Lad 1985). Over/underconfidence is *not* a component of the decomposition of the quadratic scoring rule. It is a signed difference.

The measure of over/underconfidence does not show how good the subjective probability judgments are as compared with the 'gold standard' of relative frequencies. It actually measures the coherence of the judgements. Poor calibration may result from many different incoherences and violations of the laws of probability, from probabilities that do not sum up to one, from violations of the addition rule, from probability scales that are not subdivided into equally spaced intervals, from a lack of the reliability of the judgments etc.

The psychophysics metaphor suggests that 'objective' and 'subjective' probabilities can be studied without studying the environmental context in which the events occur, and without studying the knowledge and information the subjects have about these events. The metaphor suggests that the psychophysical function can be established in an environmental and cognitive vacuum. The embedding of probability judgments into a Brunswikian Lens model provides the means to analyse the validity of environmental cues on one hand, and the subjects' cue utilization on the other hand. In the next sections we will report a series of four calibration studies in a Brunskwikian Lens model approach.

Brunswik at the baseball field

Four calibration studies on the judgement of baseball games and the probability of their outcomes were performed (Brake *et al.*, 1999). In all four studies, descriptions of 150 games randomly sampled from the 1992 Major League Baseball season were collected. The data were colleted in 1995. We give only a short description here, for more details the reader is referred to (Brake 1998; Brake *et al.* 1999).

Study 1

138 students were tested on their knowledge of baseball and the twenty students having the highest scores in the test were selected. Each of these experts were presented the descriptions of 150 baseball games that had been taken verbatim from *USA Today*. The summaries contained information about six variables: WINS (season records of the two teams playing), PITCHERS (the number of pitchers appearing in the game for each team), CLOSERS (relief pitcher with twenty or more saves that season), INNINGS (number of innings given in the summary, 3, 5, and 7 innings played), RUNS (runs scored by each

team up to the innings given), and HFA (home field advantage). For each game the subjects were asked which team would win the game and what the probability is that this team would win the game. Probabilities in the range between 0.5 and 1.0 were assessed (half-scale method). All participants completed the paper and pencil task individually.

Study 2

Seventy-five subjects were pretested on their knowledge about baseball and 29 of these were eligible for the main experiment. For technical reasons the results from 10 experts will be reported here only. They all took part in a second session in which the experimental procedure was replicated to obtain information about the reliability of the data. All participants completed the task individually in computer-controlled sessions. The full-scale method was used, that is, probability ratings between 0 and 1 were obtained.

Study 3

From 95 students who took the pretest, 37 were selected to participate in Study 3. In Study 3 (1) the *number of cues* provided in the baseball game summaries and (2) the *number* of teams (two teams or one team only) for which the information was provided varied in a 2 × 2 between subjects design. In the high predictability condition the game descriptions contained the number of innings, the number of runs scored through that inning, and the number of wins in the final season. Information about pitchers, closers, and the home team was not provided. In the low predictability condition the game descriptions contained only one cue, the number of wins in the final season. In the two team condition, the information was provided for both teams, in the one team condition, the information was provided for one team only. The full-scale method was used. The experiment was computer-controlled.

Study 4

Fifty-eight students who were highly knowledgeable about baseball but did not qualify as experts in the pretest participated in this experiment. The materials, design, and procedure for Study 4 were identical to those used in Study 3.

Task analysis

Multiple logistic regression was employed to find reference models with which the judgements of the subjects in the various experimental conditions could be compared. The information provided in Study 1 contained six predictors. The linear term for the full data set of all 150 items leads to three significant predictors and the following linear term

$$u = -1.109 + 0.813\,\text{RUNS} + 0.084\,\text{WINS} + 1.499\,\text{CLOSERS}$$

in the logistic equation

$$\text{OUTCOME} = \frac{\exp(u)}{1 + \exp(u)}.$$

Estimating the model parameters and predicting the outcomes of the games with one and the same data set (as it is usually done in Lens model studies) may lead to overfitting and to too small residuals. To avoid overfitting we used two thirds of the data to estimate the regression coefficients and one third of the data to predict the outcomes (Breiman 1996). The first set is a learning and the second a testing set. For the learning set we selected the 100 items 2, 3, 5, 6, 8, 9,..., 149, 150, and for the testing set and the remaining 50 items 1, 4, 7,..., 148. In real life, the experts in our studies clearly had a learning experience of far more than 100 games. In the test set multiple logistic regression using the estimates from the learning set resulted in three significant predictors and the following linear terms

$$u = -0.481 + 0.787 \text{ RUNS} + 0.107 \text{wins} + 1.683 \text{ CLOSERS}.$$

The estimates of the linear parameters are robust. The regression weights are practically the same as for the full set of 150 games, only the constant terms differs slightly.

The equation of the logistic linear regression model was used to obtain a *quasi-probability* for each of the 50 games in the test set. The quasi-probabilities vary between 0 and 1. When less than 0.5 they were transformed to half-scale values between 50 and 100 for reasons of better comparison with the judgements of the subjects. The mean quasi-probability is 88.27 (sd = 13.31). The mean probability ratings of the subjects in Study 1 and in Study 2 were 73.2 (sd = 6.02) and 73.3 (sd = 7.11), respectively. The subjects are much more conservative than the logistic regression model. We note that the mean probability ratings are practically identical to the values reported by Juslin *et al.* (Juslin *et al.* 2000) for their 95 selected and their 35 representative item samples, which are both 0.73. One may speculate that the subjects respond in a way that keeps the response level constant over various conditions, while the difficulty level varies over conditions, and varies across the boundary suggested by Juslin, i.e. around 0.70 to 0.75. The subjects have expectations about the difficulty of the items. They do not assume that the experimenter presents items that can only be answered at the guessing level (see also Lichtenstein and Phillips 1977).

Is the reference model overconfident when it gives a mean confidence as high as 88%? The quasi-probabilities were taken to determine the calibration of the reference model for the 50 games in the test set. The Brier scores of the multiple regressor for the various conditions are shown in Table 15.1. For the 2-teams/6-predictors condition of Study 1 the Brier score is 0.0869. This is an exceptionally good value. The ideal regressor is slightly underconfident (−0.0485). The mean Brier score of our subjects in Study 1 was 0.1521 (sd = 0.0291). The model made only four (out of fifty) false predictions. Two of

Table 15.1 Mean quasi-probabilities obtained by multiple logistic regression and mean Brier scores of the logistic model in the test set of 50 items; the model is based on 100 items and tested on the 50 remaining items

	2 Teams		1 Team	
	Prob	Brier	Prob	Brier
3/6 Predictors	88.27	0.0869		
2/3 Predictors	82.80	0.1109	76.39	0.1919
1/1 Predictors	74.30	0.2413	64.51	0.2451

these false predictions, though, were made with extreme quasi-probabilities, one with a quasi-probability of 0.007 and the other one with a quasi-probability of 0.010. The other two values were 0.68 and 0.26. We conclude that the outcomes of baseball games are clearly predictable. There are, though, a few games which are 'unpredictable' by the model we used.

In Study 3 the 3-predictors/2-teams condition leads to two significant predictors, the number of runs and the number of wins in the previous season:

$$u = -0.405 + 0.732 \text{RUNS} + 0.087 \text{WINS}.$$

The mean quasi-probability is 82.8043 (sd = 14.3958) and the mean Brier score for the quasi-probabilities is 0.1109 with a close to perfect calibration of −0.0335. The model gives seven out of 50 false predictions.

The 1-predictor/2-teams condition leads to the linear term

$$u = -0.277 + 0.083 \text{WINS}.$$

The weight for the wins is practically identical with the weight in the previous models. The mean quasi-probability is 74.2963 (sd = 12.5756). The Brier score of 0.2413 is clearly below the values in the previous conditions. There is some over-confidence, 0.0875. The model produces 17 (out 50) false predictions.

The 3-predictors/1-team condition results in two significant cues

$$u = -8.988 + 0.562 \text{RUNS} + 0.094 \text{WINS}.$$

The mean quasi-probability is 76.3876 (sd = 13.2669). The Brier score is 0.1919 with a close to perfect calibration of 0.0225.

Finally, in the 1-team/1-predictor condition we get the linear component

$$u = -5.290 + 0.063 \text{WINS}.$$

The mean quasi-probability is 64.508 (sd = 6.0475) and the Brier score is 0.2451. Table 15.2 shows the mean percentages of correct predictions made by the participants in the various experimental conditions.

For the probability ratings of the subjects in Study 3 and in Study 4, a 2 × 2 ANOVA was performed. The factors were the number of teams (one versus two) and the number of predictors (one versus three). For the data of Study 3 no significant effects were

Table 15.2 Mean percentage of correct predictions made by the participants

	2 Teams	1 Team
3/6 Predictors (Study 1)	70.08	
3/6 Predictors (Study 2)	73.33	
2/3 Predictors (Study 3)	81.11	72.06
2/3 Predictors (Study 4)	83.61	69.68
1/1 Predictors (Study 3)	71.27	60.57
1/1 Predictors (Study 4)	69.22	60.17

obtained for the number of teams and for the interaction. The number of predictors is significant (F = 7.08, 1/33 df, p = 0.012). The mean for the 1-predictor condition is 64.6, for the 3-predictors 68.9. For the data of Study 4 the ANOVA did not produce any significant differences. The grand mean was 67.7 (sd = 5.32).

The level of the mean probability judgements is nearly invariant with respect to the amount of the information available. As a between-subjects design was employed, one might expect that the sensitivity to the amount of information available would be higher in a within-subjects design.

The main result of the comparisons of the judgements of the subjects and the multiple logistic regression model is that the reference model makes better use of the cue information than our subjects do. This is the reason why the mean quasi-probabilities are much higher than the mean probability ratings of the subjects and, at the same time, the Brier score of the reference model is much better than that of the subjects. To improve the probability judgements, not calibration but cue utilization is of primary concern.

Lens model

The Lens model statistics (Cooksey 1996) of Study 3 and 4 are shown in Table 15.3. The high multiple correlation between responses and cues (R_s) indicates that participants showed substantial consistency in applying their judgement policies; this is typical of Lens model research involving domain experts. The level of achievement (correlation between criterion and response, r_a) found in this study similar to that found in other studies with comparable task predictability. The large value for the correlation between the predicted criterion and the predicted response (G) indicates that the match between the models of the judgement and the environment on the cues specified is high. The cue validities are contained in Table 15.4. The participants tend to slighlty underweight the cues.

Table 15.3 Lens model statistics for Study 3 and 4. The Lens model indices are r_a (correlation between criterion and response), R_s (multiple correlation between response and cues), G (correlation between estimated criterion and estimated response), C (correlation between the residuals for the criterion and the response). n_{cues} is the mean number of cues that were statistically significant in the subjects' equations

	2 Teams		1 Team	
	Study 3	Study 4	Study 3	Study 4
r_a	0.60	0.62	0.44	0.40
R_s	0.85	0.84	0.82	0.82
G	0.98	0.97	0.96	0.93
C	0.12	0.19	0.10	0.04
n_{cues}	2.33	2.21	2.33	2.93
r_a	0.35	0.33	0.21	0.20
R_s	0.78	0.80	0.71	0.89
G	1.00	1.00	1.00	1.00
C	0.08	0.04	0.07	0.03
n_{cues}	1.00	1.00	1.00	1.00

Table 15.4 Cue validities for Study 3 and 4. Multiple logistic regression was used to obtain the quasi-probability of the target team winning each game. These probability values were used as the dependent measure in an OLS multiple regression model to obtain the cue validities. On top three predictors (wins, innings, runs), on bottom one predictor condition (wins)

	2 Teams	1 Team
Wins	0.39	0.50
Innings	0.06	−0.15
Runs	0.78	0.61
Wins	0.99	0.99

Table 15.5 Calibration indices for Study 3 and 4. The calibration indices are PS (Brier score), CI (calibration index), DI (discrimination index), and O/Uconf (overconfidence)

	Teams		1 Team	
	Study 3	Study 4	Study 3	Study 4
Brier	0.1686	0.1674	0.2067	0.2182
CI	0.0221	0.0251	0.0147	0.0227
DI	0.1035	0.1077	0.0580	0.0545
O/Uconf	−0.0800	−0.1000	0.0179	0.0221
Brier	0.1686	0.2255	0.2461	0.2569
CI	0.0221	0.0192	0.0188	0.0283
DI	0.1035	0.0436	0.0226	0.0214
O/Uconf	−0.0800	0.0030	0.0557	0.0877

Erev, Wallsten, and Budescu (Erev *et al.* 1994), Juslin (Juslin *et al.* 2000), and (Kleiter 1996) have shown that poor reliability of the probability judgements induces poor calibration. We therefore assessed the reliability of the judgements in re-test conditions. In Study 1 we found 84% consistent predictions who would win the games from one session to the next. Each subjective probability judgement was transformed to an unbounded scale following Erev *et al.* (Erev *et al.* 1994). The mean r_{tt} is 0.90 indicating substantial reliability in probability judgements. In Study 2 the median intercorrelations are $r_{tt} = 0.78$ (half-scale) and 0.79 (full-scale). We thus observed a rather high reliability, a result, that is consistent with the Erev *et al.* analysis.

In the introduction we pointed out that in calibration studies the kind of data analysis is peculiar, and clearly differs from that in psychophysics insofar as there is no real independent variable and the data are grouped by the subject's responses. Here the psychophysics metaphor leads to the misperception of an effect resulting from the logic of data analysis. In the following section we will demonstrate that the hard–easy effect is a logical consequence of way in which the data analysis is performed.

Hard–easy effect

One of the most prominent effects in calibration studies is the *hard–easy effect: studies with difficult items lead to strong overconfidence, studies with easy items lead to weak overconfidence, good calibration or even underconfidence.* Several attempts were made to explain the psychological processes underlying the effect. In our attempt to explain the hard–easy effect we will use points that were raised by Gigerenzer, Hoffrage, and Kleinbölting (Gigerenzer *et al.* 1991), by Klayman (Klayman *et al.* 1999), and Juslin (Juslin *et al.* 2000). We will argue that the hard–easy effect is not caused by psychological processes, but is a logical consequence of item selection.

To show this we first imagine that the subject's responses are categorized as 'correct' and 'incorrect' and arranged in a subject × response matrix **A**. The matrix has n rows, one for each subject, and m columns, one for each item. The entry in cell ij is 1 or 0, $a_{ij} = 1$ (correct) or $a_{ij} = 0$ (incorrect). The row means $a_{i.}$ and the column means $a_{.j}$ may be used to estimate the the amount of knowledge or expertise subject i has, and the easiness of item j, respectively. It is reasonable to assume that subjects having equal $a_{i.}$ have the same amount of knowledge and that items having equal $a_{.j}$ have the same easiness. Such items build a homogeneous scale. This principle is well known in latent trait analysis and in item response models.

Binary items that are answered incorrectly 50% or more of the time, are predominantly answered in the 'wrong direction'. The information provided favours one alternative but the other alternative is actually true. We all know that this can easily happen in sports events. Even experts (and especially experts!) will be wrong here. These items do not measure expertise or knowledge. They do not belong to a scale that assesses baseball knowledge. They correlate negatively with the total score of an expert. A psychometric test would discard such items. In the field of psychological testing there exists a rich repertoire of models that measure item parameters, person parameters, or the homogeneity of scales. Partial credit models or Rasch models for the analysis of rating scales (Wright and Masters 1982) would allow a much better characterization of the items contained in a calibration study. An item response model identifies such items as not belonging to a scale measuring a latent trait. Items of this sort were called 'contrary items' by Klayman *et al.* (1999) or 'trick items' by Gigerenzer *et al.* (1991). We will speak of *counter-intuitive* items. Few experiments investigated systematically the properties of the items used in calibration studies. An exception is Lichtenstein and Phillips (Lichtenstein and Phillips 1977).

The *easiness/difficulty* of an item is obviously defined as follows: an item is easy if it is predicted correctly by a high proportion of subjects, and it is difficult if only a small proportion of subjects predict it correctly. An easy item is expected to obtain a probability rating close to 1.0, a difficult item is expected to obtain ratings close to the guessing level 0.5. We introduce the following definition for the calibration of an *item*:

An item is said to be well calibrated if the observed proportion of its correct predictions (or its easiness) and its mean uncertainty rating are equal. We call an item counter-intuitive if it is predicted correctly by less than fifty percent of the subjects and at the same time obtains a mean uncertainty rating greater than 0.5.

The scatter diagrams in Fig. 15.1 and Fig. 15.2 plot the items' easiness on the abscissa against the items' mean uncertainty ratings on the ordinate. In Study 1 (Fig. 15.1) 36 items

THE PSYCHOPHYSICS METAPHOR IN CALIBRATION RESEARCH   251

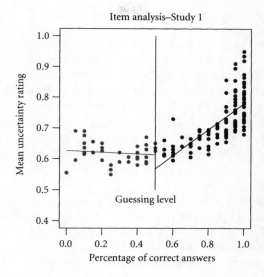

Figure 15.1 Analysis of the items of Study 1. Items answered by less than 50% of the subjects correctly are below the guessing level and on the left half of the diagram. Items answered by more than 50% of the subjects correctly are above the guessing level and plotted on the right half of the diagram. Also plotted are the regression lines for the two sets of items.

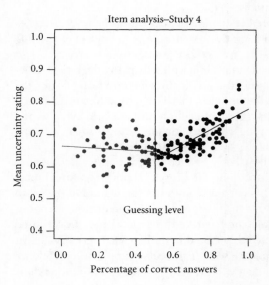

Figure 15.2 Analysis of the items of Study 4. Items answered by less than 50% of the subjects correctly are below the guessing level and on the left half of the diagram. Items answered by more than 50% of the subjects correctly are above the guessing level and plotted on the right half of the diagram. Also plotted are the regression lines for the two sets of items.

are left of the guessing level of 50% on the abscissa. They are randomly spread out. The correlation between the easiness and the mean ratings is $r = -0.116$. The intercept of the fitted regression line is 0.626, its slope is -0.026, and the mean rating is 0.619 (sd = 0.035). 114 items are above the guessing level, for these $r = 0.701$, the intercept is 0.357, the slope is 0.423, and the mean is 0.727. As the slope of the regression line is less than 1.0, the relationship shows underconfidence. Items that are predicted correctly by 90% of the subjects or more obtain only mean ratings of about 80%. Difficult to predict games produce erratic random ratings, easy to predict games obtain underconfident probability ratings.

In Study 3 fifty items are below the guessing level. For these items $r = -0.07$, the intercept is 0.651, the slope is -0.027, and the mean rating 0.643 (practically equivalent to the intercept). One hundred items are above the guessing level, the correlation is $r = 0.674$, the intercept is 0.509, the slope is 0.229, and the mean rating is 0.678.

In Study 4 (Fig. 15.2) 53 items fall into the area left of the guessing level of 50% on the abscissa. The correlation between the easiness and the mean reatings is $r = -0.092$. The intercept of the regression line is 0.665, and its slope is -0.036. The probability ratings are distributed around the mean of 0.65 (sd = 0.047) and the regression line is practically horizontal. The mean rating of the items right of the guessing level (97 items, right of the guessing point 0.5 in Fig. 15.2) show a clear sensitivity to the easiness. The correlation is $r = 0.723$, the intercept of the regression line is 0.464, and the slope is 0.315 (standard error = 0.031).

The pattern of these results is very uniform. Counter-intuitive items show no systematic relationship with the mean probability rating they obtain, intuitive items show a relationship that is regressed toward the mean indicating underconfidence. For items for which we have partial knowledge we are underconfident, for items we do not know we are erratic and as a consequence very overconfident. Unfortunately we do not know which items we know and which items we don't. The hard–easy effect is a consequence of the mixture of items from these two pools.

From time to time we all make false predictions. We predict HEAD when TAIL becomes true. When making a false prediction our cognitive system may be in one out of two different epistemic states, guessing or believing. We say false guesses (of a binary event) go along with a 50% degree of belief. False beliefs go along with affirmative degrees of belief that are greater than 50%. We often make a false prediction when we really favour the wrong alternative, when we have a degree of belief that is clearly different from the guessing level. From the viewpoint of the standard calibration paradigm every wrong prediction should go along with a 50% rating. By definition *any wrong prediction getting a probability rating over 50% is overconfident.*

To escape overconfidence we should be aware of which proposition we don't know, and we should assign to each a guessing probability. To be well calibrated requires one to have no false beliefs. To believe in HEAD when TAIL comes true by definition means to assess a probability greater than 0.5 for HEAD and if *ex post facto* this predictions turns out to be false, then the assessment necessarily contributes to overconfidence. We favoured the false event. If we are perfectly calibrated this should never happen. Every instance in which it happens increases our overconfidence score. False beliefs imply overconfidence and, as a final consequence, violate rationality. Such a rationality criterion, though, is much too strong. Having false beliefs does by no means violate rationality.

If over- and underconfidence is a property that is so closely associated with properties of the items, it is a direct consequence that the item selection is crucial for the over- and underconfidence of the subjects. In a selection of predominantly difficult and counter-intuitive items we sample most items from the left hand side of the scatter diagram of Fig. 15.2 or Fig. 15.1. As a consequence we will observe overconfidence. If we have a selection of predominantly easy and intuitive items we sample from the right hand side of the scatter diagrams and obtain a selection of items that leads to underconfidence (remember the regressed regression line). A balanced cocktail will lead to good calibration. The hard–easy effect is a direct consequence of the composition of the cocktail.

This 'prediction' can be 'tested' empirically with our data (it is actually a *logical* relationship that can be illustrated by the data). In Study 4 the median of the percentage correct is 61.2. We analyse the items below (difficult items) and above the median (easy items) separately. The mean overconfidence of the 75 difficult items is 0.14 (sd = 0.12), the mean of the 75 easy items is −0.12 (sd = 0.10) (positive values indicate overconfidence and negative ones underconfidence). All (!) 58 subjects show higher mean calibration scores for the difficult than for the easy items.

In Study 3 similar results are obtained. The median uncertainty rating is 0.622, the mean overconfidence of the easy items is −0.17 (sd = 0.09) indicating underconfidence, the mean of the difficult items is 0.16 (sd = 0.11) indicating overconfidence. Again all 37 subjects show this pattern. In Study 1 the median percentage correct is 0.85. The mean of the easy items is −0.20 (sd = 0.10) indicating considerable underconfidence, the mean of the difficult items is 0.045 (sd = 0.06) indicating close to perfect calibration. We note that the 'difficult' items in this study are much easier than those in the other conditions. This explains the close to perfect calibration in this group.

Discussion

Juslin *et al.* (2000) have shown that the hard–easy effect is a consequence of the way in which the data are analysed. In a meta-analysis they investigated 130 studies in the literature. They correlated the proportion of correct answers and the degree of over/underconfidence and found coefficients of 0.84 for selected and 0.76 for representative samples, respectively. Our item analysis fits perfectly well in their meta-analysis, though not on the level of different studies but on the level of the items. Juslin *et al.* attribute the hard–easy effect to scale–end effects, linear dependency, and regression effects. We relate the effect to the inhomogenous counter-intuitive items. The psychophysics metaphor has led calibration research to think about 'subjective' and 'objective' probability in a way that can be terribly misleading—Juslin *et al.* speak of 'naive empiricism' and 'dogmatism'.

Let us close with a remark concerning the relationship between frequencies and probability judgements. In some steps of the statistical analysis we employed quasi-probabilities. These were values between 0 and 1 that were derived from multiple logistic regression. The values were predictors in the [0, 1] interval of the binary 0/1-criterion given the information provided by the instantiated cues. In the logistic model, *the regression weights encode conditional probabilities*. It is plausible to assume that the cognitive processes by which subjects arrive at probability judgements often use a similar weighting procedure. A regression-like encoding is a compact and simple method to derive probabilistic predictions. The weights, though, do not encode relative frequencies. We cannot, for example, translate the intercept and the slope of a regression line into relative frequencies. A regression-like encoding, though, can easily be used to make single case inferences. Thus, the uncertainty processing can be done efficiently on the basis of a non-frequentistic weighting scheme. One disadvantage of a regression-like encoding is that it is directed and difficult to invert. But exactly this property seems to make probabilistic inference tasks difficult for the human problem solver.

Acknowledgement

This research was supported by the SFB F012 from the Fonds zur Förderung der Wissenschaften (Vienna, Austria).

References

Bernardo, J. M. & Smith, A. F. M. (1994). *Bayesian theory*. Chichester: Wiley.

Blattenberg, G. & Lad, F. (1985). Separating the Brier score into calibration and refinement components: a graphical exposition. *The American Statistician, 39*: 26–32.

Brake, G. L. (1998). *Calibration of probability judgments: effects of number of focal hypotheses and predictability of the environment*. Ph.D. Thesis, Bowling Green State University, Department of Psychology.

Brake, G. L., Doherty, M. E. & Kleiter, G. D. (1999). *An experimental assessment of the hard–easy effect*. Technical report, Department of Psychology, Bowling Green State University.

Breiman, L. (1996). Baggin predictors. *Machine Learning, 24*: 123–140.

Cooksey, R. W. (1996). *Judgment analysis*. San Diego: Academic Press.

de Finetti, B. (1974). *Theory of probability, vol. I*. Wiley, London.

Doherty, M. E., Schiavo, M. D., Tweney, R. D. & Mynatt, C. (1979). Pseudodiagnosticity. *Acta Psychologica, 43*: 111–121.

Erev, I., Wallsten, T. S. & Budescu, D. V. (1994). Simultaneous over- and under-confidence. the role of error in judgment processes. *Psychological Review, 101*: 519–527.

Gigerenzer, G., Hoffrage, U. & Kleinbölting, H. (1991). Probabilistic mental models: A Brunswikian theory of confidence. *Psychological Review, 98*: 506–528.

Juslin, P., Winman, A. & Olsson, H. (2000). Naive empiricism and dogmatism in confidence research: A critical examination of the hard–easy effect. *Psychological Review, 107*: 384–396.

Keren, G. (1991). Calibration and probability judgments: conceptual and methodological issues. *Acta Psychologica, 77*: 217–273.

Klayman, J., Soll, J. B., Gonzalez-Vallejo, C. & Barlas, S. (1999). Overconfidence: It depends on how, what and whom you ask. *Organizational Behaviour and Human Decision Processes, 79*: 216–247.

Kleiter, G. D. (1996). *A hidden probability model of overconfidence and hyperprecision. Comments on Erev, Wallsten & Budescu*. Technical report, Institut für Psychologie, Universität Salzburg.

Krantz, D. H., Luce, R. D., Suppes, P. & Tversky, A. (1971). *Foundations of measurement vol I*. New York: Academic Press.

Lad, F. (1984). The calibration question. *British Journal for the Philsophy of Science, 35*: 213–221.

Lad, F. (1996). *Operational subjective statistical methods*. New York: Wiley.

Lakoff, G. & Núñez, R. (1997). The metaphorical structure of mathematics: Sketching out cognitive foundations for a mind-based mathematics. In L. English (ed.) *Mathematical reasoning: analogies, metaphors, and images* (pp. 21–89). Hillsdale, NJ: Erlbaum.

Lichtenstein, S., Fischhoff, B. & Phillips, L. D. (1982). Calibration of probabilities: The state of art to 1980. In D. Kahneman P. Slovic & A. Tversky (eds), *Judgment under uncertainty: heuristics and biases* (pp. 306–334). Cambridge: Cambridge University Press.

Lichtenstein, S. & Phillips, L. D. (1977). Calibration of probabilities: The state of the art. In H. Jungermann & G. de Zeeuw, (eds) *Decision making and change in human affairs* (pp. 275–324) Dordrecht: Reidel.

Núñez, R. E. & Lakoff, G. (1988). What did Weierstrass really define? the cognitive structure of natural and ε–δ continuity. *Mathematical Cognition*, 4: 85–101.

Staël von Holstein, C.-A. S. (1970). *Assessment and evaluation of subjective probability distributions*. Ph.D. Thesis, EFI, The Economic Research Institute at the Stockholm School of Economics, Stockholm.

Wright, B. D. & Masters, G. N. (1982). *Rating scale analysis*. Chicago: Mesa Press.

Part III
PRACTICAL IMPLICATIONS

CHAPTER 16

FREQUENCY EFFECTS IN CONSUMER DECISION MAKING

JOSEPH W. ALBA

Abstract

The commercial world provides an opportunity to count the frequency of occurrence of a variety of product-related stimuli. Firms signal their advantage over competitors through repeated exposure and by enumerating their points of differentiation; consumers, in turn, encode frequency of occurrence and incorporate numerosity information into their product judgements and purchase decisions. This chapter reviews research regarding the potential influence of frequency information in the competitive marketplace. Consistent with findings obtained in other contexts, consumer research demonstrates how misperception of true frequency can affect decision variables. Consumer research further demonstrates that frequency cues can be a determining influence on behaviour irrespective of accuracy of encoding, even when pitted against more diagnostic product cues. Unsurprisingly, however, this research also suggests that the influence of frequency information varies as a function of a variety of factors internal and external to the consumer.

Frequency information has long been implicated in consumer judgement. For example, economists have argued that consumers infer the quality of a brand from the frequency of its advertising, because the highest long-term incentives for advertising exist among those firms that produce the highest quality products (Nelson 1974). Marketers note that consumers attend to the frequency of other consumers' behaviour. In non-monopolistic markets, the frequency with which brands or products are observed in a competitive market may signal product attractiveness. Thus, market share becomes a driver of product purchase, except in cases in which the consumer seeks individuality (and therefore frequency drives non-purchase). Similarly, the rate at which a new product diffuses through a population may vary as a function of perceived consensus. As an innovation is observed more frequently in the environment, it may be viewed as possessing less risk of purchase (if, as in the case of VHS versus Beta, the winner of a technology battle is dependent on which technology attracts the most adopters) and greater value (if, as in the case of fax machines, widespread ownership raises the utility of the product).

Despite the apparent role that frequency information plays in the marketplace, direct measurement of frequency knowledge and its influence on decision making is a relatively new development in consumer research. The remainder of this chapter reviews the progress made thus far. It should become evident that frequency information can be an important element in the judgement process, especially if an expansive view of frequency is adopted. Consistent with tradition, frequency effects in consumer settings may be examined in terms of the number of occurrences of a particular event, such as when estimating the number of occasions on which a consumer has engaged in a particular behaviour (e.g. Menon 1993; Menon et al. 1995; Nunes 2000). However, frequency may also be defined at a more abstract level when referring to the number of times a broader category has been instantiated (e.g. Alba et al. 1980). This conception of frequency is particularly applicable in decision contexts. For example, frequency is used in this chapter to refer to the number of arguments that support or refute a particular hypothesis, the number of attractive or unattractive features of an object, and the number of times one object is deemed superior to an alternative.

The review begins with consumer research that adheres most closely to the traditional paradigm by considering the frequency with which particular prices are perceived to occur in the marketplace. The discussion then broadens to include more abstract stimuli as they pertain to persuasion and consumer choice.

Price promotions and the accuracy of encoding

Although many stimuli are amenable to counting, it is not surprising that a focal concern of consumer research has involved the accuracy with which consumers encode the frequency of prices and price 'deals'. Research suggests that frequent but modest price reductions lead to higher perceptions of value and higher rates of purchase (Buyukkurt 1986; Hoch et al. 1994). Moreover, the price consumers are willing to pay is directly related to their perceptions of deal frequency (Krishna 1991). Consequently, it is important for firms to understand how consumers are perceiving their promotional activities and for consumers to appreciate the frequency and regularity of the firms' behaviour. When consumers can anticipate deals, they can plan their purchases in a way that minimizes both the average price paid and their own stock-keeping costs (Krishna 1994). Preliminary research along these lines has produced some intuitively appealing results. Consistent with findings from other domains, consumers overestimate the occurrence of low frequency deals and underestimate the frequency of high frequency deals; consistent with research on behavioural frequency, consumers perform best when firms deal on a frequent and regular basis (Krishna 1991; Krishna et al. 1991).

An attractive aspect of investigating frequency effects in an applied consumer setting is that it raises issues of pragmatic importance. For example, because competitive markets typically contain numerous and aggressive members, the issue of interference looms large. Accuracy for any given brand's pricing profile is a function of its own promotional activity as well as the activities of its competitors. Consumers are most accurate about their preferred brand but nonetheless will be biased in the direction of the frequency of a competitor. The biasing effect of a competitor itself appears to be moderated by the competitor's deal schedule. The degree to which accuracy for a target brand is degraded is a function of the regularity in the competitor's behaviour such that accuracy is greatest when competitive action is most regular (Krishna 1991).

A second pragmatic issue not likely to be investigated in traditional paradigms concerns the nature of the deal itself. The simplest deal pattern a firm can employ would be to deal on a regular schedule *and* at the same discount level each time. Everyday experience informs us that although such patterns are occasionally encountered, the size of the discount is likely to vary. Krishna and Johar (1996) have investigated the simplest variation on a constant discount pattern and report effects that are consequential to both firms and consumers. Their paradigm allows firms to deal at two discrete discount levels. Results show that deal frequency is not uniformly encoded at each level; instead, the frequency of the smaller of the two price discounts is encoded less accurately. The ultimate outcome is that total deal frequency is underestimated (relative to a control firm that deals at a uniform depth), and average deal size is overestimated. The latter error results from the former. The deeper discount is more prominent and is overweighted in the calculation of average deal price. Krishna and Johar view this result as a direct consequence of an arithmetic combination of perceived deal frequency at each level of discount and the more accurately encoded true levels of price discount. The effect on overall average price (and therefore consumers' willingness to pay it) is negligible because the effects negate each other. That is, mean deal size is overestimated but mean deal frequency is underestimated. Clearly, however, this research only scratches the surface of potential interference effects. Firms deal at multiple levels and in noisy environments. Opportunity for misperception is high, and the consequences for consumer welfare are significant.

The remainder of this chapter places even greater emphasis on consumer welfare by shifting attention from the accuracy of encoding to the use of frequency information in decision making. Decision research assumes that frequency is encoded at a reasonably accurate level and therefore focuses on how frequency cues influence judgement of the relative attractiveness of competing options. The impact of frequency information is not obvious because frequency information does not exist in isolation. In critical tests, it competes against other decision-relevant information.

Persuasion

In persuasion contexts it is important to distinguish 'frequency' from mere information 'mass'. Persuasive messages that contain a large number of arguments in favour of a position also contain a large amount of information. If allowed to covary, the unique effect of frequency cannot be isolated. Consider, for example, research by Chaiken (1980), which manipulated the number of arguments favouring a position. Results indicated that motivated subjects became more persuaded as the number of arguments in favour of a position increased. Inasmuch as all the arguments were cogent, one cannot conclude that subjects were responding to frequency, per se. The message that contained more arguments was simply a more persuasive message. Less normative behaviour is demonstrated in judgements influenced by the mass of available evidence when mass is non-diagnostic of the criterion one wishes to attain (Josephs *et al.* 1994).

Non-normative behaviour that is somewhat closer to a frequency rather than quantity explanation can be observed in the use of the numerosity heuristic (Pelham *et al.* 1994). In this instance, people make judgements that are consistent with the number of pertinent 'units' of information rather than the information contained in those units. The paradigm employed to produce a numerosity effect allows for a stronger inference regarding the role of frequency because the total value of the information can be varied independently of the number of units expressing it. The numerosity heuristic is reminiscent of earlier research by Estes (1976), who described conditions under which decision makers would choose an option associated with a higher number of 'wins' over an option with a larger proportion of wins. More recent research shows that when presented with two options that possess identical ratios of success (e.g. 1/10 and 10/100), decision makers are biased toward the option that contains the larger absolute number of winning instances (Denes-Raj & Epstein 1994).

These studies provide ample reason to believe that numerosity may play a dysfunctional role in decision making, although only Estes demonstrated the effect in a choice context. In the other studies, individuals were shown to be biased by quantity or numerosity cues, but little evidence was provided for the proposition that decisions can be guided by numerosity in the face of countervailing evidence for a more preferable option. Classic decision paradigms, discussed next, offer such a test.

Relative frequency

The dominant paradigm in traditional decision research involves choice from among a set of alternatives that are described along a set of common dimensions. In such settings, a wide variety of decision heuristics is possible. A frequency rule itself may take on different variations but in choice environments typically involves a count of the number of dimensions favouring one option over another (Svenson 1979). It is interesting to note that the use of such rules has been observed at non-trivial levels in decision research (e.g. Bettman & Park 1980) and, moreover, that such rules can be attractive from an effort–accuracy perspective. However, their attractiveness is highly situation-specific. The effort required to execute a frequency rule depends on the number of options being compared. Because the rule operates by taking pairs of items and declaring a winner based on the number of times one option outperforms another, effort is a function of the number of available options. When the options are numerous and no other screening mechanism is employed, the effort–accuracy trade-off is not favourable (Payne et al. 1988). Effort can be reduced by performing an initial screening to reduce the number of alternatives, and therefore it is unsurprising that frequency rules are more likely to be observed in later phases of the decision process (Bettman & Park 1980) or in binary choice situations (Russo & Dosher 1983).

Regardless of the effort required, simulation experiments suggest that frequency rules can return a surprisingly high degree of accuracy given the degree of data reduction (see Bordley 1985; Russo & Dosher 1983). A strict frequency rule treats all dimensions of comparison as equally weighted and ignores any magnitude differences that exist among options on those dimensions. For example, when comparing two automobiles, a frequency rule will select as the winner the brand that outperforms the competitor on the greatest number of dimensions. Superior performance on gas mileage carries no more influence than superior performance on the number of cup holders possessed by each brand. Even when attributes are equally important to the decision maker, a large relative advantage on one dimension (e.g. gas mileage) cannot compensate for a narrow disadvantage on another dimension (e.g. repair record).

Of course, the frequency rule will be most attractive when frequency and substance are correlated. A problem for consumers arises when companies or happenstance create comparisons that do not reflect a positive correlation. The remaining discussion examines such situations.

Peripheral processing

Companies may exploit the use simplifying heuristics to prompt decisions that favour the company, particularly when the consumer's vigilance is low. For example, one popular magazine has been known to offer many extra incentives for subscribing to its publication. These incentives are often a combination of typical features of the product and some premiums (read: trinkets), both of which are described in terms of their specific characteristics to enhance the perceived number of benefits associated with purchase. The offer may appear tempting under typical TV viewing conditions but far less so on further inspection.

Experimental evidence is provided by Petty and Cacioppo (1984). In a classic persuasion setting, they provided subjects with either 3 or 9 arguments in favour of an issue. The arguments were either uniformly cogent or specious, and subjects were either motivated or unmotivated to process the information deeply. Results were consistent with the large body of research dealing with the distinction between central and peripheral processing. When involvement in the message was high, persuasion was driven by the quality but not the quantity of the arguments; when involvement was low, persuasion increased with the number of arguments but not their quality. The results from the low involvement condition are presumably due to a failure to assess the quality of the message and a reliance on a plausible indicator of the validity of a position. The number of arguments that can be made in favour of a position should covary with the correctness of the position, *ceteris paribus*. Note that the independent manipulation of argument quality and quantity avoids the interpretational problem described earlier in studies that report a relationship between persuasion and the amount of information presented. In that research, all information was cogent and therefore the independent effects of quantity and quality could not be assessed. The experiment by Petty and Cacioppo illustrates how frequency information can influence judgements in the presence of more diagnostic data.

Firms do not normally create messages composed primarily of specious arguments because a lack of involvement cannot be assumed to exist across consumers or opportunities to view the message. However, deceptive firms may create false frequency-like impressions via a more subtle technique. A piecemeal advertisement refers to a format in which a firm claims superiority over one competitor on dimension A, a second competitor on dimension B, a third competitor on dimension C, and so on. The implication is that the firm's offering is frequently superior to its competitors when the truth may be that it performs next-to-last on each dimension, with the worst performer being the explicitly identified competitor. Research suggests that consumers may be sensitive to such deception only under rare circumstances that inspire extremely high levels of vigilance (Muthukrishnan *et al.* 2001).

Minority configurations

Decision research makes a clear distinction between judgement of individual alternatives and choice among alternatives. Most of the research described thus far has involved judgement. With the exception of deceptive piecemeal messages, prior research suggests that frequency information may bias judgement but offers little direct evidence that consumers will make truly suboptimal decisions. For example, in the experiment by Petty and Cacioppo, subjects were persuaded by many weak arguments, but there was no criterion for judgement accuracy. In contrast, consider the consequences of decisions that require choices to be made among competing alternatives. Russo and Dosher (1983) required subjects to make a large number of binary choices. The decision heuristic observed most often, particularly as the number of comparison dimensions increased, was a frequency heuristic (referred to as the 'majority of confirming dimensions' or MCD heuristic by Russo and Dosher). Based on predetermined utilities associated with each option, Russo and Dosher computed the proportion of correct choices made by each subject. Results showed that choice accuracy was largely determined by the configuration of dimensional superiority. When the majority of dimensions favoured the truly superior alternative, accuracy was quite high; when the majority of dimensions favoured the inferior alternative, accuracy was quite low. Among subjects identified as MCD users, accuracy on these latter cases was often well below chance.

Russo and Dosher did not conspire to mislead subjects. Their stimuli reflected a factorial combination of utilities and stimulus configurations. Thus, it was also the case that any true correlation in the real world was avoided. The problem for subjects was simply that they encountered a non-trivial number of 'minority configurations', in which the truly superior alternative outperformed the other alternative on only a minority of the dimensions. Consequently, overall decision accuracy was degraded to levels that would be considered unacceptable in most environments.

One might argue that the poor performance exhibited by these subjects is attributable to the complexity of the task. Subjects were asked to make numerous decisions, some of which involved many attribute dimensions. Related research also has shown how attribute frequency may drive choice when decision makers are constrained by motivation, time pressure, or ability to assess the implications of the attributes (Alba & Marmorstein 1987). However, other research has shown that frequency bias is not restricted to unfriendly environments. The true seductiveness of frequency cues is revealed by evidence of their influence on choice when only a single decision is required, the decision maker is motivated, unconstrained by time, and knowledgeable, and the trade-offs among attributes are easily made.

Consider again the task investigated by Russo and Dosher. Subjects were required to make a choice among pairs of applicants for a college scholarship based on the applicants' SAT scores, high school GPA, and family income. Such a task requires subjects to assess the importance of each dimension, the relative differences between the options on each dimension, and the appropriate trade-offs among attributes that are each described along qualitatively different dimensions. Depending on the exact values and configuration, even a single decision may require considerable thought. In contrast, consider the context examined by Alba et al. (1994). In the base condition, subjects were provided with prices of 60 brands at two competing grocery stores. The total basket price of the items at each store was the same. However, one store (the frequency store) was less expensive on 40 of the items. The other store was less expensive on the remaining 20 items, which meant that the magnitude of its price advantage on those items was twice as large as the magnitude of its disadvantage on the 40 items on which it was more expensive. Subjects were asked to view the prices at each store, which were aligned on the same page, and then estimate the total prices. Thus, unlike prior research, no 'trade-offs' were required. All 'attributes' were expressed on a common dollar metric, and subjects were merely asked to engage in an arithmetic task. Despite the lack of ambiguity, a robust frequency effect was observed. Across several replications, the store with the greater number of price advantages was routinely judged to have the lower aggregate price. Given the stimuli, there was no reason to expect such a result. Although the frequency cue was clearly observable, so too was the countervailing magnitude cue. Even cursory attention to the absolute differences among a randomly chosen subset of the items would have revealed that the low frequency store had deeper discounts on the items that favoured it. And, inasmuch as absolute dollar differences are salient to consumers, a plausible hypothesis was that the magnitude information would dominate.

This frequency effect is noteworthy in light of attempts to eliminate it. For example, one may argue that stimulus ambiguity existed in the form of information density. Even when ample time is provided, 60 pairs of prices might tax one's computational abilities (although, as noted, computation of a subset of the list would have revealed its true structure). However, a significant frequency bias was observed when (1) the time allotted to view the list exceeded the average time that subjects spent viewing it, (2) the number of items was reduced to nine, (3) subjects were sensitized to the price structure, and (4) the precise $+/-$ dollar difference was computed for each item and presented adjacent to the prices at the two stores. In addition, the motivation explanation of Petty and Cacioppo was investigated by providing subjects with incentives to search the information more thoroughly. In one experiment, subjects were shown the 60 items in six-item blocks. Each block had the same configuration as the overall list (i.e. a 2:1 frequency ratio and an equal overall price across the items). Subjects were allowed to search as many blocks as they wished before declaring one store to be less expensive than the other. They also were given a $3.00 stake at the start and were told that they would lose 30 cents for each block searched but that an incorrect answer at any time would lead to loss of the entire remaining stake. Results showed that subjects viewed an average of two of the ten blocks before concluding that the store with the frequency advantage was less expensive.

Perhaps most surprising in light of prior research is that the frequency-advantaged store was deemed less expensive even when pitted against prior beliefs about which store was less expensive. It is rare in psychological research to find a dominance of data over prior beliefs. Indeed, some of the most popular effects in social and cognitive psychology attest to the persistence of prior beliefs in the face of unambiguous and contrary data (Edwards & Smith 1996; Gilovich 1983). In the present research, prior beliefs were both measured (via the use of real brand names) and manipulated (using fictitious names along with a description of store size and service philosophy). Although prior beliefs did reduce the size of the bias to a slight degree, the frequency bias was never eliminated. Thus, the frequency store was deemed less expensive in the face of the combined effects of prior beliefs and the data-based magnitude advantage of the other store. The conflict between prior beliefs and the frequency cue did not result in deeper processing of the data (Maheswaran & Chaiken 1991), which would have revealed the magnitude differences between the stores, nor did prior beliefs guide confirmatory examination of the data or serve as a simplifying heuristic, as might be expected in difficult learning environments (Hoch & Deighton 1989). Instead, a data-based frequency cue dominated all other information.

Magnitude effects

Are there no boundaries on this price-frequency effect? In fact, the basic outcome reported by Alba *et al.* (1994) can be reversed by a very simple manipulation. Alba *et al.* (1999) employed the same basic paradigm used by Alba *et al.* (1994) but altered the context to correspond more closely to the previously described research on deal frequency. Thus, instead of presenting the prices of multiple products at two competing stores, they presented the prices of competing brands of a single item (or competing stores on a single brand) over multiple time periods. Within any time period, each brand was presented at either its regular price or normal discount price. One brand was discounted more frequently than the other but at a shallower level; the other brand was discounted less frequently but at a deeper level. The regular price of each brand was the same and, as in Alba *et al.* (1994), the average price of each brand across time periods was identical. In sharp contrast to Alba *et al.* (1994), Alba *et al.* (1999) found that the less frequent but more deeply discounted alternative was deemed less expensive. This effect was robust across depth/frequency ratios, data presentation format (sequential versus concurrent), dependent price measure (total versus average), and salience of the discount signal. The latter is especially important. A bias in favour of the depth brand could obtain from a simple misestimation of frequency. That is, if the magnitude of discount for each brand were accurately encoded, the depth brand would be favoured if subjects underestimated the frequency of the frequency brand. Although Alba *et al.* (1999) found that the depths of discount were accurately encoded (as they should have been given that the discount price of each brand never varied) and the frequencies of discount were indeed misestimated, a simple arithmetic explanation could not account for the depth bias. Across and within experiments, the depth effect was consistently observed despite large changes in the accuracy of the perceived number of discounts for each brand.

A final manipulation revealed the key moderating factor. When the sale prices of the brands were allowed to vary from period to period, the frequency bias returned. Note that a varying sale price corresponds to the stimulus structure used in the original studies (Alba *et al.* 1994). Such a structure possesses considerably more complexity because the price difference between competing brands is unique at every point of comparison. Hence, a mental accounting of the absolute differences becomes arduous and the frequency heuristic becomes correspondingly appealing. Although it is understandable why consumers might be less tempted to employ a frequency heuristic when the price structure is simple, it is less clear why the frequency bias would be replaced with a magnitude bias. One possibility is that subjects anchored on the more salient discount price of the depth brand and failed to adjust sufficiently. This explanation is interesting in light of Krishna and Johar's (1996) research described earlier. They found that an arithmetic combination of perceived deal frequency and discount provided a good approximation of subjects' estimates of the average deal price. Nonetheless, they were unable to rule out a heuristic process wherein the lower deal price is used as an anchor and upward adjustment is insufficient. Alba *et al.* (1999) ruled out an arithmetic explanation but could not rule in an anchoring explanation. The processes are not mutually exclusive, and either may be used depending on the processing constraints imposed by the decision environment.

Memory effects

Much research on frequency counting treats frequency as a memory phenomenon by measuring the extent to which people can recall the frequency with which an item was previously encountered. In the choice experiments discussed thus far, memory played a minor role. In the research of Dosher and Russo (1983), memory was not a factor, inasmuch as subjects made their choices in a stimulus-based setting in which all information was plainly available. In the pricing research of Alba *et al.* (1994, 1999), subjects essentially were presented with an arithmetic task in which comparison of the choice alternatives was conducted 'online' and observation of a frequency vs. magnitude effect was driven by the extent to which subjects attended to the magnitude information. Thus, an unanswered question concerns the impact frequency information has on a decision when the constraint is not the complexity of the information but rather the memorability of the frequency cue vis-à-vis other, perhaps more diagnostic, information.

In general, memory has been an important issue in the study of consumer decision making (e.g. Alba *et al.* 1991), but investigation of memory has been narrow in scope. Understandably, researchers concerned either with advertising effectiveness or memory-based decision making have limited their attention to recall or recognition of message information or product attributes. Most decision rules require comparison among options along substantive product dimensions. Thus, the consumer's ability to recall the dimensions and the values of those dimensions is critical. When memory for specific information is poor, decision makers may engage in 'judgement-referral', insofar as recalled global evaluations are more memorable than specific facts *and* can discriminate between competing options (see Alba *et al.* 1991). Frequency information is noteworthy in this regard. An intriguing aspect of frequency information from a traditional cognitive perspective is the ease with which it is encoded (Hasher & Zacks 1984). Less attention in any domain has been devoted to how well it is retained. Based on some models of persuasion (see Petty & Cacioppo 1986), the frequency cue may be expected to be forgotten quickly because it is not typically the recipient of elaboration. On the other hand, frequency is unique in the sense that it is easily learned and devoid of the semantic detail that is known to decay rapidly in memory-based decision contexts. If the frequency cue is memorable, it may affect decision making long after memory for attribute values has faded.

One study has examined the relative retrievability of frequency and attribute information in a consumer decision context, albeit at relatively short retention intervals. Alba and Marmorstein (1987) manipulated the rate of presentation of sequentially presented attributes of competing brands. Consistent with prior memory research, subsequent memory tests showed that study time strongly influenced attribute recall but had no effect on estimation of the number of attractive attributes of each brand.

This result, however, may be viewed more as a learning effect than a memory effect. Thus, a subsequent study by Alba *et al.* (1992) is informative. Subjects viewed a description of either a Strong or Weak brand of television. The Strong brand was described as performing well on three important television dimensions (e.g. picture quality). The Weak brand performed poorly on these dimensions but also possessed several relatively unimportant unique features (e.g. earphone jack). Directly thereafter or 48 hours later, subjects were presented with a Comparison brand and were asked to make a choice. The Comparison brand was of intermediate attractiveness, possessing four total features and performing moderately well on the three important dimensions. Results showed that when the retention interval was brief, subjects shown the Strong brand preferred it over the Comparison brand, whereas subjects shown the Weak (frequency) brand preferred the intermediate Comparison brand. These no-delay conditions serve as controls and indicate that subjects had learned the information sufficiently well to make appropriate decisions when memory for the attributes was high. The delay conditions showed an opposite pattern. Subjects preferred the Comparison brand over the Strong brand and the Weak brand over the Comparison brand. The former result is consistent with a known tendency to be risk-averse when memory for a previously encountered option is poor and an acceptable option is available (Biehal & Chakravarti 1983). Thus, lacking confidence in their memory for the precise attributes of the Strong brand, subjects chose the acceptable Comparison brand. This bias highlights the results from the remaining condition. Despite an inclination to select a satisfactory brand that is available, subjects opted for the Weak (frequency) brand encountered two days earlier. Post-decision rationales confirmed that the Weak brand was preferred over the Comparison brand because it was recalled as possessing a larger *number* of features.

A follow-up study showed that this outcome was not attributable to subject laziness, which might prompt reliance on a memorable frequency cue over the more arduous process of retrieving specific facts. An incentive to produce a correct response was created by making subjects accountable for their choices at the time of their decisions. When subjects believed that they would need to justify their choices, the frequency effect was exacerbated. Attribute numerosity was apparently viewed as a plausible rationale for product choice. Thus, contrary to some views of simple persuasion cues, the effect of the frequency cue can be powerful and enduring. The memorability and long-term persuasiveness of a cue is a function not only of the degree to which it is rehearsed and elaborated, but also of its inherent nature. Simple numerosity, which lacks specific semantic detail, appears to endure—at least when interference from other numerosity cues is low.

Conclusion

In some cases, consumers need to understand the frequency of occurrence of an event for its own sake. In these instances, absolute accuracy is important. The evidence thus far suggests that encoding of product-related frequency information is error-prone in ways that may lead to inappropriate judgements. In different situations, frequency information may supplement or compete against other information in the decision process. Research on decision making initially viewed the frequency heuristic in its classic sense of a decision shortcut, and attention therefore was directed at understanding the effort–accuracy trade-off involved. More recently, evidence has accumulated in favour of the notion that frequency is persuasive not only because it circumvents more extensive thought but because it is a seductive cue that is perceived by decision makers to offer a plausible basis for decision making (cf. Pelham *et al.* 1994). In this chapter, frequency biases were described in situations in which processing was constrained as well as in situations in which ample opportunity and motivation were provided to arrive at a correct judgement. Boundary conditions were also identified. Future research must determine whether the more prevalent reporting of a frequency bias reflects a frequency bias on the part of the investigator.

References

Alba, J. W., Broniarczyk, S. M., Shimp, T. A. & Urbany, J. E. (1994). The influence of prior beliefs, frequency cues, and magnitude cues on consumers' perceptions of comparative price data. *Journal of Consumer Research*, 21:219–235.

Alba, J. W., Chromiak, W., Hasher, L. & Attig, M. S. (1980). Automatic encoding of category size information. *Journal of Experimental Psychology: Human Learning and Memory*, 6:370–378.

Alba, J. W., Hutchinson, J. W. & Lynch, J. G., Jr. (1991). Memory and decision making. In T. S. Robertson and H. H. Kassarjian (eds) *Handbook of consumer behavior* (pp. 1–49). Englewood Cliffs, NJ: Prentice-Hall.

Alba, J. W. & Marmorstein, H. (1987). The effects of frequency knowledge on consumer decision making. *Journal of Consumer Research*, 14:14–25.

Alba, J. W., Marmorstein, H. & Chattopadhyay, A. (1992). Transitions in preference over time: The effects of memory on message persuasiveness. *Journal of Marketing Research*, 29:406–416.

Alba, J. W., Mela, C. F., Shimp, T. A. & Urbany, J. E. (1999). The effect of discount frequency and depth on consumer price judgements. *Journal of Consumer Research*, 26:99–114.

Bettman, J. R. & Park, C. W. (1980). Effects of prior knowledge and experience and phase of the choice process on consumer decision processes: A protocol analysis. *Journal of Consumer Research*, 7:234–248.

Biehal, G. & Chakravarti, D. (1983). Information accessibility as a moderator of consumer choice. *Journal of Consumer Research*, 10:1–14.

Bordley, R. F. (1985). Systems simulations comparing different decision rules. *Behavioral Science*, 30:230–239.

Buyukkurt, B. K. (1986). Integration of serially sampled price information: Modeling and some findings. *Journal of Consumer Research*, 13:357–373.

Chaiken, S. (1980). Heuristic versus systematic information processing and use of source versus image cues in persuasion. *Journal of Personality and Social Psychology*, 39:752–766.

Denes-Raj, V. & Epstein, S. (1994). Conflict between intuitive and rational processing: When people behave against their better judgement. *Journal of Personality and Social Psychology*, 66:819–829.

Edwards, K. & Smith, E. E. (1996). A disconfirmation bias in the evaluation of arguments. *Journal of Personality and Social Psychology*, 71:5–24.

Estes, W. K. (1976). The cognitive side of probability learning. *Psychological Review*, 83:37–64.

Gilovich, T. (1983). Biased evaluation and persistence in gambling. *Journal of Personality and Social Psychology*, 44:1110–1126.

Hasher, L. & Zacks, R. T. (1984). Automatic processing of fundamental information: The case of frequency of occurrence. *American Psychologist*, 39:1372–1389.

Hoch, S. J. & Deighton, J. (1989). Managing what consumers learn from product experience. *Journal of Marketing*, 53:1–20.

Hoch, S. J., Dreze, X. & Purk, M. E. (1994). EDLP, Hi-Lo, and Margin Arithmetic. *Journal of Marketing, 58*:16–27.

Josephs, R. A., Giesler, B. & Silvera, D. H. (1994). Judgements by quantity. *Journal of Experimental Psychology: General, 123*:21–32.

Krishna, A. (1991). Effect of dealing patterns on consumer perceptions of deal frequency and willingness to pay. *Journal of Marketing Research, 28*:441–451.

Krishna, A. (1994). The impact of dealing patterns on purchase behaviour. *Marketing Science, 13*:351–373.

Krishna, A., Currim, I. S. & Shoemaker, R. W. (1991). Consumer perceptions of promotional activity. *Journal of Marketing, 55*:4–16.

Krishna, A. & Johar, G. V. (1996). Consumer perceptions of deals: Biasing effects of varying deal prices. *Journal of Experimental Psychology: Applied, 2*:187–206.

Maheswaran, D. & Chaiken, S. (1991). Promoting systematic processing in low-motivation settings: Effect of incongruent information on processing and judgement. *Journal of Personality and Social Psychology, 61*:13–25.

Menon, G. (1993). The effects of accessibility of information in memory on judgements of behavioral frequencies. *Journal of Consumer Research, 20*:431–440.

Menon, G., Raghubir, P. & Schwarz, N. (1995). Behavioral frequency judgements: An accessibility-diagnosticity framework. *Journal of Consumer Research, 22*:212–228.

Muthukrishnan, A. V., Warlop, L. & Alba, J. W. (2001). The piecemeal approach to comparative advertising. *Marketing Letters, 12*:63–73.

Nelson, P. (1974). Advertising as information. *Journal of Political Economy, 82*:729–754.

Nunes, J. C. (2000). A cognitive model of people's usage estimates. *Journal of Marketing Research, 37*:397–409.

Payne, J. W., Bettman, J. R. & Johnson, E. J. (1988). Adaptive strategy selection in decision making. *Journal of Experimental Psychology: Learning, Memory, and Cognition, 14*:534–552.

Pelham, B. W., Sumarta, T. T. & Myaskovsky, L. (1994). The easy path from many to much: The numerosity heuristic. *Cognitive Psychology, 26*:103–133.

Petty, R. E. & Cacioppo, J. T. (1984). The effects of involvement on responses to argument quantity and quality: Central and peripheral routes to persuasion. *Journal of Personality and Social Psychology, 46*:69–81.

Petty, R. E. & Cacioppo, J. T. (1986). *Communication and persuasion: Central and peripheral routes to attitude change.* New York: Springer-Verlag.

Russo, J. E. & Dosher, B. A. (1983). Strategies for multiattribute binary choice. *Journal of Experimental Psychology: Learning, Memory, and Cognition, 9*:676–696.

Svenson, O. (1979). Process descriptions of decision making. *Organizational Behaviour and Human Performance, 23*:86–112.

CHAPTER 17

FREE WORD ASSOCIATIONS AND THE FREQUENCY OF CO-OCCURRENCE IN LANGUAGE USE

MANFRED WETTLER

Abstract

According to the contiguity theory of associative learning free word associations should be determined by the frequency with which words co-occur in language use. We have counted the co-occurrences between words in large machine readable texts, and have found that the strength of the associative relationship between two words which is observed in the free association experiment can be predicted on the basis of their co-occurrences. For German as well as for English the agreement between the predicted and the observed primary responses is only slightly weaker than the average agreement between the responses of a single participant and the primary response. It is shown how these results can be used in marketing research.

Introduction

A free associative response is the first word that occurs to a person after they have perceived another word, the so-called associative stimulus. Since the first systematic study of word associations by Galton (1880) this concept has played an important role both in theoretical and various branches of applied psychology. According to William James, associations are the 'mechanical conditions on which thought depends' (James 1890, vol. 1, p. 553). In psychoanalysis, free word associations are used as a tool to disclose the complexities of our memory (Jung & Riklin 1906), and in contemporary consumer research, associations are collected in order to predict the behaviour of potential customers (Kroeber-Riel 1988).

According to traditional learning theories, associations are learnt by the law of contiguity:

Objects once experienced together tend to become associated in the imagination, so that when any one of them is thought of, the others are likely to be thought of also, in the same order of sequence or coexistence as before. (James 1890, vol. 1, p. 561)

The generality of this law, which has been the cornerstone of most theories of learning, has been called into question by cognitive psychologists, who argue that many aspects of verbal behaviour—embedded constructions for example—cannot possibly be explained as the result of learning by contiguity (Chomsky 1959; Lashley 1951). According to Clark (1970) free word associations are the result of a series of symbolic processes. First, the stimulus word is encoded semantically. Then the resulting semantic representation will be transformed by changing one or several of its features, and finally the response word which corresponds to the new semantic representation will be generated. However, as long as we do not dispose of a system of rules which assign semantic descriptions to natural language words, semantically based theories about the generation of free word associations are not testable.

During the last 15 years the associationist approach has increased in popularity in the psychology of learning, psycholinguistics and computational linguistics. An increasing number of studies has shown that a wide variety of empirical observations about the processing of verbal material can be predicted on the basis of statistics about co-occurrences of words (see e.g. Christiansen & Chater 1999). There appears to be converging evidence that different domains of verbal behaviour whose occurrence has been explained by cognitive psychologists as a result of symbolic processes might be better explained by associative learning.

The next section will show which predictions about the relation between the strength of free word associations and the frequency of co-occurrence of the stimulus and the response word in language use can be derived from the stimulus sampling theory of Estes (1950). We will describe a method to test these predictions and show that they prove correct. Examples will be given to demonstrate how the possibility of predicting human word associations can be used in consumer research. Later sections will link our work to other approaches which relate co-occurrences between words to word associations as well as to other observations about verbal behaviour.

Word associations and co-occurrences

Associative learning of free word associations

According to the stimulus sampling theory of Estes (1950), the strength of an association from a situation j to an outcome i increases by a constant fraction of the maximal possible increment whenever i and j co-occur. If associative strengths range between 0 and 1 we have:

$$a_{i,j(t)} = a_{i,j(t-1)} + \theta(1 - a_{i,j(t-1)}), \quad \text{if}(i \& j). \tag{1}$$

where $a_{i,j(t-1)}$ denotes the associative strength before and $a_{i,j(t)}$ after the co-occurrence of i and j. θ, the learning rate, equals the fraction of the features of situation j which are sampled at time t.

If i does not occur in situation j, then the strength of the association from j to i diminishes by the same fraction of the existing association strength:

$$a_{i,j(t)} = a_{i,j(t-1)}(1 - \theta), \quad \text{if}(i \& \neg j) \tag{2}$$

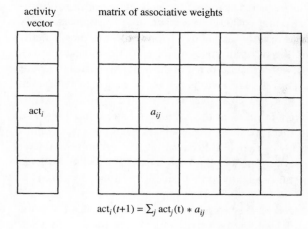

Figure 17.1 The co-occurrence net. Each row and each column of the quadratic matrix stands for a word of the vocabulary. The value in each cell gives the strength of the association from the row to the column word. The activity vector on the left describes the actual state of the net. The multiplication of the activity vector with the matrix of weights produces the associative response to the set of previously activated words.

With increasing t $a_{i,j}(t)$ will approach the conditional probability of i given j.

$$a_{i,j} = p(i|j) \qquad (3)$$

We predict that a stimulus word j is responded to by that word i for which $p(i|j)$ reaches a maximum.

The associative relations within a set of words can be described by the matrix shown in Fig. 17.1. The rows and the columns of this matrix represent the words of the vocabulary. The values in the cells are the conditional probabilities that the row-word occurs, given the column-word.

Each word has an activity value. These activity values are given in the so-called activity vector which is shown in Fig. 17.1 to the left of the association matrix. When a stimulus is presented in the free association task, the vector element of the stimulus word becomes activated by a constant amount. The production of an associative response can be described as the multiplication of this activity vector with the association matrix. In the free association experiment, where one element of the stimulus vector has the value one and all other elements the value zero, the activity values of the resulting vector are equal to the conditional probability of the response given the stimulus word.

In the general case which will be considered later in the section on sales promotion, the activity of a word i after multiplication of the input vector with the association matrix will be:

$$\mathrm{act}_{i(t+1)} = \sum_j \mathrm{act}_{j(t)} * a_{ij} \qquad (4)$$

Method

To test this prediction, free word associations produced by experimental participants were compared with counts of contiguities between words in language use. This comparison was made for word associations in German (Wettler et al. 1993) as well as in English (Wettler & Rapp 1991; Wettler et al. unpublished). The method and results regarding English associations, which will be given below, stem from this later study.

Word associations of German students were taken from the association norms by Russell (1970), which list the associative responses of 331 participants to 100 stimulus words. Different machine-readable collections of texts containing 21 million words, mostly newspaper texts, were used for counting the co-occurrences between 65,356 different words. This vocabulary comprises the 2,012 words that occurred in the study by Russell (1970) as stimulus or as response and, in addition, all 63,344 words that occurred at least 10 times in the texts used. Punctuation marks and special characters were counted as words.

The English word associations were taken from the *Edinburgh Association Thesaurus* (Kiss et al. 1973; Kiss 1975). From these we chose the associations to the 100 stimulus words that had been used in the study by Kent and Rosanoff (1910) and in several later studies. For the counting of the co-occurrences in English we used *The British National Corpus* (Burnard 1995), which consists of slightly more than 100 million word tokens of modern British English.

The frequencies of the co-occurrences have been counted with the so-called window technique. This technique involves a window of fixed length being moved from left to right step by step across the text. At each step the word at the left side of the window is covered and the next word at the right side uncovered. After each step the co-occurrences of the words within the window are counted. The sum of these co-occurrences gives the basis for estimating the strength of the associative relations between the co-occurring words.

Within this general method different factors may be varied. One such factor is the length of the window, which is usually defined by the number of words it contains. The longer the window the greater the distance between the words considered to co-occur and the higher the total number of co-occurrences.

A second factor that can influence the co-occurrence counts is the counting algorithm. Usually one of two algorithms used. In the first one all word pairs that can be formed from the words within the window are counted as co-occurrences. This means that with a window of length of n, $n(n-1)/2$ co-occurrences are counted at each step. Immediate adjacency between two words leads to $(n-1)$ co-occurrence counts. In the second method only the co-occurrences in relation to one word within the window are counted in each step, e.g. the word at the left end of the window. Thus in each step $(n-1)$ co-occurrences are counted, and between two words which co-occur within the distance of $(n-2)$ other words only one co-occurrence is counted, irrespective of their distance. Our study used the first of these two methods.

Further factors that influence the counting of co-occurrences using the window technique are the type and the length of the text, if inflected word forms are replaced by the root forms or not, and if the word order is to be taken into account. For German the length of the window was set to 12 words and for English to 20 words. Inflected word forms were always replaced by their root forms and word order was neglected.

Results

Preliminary analyses of participants' associative responses showed that they preferably produced words with medium word frequencies. This might be due to a tendency to produce responses of similar word frequencies to the stimulus words. In order to account for this response bias, the values computed by equation (3) have been corrected for word frequency. For this the values calculated according to equation (3) were transformed so that the word frequencies of the predicted responses correspond to the word frequencies produced by the participants in the free association experiment.

Which measure should be used to test the accuracy of our prediction? Other studies on the relation between co-occurrences and free word associations by Spence and Owens (1990) and by Lund *et al.* (1996), which will be described later, considered all stimulus response pairs produced by at least one participant in the Palermo and Jenkins study and computed the Spearman correlation between number of co-occurrences and the frequency with which the response word was given as an associative response to the stimulus. This procedure does not take into account possible response words not observed in the experiment. If we assume that all response words produced co-occur once with the corresponding stimulus word, and that all other words never co-occur with the stimulus this procedure would yield a zero correlation, because it considers only observed stimulus response pairs. Nevertheless the list of observed responses to each stimulus could be predicted perfectly. The correlation over all possible word pairs of the 100 stimulus words with all other words in the vocabulary would not be useful either: in our study of German word associations this would total more than 6.3 million word pairs. For more that 90% of these word pairs, the two words never co-occur in the texts used, i.e. they have zero co-occurrence.

To compare the predicted associative responses with the observed ones quantitatively we examined how many primary responses had been correctly predicted. The primary response to a stimulus word is the most frequent response word. In the study of German we found that for 20 of 100 stimulus words the predicted corresponded to the primary response in the study of Russell (1970). In comparison the test participants in the study by Russell (1970) produced on average 22.5 primary responses. The difference between the predicted and observed primary responses is thus only marginally larger than the difference between the responses of an average test person and the observed primary responses.

Similar good agreement levels between the predicted and observed associations were found when we counted, for each produced associative response, the number of other participants in the association experiment who gave the same response. On average the associative response of a participant to a stimulus word was given by 28.8 of the 330 other participants. The predicted associative responses from the simulation were given by an average of 22.9 test participants.

The results of the study of English associations are similar: for 29 of the 100 stimulus words our model produced the primary associative response. The average participant in the free association experiment produced 28.3 primary responses. Overall, 64 of the 100 predicted primary responses were produced by at least one participant. In comparison, an average participant in the study of Kiss (1975) produced 72 responses which were also given by at least one other participant. These results show that in German as well as in English free word associations can be predicted on the basis of co-occurrences of stimulus and response words. Thus the behaviour of participants in the free association task can be explained by associative learning of the contiguities between words. These observations do not disprove the operation of semantic structures during the free association task, but our model makes this assumption seem unnecessary.

The role of word associations in sales promotion

The possibility of computing the associations to any stimulus words can be used to predict the communicative effect of advertisements. When customers have the choice between different products of equal or similar quality and price they will decide on the basis of intuitive impressions, i.e. on the basis of associations that the products evoke. The aim of advertisements is the creation and the reinforcement of associations of the product to other positively evaluated concepts and the inhibition of existing associations to concepts which might impede the purchase of a product.

Figure 17.2 illustrates this process with some concepts which are relevant for the promotion of coffee. The rectangle at the bottom left shows the strengths of the associations between 'coffee' and six related concepts. The width of the arrows is proportional to the logarithms of the computed association strength. Some of these concepts have positive connotations (pleasure, flavour) and some have negative ones (nervous, work).

The starting point for the advertising of brand name products is their positioning. In the example shown in Fig. 17.2 some of the positive concepts associated with coffee are chosen, and others that are to be associated to the product are added. These concepts can be found in the central rectangle. In this fictive example the coffee brand is given a sociable and friendly image. When positioning a product, care should be taken that the concepts used are coherent. Coffee, for example, should not be described as a drink

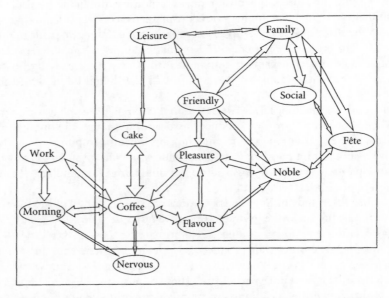

Figure 17.2 The development of a treatment for the promotion of coffee. The width of the arrows is proportional to the logarithm of the computed associative strength between the two connected concepts. The distances between the concept nodes have no significance. The rectangle at the lower left includes different concepts which are associated with coffee. The rectangle in the middle includes the concepts of the positioning. The upper rectangle shows the concepts which are used in the treatment.

that wakes the consumer at breakfast and at the same time creates a cosy atmosphere in the afternoon. Such incoherences can be found when analysing the associations between the concepts used in positioning.

On the basis of positioning the so-called treatments are developed. Depending on the media used for the advertisement they can be TV commercials, pictures or texts. The treatment aims to strengthen the associations made by the potential consumer between the product and the concepts used in the positioning. In the example shown this is realized in a scene that shows a middle class family celebrating together. This should have the effect of strengthening the association to 'friendly and social'.

Analyses of the associative structure of positioning and advertisements are used by various brand name product manufacturers to predict the success of future advertising campaigns. They make it possible to dispense with time-consuming and cost intensive surveys by market research institutes (Wettler & Rapp 1993; Wettler *et al.* 1998).

A procedure similar to that used for the calculation of associations to individual stimulus words is used to predict the communicative effect of possible advertisements. The text whose effect is to be calculated is described by a vector whose length represents the number of words in the vocabulary used (usually we use a vocabulary with about 4,000 word stems for marketing studies). Each value in the vector represents the frequency of the corresponding word in the text. This input vector is multiplied with the associations matrix. The resulting vector describes the predicted communicative effect of the text. The higher the value of a vector element the more strongly this word will be evoked by the input text. This allows the calculation of similarity between the communicative effect evoked by treatment and positioning.

We will now analyse two TV commercials for German coffee brands. They were broadcast at the same time, with different degrees of success. The first commercial is from a company that has since been bought by its main competitor. It contains nine cuts:

A sack of coffee is opened and the coffee beans are poured into a coffee roasting machine.
A baker dusts flour onto a loaf of bread.
A hand cuts grooves into a loaf of unbaked bread.
A coffee roasting machine is turned on.
A baker puts a loaf of bread into the oven using a wooden spatula.
The coffee roaster takes a handful of beans out of the roasting machine and smells it.
The baker takes the bread out of the oven.
The coffee roaster opens the drawer of a traditional household coffee grinder, takes a handful of ground coffee out and smells it.
The baker and the coffee roaster drink a cup of coffee together and make jokes.
During the commercial the following text can be heard from off:
Our partners and us, our freshness, freshly baked, fresh beans, freshly ground, freshly baked bread also tastes better, a feeling for coffee.

The combination of baker and coffee roaster in one depiction alludes to the fact that this brand of coffee has been sold at bakeries where the coffee was ground freshly in the presence of the buyer.

Figure 17.3 shows the structure of the associative connections between the concepts used in the commercial. In the figure two distinct overlapping clusters of strongly associated concepts can be seen. The upper cluster contains the concepts baker, flour, bread

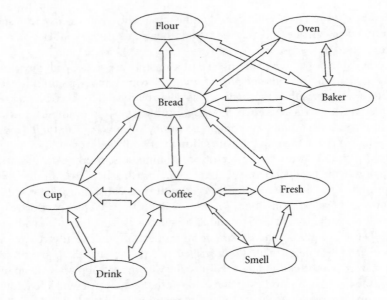

Figure 17.3 Associative relations between the concepts of the baker spot.

and baking, the lower contains the concepts coffee, drink, cup, and bread. The upper cluster has to do with the production of bread and the lower cluster with the consumption of coffee. There are few associative connections between the two clusters. Flour, which is dry, white, and rather odourless, is a weak concept to strengthen the hedonistic aspect of coffee consumption.

The second commercial was broadcast at about the same time and advertises Dallmayr coffee. It was a very successful advertising campaign. It also consists of 9 cuts:

A gentleman's house from the eighteenth century. Camera to entrance door.
Inside a shop with a fruit stand.
A chocolate counter; a hand removes a chocolate using tongs.
Picture of a female sales assistant in uniform.
The cake counter: a sales assistant passes a customer cakes.
The coffee counter: a friendly, smiling sales assistant fills a packet with coffee beans and passes it to a customer.
An old-fashioned coffee dispenser; coffee beans come out of it.
A cup of steaming hot coffee.
A packet of coffee falls over, spilling coffee beans onto the counter.
During the commercial the following text is heard from off:
Luckily, Dallmayr is not only in Munich. But that is where it comes from. From the house with the same name as the coffee. There is a pleasant aroma (german: Duft) when the coffee beans are freshly roasted. I like to drink the first cup immediately. Perfect top quality coffee.

Figure 17.4 shows the structure of the associative connections between the concepts used in the Dallmayr commercial. It is noticeable that a large number of concepts associated

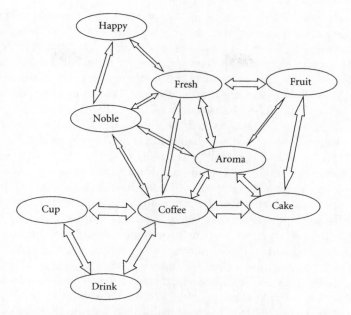

Figure 17.4 Associative relations between the concepts of the Dallmayr spot.

with coffee are also part of the commercial. Although the strength of the associative connection between coffee and the concepts used in the commercial is no greater than those in the first commercial, the second one has better conceptual coherence, which leads to a stronger association with taste—both literal and figurative—and other positively connotated feelings which are desirable in the promotion of coffee.

Co-occurrences, co-relations, and semantic organization

Free word associations can be predicted on the basis of the frequencies with which words co-occur in language use. How does this result relate to other studies which compare the distribution of words to word associations and other types of verbal behaviour?

These studies fall into two groups according to the type of distributional relation that they examine. Studies in the first group look at co-occurrences between words, as we have done.

Spence and Owens (1990) investigated the relationship between free word associations and the co-occurrences of words in the Brown corpus, a collection of texts which contains slightly more than one million words. The free word associations were taken from the word association norms of Palermo and Jenkins (1964) which list the associations of slightly less than 1000 participants to 200 stimulus words. From these 200 stimuli the authors choose the 47 most frequent concrete nouns whose primary response has a frequency of more than 100. Each of these 47 stimuli was paired with the concrete noun which was given most frequently as an associative response to this stimulus word in the Palermo and Jenkins study. Co-occurrence of a stimulus–response pair was defined as the common appearance of the two words within a window of 250 characters. They found a correlation of 0.42 between the number of co-occurrences and the frequency that the second word of a pair was produced as an associative response to the first word. Moreover, the stimuli and their primary responses had a significantly larger number of lexical co-occurrences than a control sample composed of the same stimuli matched with equally frequent but unrelated words. This difference remained significant with window sizes of up to 1000 characters.

Church and Hanks (1990) computed the contexts of all stimulus words of the Palermo and Jenkins (1964) word association norms in 60 million words of text, largely Associated Press news coverage. Their predictions of free word associations used an asymmetric measure by counting the number of times that a word x is followed by y within a window of five words. The predicted associative strength from a stimulus word x to a response word y was computed as the dual logarithm of the probability of y following x divided by the product of observing x and y independently. Church and Hanks (1990) give several examples where this method successfully predicts associations from the Palermo and Jenkins (1964) norms. However, it does not include a systematic comparison of the computed associative relations with the associations which have been observed in the Palermo and Jenkins study; therefore a quantitative evaluation of the results of Church and Hanks is not possible. The studies of Spence and Owens (1990) and Church and Hanks (1990) are in agreement with our results that free word associations are a function of the common occurrence of the stimulus and the response word in language use.

A second type of distributional relatedness, which is sometimes called semantic similarity, is given when words appear in similar contexts. An example is the word pair coffee and tea. Coffee co-occurs with the words hot and drink. The same is true of the word tea. Thus coffee and tea have common contexts; they are semantically similar. The differentiation between relatedness by co-occurrence and relatedness by common context also played an important role in pre-Chomskian linguistics. The first type of relation (co-occurrence) is called syntagmatic, and the second type, (common context), paradigmatic. Usually paradigmatically related words have the same part of speech and syntagmatically related words have different parts of speech. According to Ferdinand de Saussure (1916) and to the Prague school (see e.g. Jakobson 1990) syntagmatic and paradigmatic relatedness are the two basic principles on which all linguistic structures are founded.

Lund et al. (1996) have compared associative relatedness with contextual similarity. They partitioned the stimulus response pairs observed in the study of Palermo and Jenkins (1964) into two groups. The first group includes pairs of words which occur in similar contexts. These contextual similarities were computed on the basis of so-called context vectors. For this they used a corpus of 160 million words of English texts which were gathered from Usenet. For each of the 70,000 most frequent word types in this text a context vector with 140,000 values was computed. The values of the first 70,000 elements of the vector give the frequencies at which the word is preceded by each of the 70,000 words of the vocabulary, and the values of the second half of the vector give the frequencies at which this word is followed by each of the 70,000 words of the vocabulary. The contextual similarity between two words has been computed as a function of the dot product of their context vectors. For each stimulus word all other words of the vocabulary were ranked according to their semantic similarity. The fifty words with the smallest semantic distances were considered as 'semantic neighbours' of the stimulus word and all other words as 'non-neighbours'. The correlation between the frequency that a word was given as a response to a stimulus in the Palermo and Jenkins study and the number of co-occurrences between these two words was computed for the five most frequent associative responses to each stimulus word. For semantic neighbours the correlation between the number of co-occurrences and associative relatedness is 0.48, and for non-neighbours it is 0.05. Lund *et al.* (1996) concluded that the correlation between co-occurrences and associative relatedness only holds for semantically related pairs of words, i.e. paradigmatically related words.

This conclusion is supported by a study by Landauer and Dumais (1997), who computed the similarities between 60,000 different words on the basis of the number of articles in Grolier's *Academic American Encyclopaedia* in which each word pair appears. They could show that their model predicts the performance of human participants in a synonym test as well as the results of the experiments of Till, Mross and Kintsch (1988) on semantic priming. Another study by Lund *et al.* (1995) has found that similarity of context between prime and target predicts the amount of semantic priming.

These results seem to show that lexical organization and associative relatedness between words is determined by paradigmatic rather than by syntagmatic relatedness. This conclusion is contradictory to our own results, which have shown that free word associations can be predicted on the basis of co-occurrences.

In order to find the causes of the differences between the results of Lund *et al.* (1996) and our own results we analysed the predicted and observed primary responses in our study on German word associations in detail. Primary responses to the 100 stimulus words were grouped into paradigmatic responses, where the response word belongs to the same part of speech as the stimulus, and syntagmatic responses, where stimulus and response belong to different parts of speech. Eighty-five primary responses are paradigmatic and 15 are syntagmatic. Our model predicts 20 paradigmatic primary responses correctly but none of the syntagmatic primary responses; so our results seem to confirm the thesis of Lund *et al.* (1996) that word associations are determined by paradigmatic rather than by syntagmatic relatedness.

This is surprising since syntagmatic relations have been defined as co-occurrence relations, and our predictions are based on the number of co-occurrences between stimulus and response word. However, the fact that paradigmatic primary responses are better predicted than syntagmatic primary responses might be due to the fact that the paradigmatic primary responses were produced by a considerably higher number of participants than the syntagmatic primary responses. In the study on German word associations by Russell (1970) all but one of the 15 syntagmatic primary responses were produced by less than 100 of the 331 participants. On the other hand 25 of the 85 paradigmatic primary responses were produced by more than 100 participants. Out of these 25 frequently given paradigmatic primary responses 11 were predicted correctly by our model, but only 9 of the 60 paradigmatic primary responses which were produced by less than 100 participants.

Primary responses are better predicted the higher the proportion of participants who produce them. On the average paradigmatic primary responses are produced by a higher proportion of participants than syntagmatic primary responses. Therefore they can be better predicted.

Our results show that the assumption of McNeill (1966) and others that paradigmatically related words 'rarely appear together' is wrong. This observation casts a new light on the syntagmatic–paradigmatic dichotomy which has played an important role in the history of linguistics (Jakobson 1990) and in theories on mental development (Nelson 1977; Piaget 1970). When we look at the contiguities between words within a distance of 10 to 20 words, then semantically similar words also tend to co-occur. Thus the syntagmatic–paradigmatic dichotomy does not coincide with relatedness by co-occurrence versus relatedness by common context.

This is a consequence of different definitions of the term context. In distributional linguistics the differentiation between syntagmatic (co-occurrence) and paradigmatic (common context) relations is restricted to immediate succession: in modern terms, a two word window. Due to syntactical constraints, words that follow each other immediately usually have different parts of speech and words that occur in the same context are of the same part of speech. Thus syntagmatic relations cannot simultaneously be paradigmatic.

The models employed in cognitive psychology use larger window sizes in order to capture relations between words which cannot, because of grammatical restrictions, be uttered in immediate succession. Spence and Owens used a search window of 250 characters (about 50 words), Landauer and Dumais (1997) counted the co-occurrences within articles with a maximal length of 151 words and Lund et al. (1996) used a window of 10 words. In our studies we used window sizes between 10 and 20 words.

If the window size is larger than two, co-occurring words necessarily must have contextual overlap: when we have a chain—A B C D—and a window size of four or more then A and C co-occur and have both B and D as common contexts. However, this dependency is not reciprocal: there might be pairs of words which have similar contexts but do not co-occur. Candidates for such pairs are synonyms which are not used together usually. 'pesticides' and 'plant-protection agents' might be an example or 'revolutionary' and 'terrorist'.

There are two ways to account for the effects of semantic relatedness (common context) on verbal behaviour. The first approach was adopted by Landauer and Dumais (1997) and by Burgess (1998), who postulate that the representation in the lexicon reflects similarity relations. This means that the computation of similarities occurs before the representation is built up. However, it also may be that it is the co-occurrences that are represented in memory, and that the computation of the similarities is done later when a task makes it necessary.

In addition to the direct associations (the computation of which was described in the section on methods) a co-occurrence matrix can be used to compute indirect or second order associations. For this purpose the activity vector resulting from the first application of equation (4) is multiplied with the co-occurrence matrix. The result of this second multiplication is the vector of indirect associations to the stimulus words.

$$\text{act}_{k(t+2)} = \Sigma_j \text{act}_{j(t+1)} * a_{kj} = \Sigma_j (\Sigma_i \text{act}_{i(t)} * a_{ij}) * a_{kj} \tag{5}$$

In the case of the association experiment where one stimulus word i has an activity value of one and all other words have zero activities, the strength of the indirect association from i to k is:

$$\text{act}_{k(t+2)} = \Sigma_j a_{ij} * a_{kj} \tag{6}$$

This is the dot-product of the context vectors of i and k, i.e. the same measure that Landauer and Dumais (1997) and Lund *et al.* (1996) used for the computation of the semantic similarities (see Jones & Furnas 1987).

The assumption that the mental lexicon represents co-occurrences and not contextual similarities does not imply that we do not compute co-relations. For many tasks, categorization, for instance, the similarities of the contexts of different words have to be calculated and compared. The question is when these computations are done. We assume that this happens at the time when a task, a similarity judgement for instance, has to be performed. Landauer and Dumais (1997) and Burgess (1998) assume that the computed similarities are already part of the lexicon. The computations which have to be done are the same in both theories. However, similarity models of the lexicon are unable to explain the reproduction of relations which have been lost when the paradigmatic representation was constructed. Syntagmatic associations are an example of this.

References

Burgess, C. (1998). From simple associations to the building blocks of language: Modeling meaning in memory with the HAL model. *Behaviour Research Methods, Instruments, and Computers*, 30:188–198.

Burnard, L. (1995). *Users reference guide for the British National Corpus, Version 1.0*. Oxford: Oxford University Computing Services.

Chomsky, N. (1959). Verbal Behaviour (A review of Skinners book). *Language*, 35:26–58.

Christiansen, M. H. & Chater, N. (1999). Connectionist natural language processing: The state of the art. *Cognitive Science*, 23:417–437.

Church, K. W. & Hanks, P. (1990). Word association norms, mutual information, and lexicography. *Computational Linguistics*, 16:22–29.

Clark, H. H. (1970). Word associations and linguistic theory. In J. Lyons (ed.) *New horizons in linguistics* (pp. 271–286). Harmondsworth: Penguin.

Estes, W. K. (1950). Toward a statistical theory of learning. *Psychological Review*, 57:94–107.

Galton, F. (1880). Psychometric experiments. *Brain*, 2:149–162.

Jakobson, R. (1990). *On language*. Cambridge MA: Harvard University Press.

James, W. (1890). *The principles of psychology*. New York: Dover.

Jones, W. P. & Furnas, G. W. (1987). Pictures of relevance: a geometric analysis of similarity measures. *Journal of the American Society for Information Science*, 38:420–442.

Jung, C. G. & Riklin, F. (1906). Experimentelle Untersuchungen über die Assoziationen Gesunder. In C. G. Jung (ed.) *Diagnostische Assoziationsstudien* (pp. 7–145). Leipzig: J. A. Barth.

Kent, G. H. & Rosanoff, A. J. (1910). A study of association in insanity. *American Journal of Insanity*, 67:37–96:317–390.

Kiss, G. R. (1975). An associative thesaurus of English: Structural analysis of a large relevance network. In A. Kennedy & A. Wilkes (eds) *Studies in long term memory* (pp. 103–121). London: Wiley.

Kiss, G. R., Armstrong, C., Milroy, R. & Piper, J. (1973). An associative thesaurus of English and ist computer analysis. In A. J. Aitken, R. W. Bailey & N. Hamilton-Smith (eds) *The computer and literary studies* (pp. 153–166). Edinburgh: Edinburgh University Press.

Kroeber-Riel, W. (1988). *Strategie und Technik der Werbung*. Kohlhammer, Stuttgart.

Landauer, T. K. & Dumais, S. T. (1997). A solution to Plato's problem: the latent sematic analysis theory of acquisition, induction, and representation of knowledge. *Psychological Review, 104*:211–240.

Lashley, K. (1951). The problem of serial order in behaviour. In L. A. Jeffreys (ed.) *Cerebral mechanisms in behaviour: The Hixton symposium*. New York: Hafner Publishing.

Lund, K., Burgess, C. & Atchley, R. A. (1995). Semantic and associative priming in high-dimensional semantic space (pp. 660–665). *Proceedings of the Cognitive Science Society*. Hillsdale, NJ: Erlbaum.

Lund, K., Burgess, C. & Audet, C. (1996). Dissociating semantic and associative word relationships using high-dimensional semantic space (pp. 603–608). *Proceedings of the Cognitive Science Society*. Hillsdale, NJ: Erlbaum.

McNeill, D. (1966). A study of word association. *Journal of Verbal Learning and Verbal Behaviour, 5*:548–557.

Nelson, K. (1977). The syntagmatic-paradigmatic shift revisited: A review of research and theory. *Psychol. Bulletin 84:93–116*.

Palermo, D. & Jenkins, J. (1964). *Word association norms*. Minneapolis, MN: University of Minnesota Press.

Piaget, J. (1970). Piaget's theory. In P. H. Mussen (ed.) *Carmichael's manual of child psychology*, vol. 1, (pp. 708–732). New York: Wiley.

Russell, W. A. (1970). The complete German language norms for responses to 100 words from the Kent-Rosanoff association test. In L. Postman & G. Keppel (eds) *Norms of Word Association* (pp. 53–94). New York: Academic Press.

Saussure, F. de (1916). *Cours de Linguistique Générale*. Lausanne: Payot.

Spence, D. P. & Owens, K. C. (1990). Lexical co-occurrence and association strength. *Journal of Psycholinguistic Research, 19*:317–330.

Till, R. E., Mross, E. F. & Kintsch, W. (1988). Time course of priming for associate and inference words in a discourse context. *Memory and Cognition, 16*:283–298.

Wettler, M. & Rapp, R. (1991). Prediction of free word associations based on Hebbian learning. *Proceedings of the International Joint Conference on Neural Networks Singapore*, vol. 1:25–29.

Wettler, M. & Rapp, R. (1993). Associative text analysis of advertisements. *Marketing and Research Today, 21*:241–246.

Wettler, M. Rapp, R. & Ferber, R. (1993). Freie Assoziationen und Kontiguitäten von Wörtern in Texten. *Zeitschrift für Psychologie, 201*:99–108.

Wettler, M. Rapp, R. & Sedlmeier, P. Free word associations correspond to contiguities between words in texts (unpublished).

Wettler, M., Weber, A., Böhnisch, M., Rapp, R. & Marten, U. (1998). Analyse associative de texte: une méthode pour anticiper l'effet communicatif de la publicité. *Recherche et Applications en Marketing, 13*:69–81.

CHAPTER 18

TECHNOLOGY NEEDS PSYCHOLOGY: HOW NATURAL FREQUENCIES FOSTER INSIGHT IN MEDICAL AND LEGAL EXPERTS

RALPH HERTWIG AND ULRICH HOFFRAGE

Abstract

Modern diagnostic technologies such as cancer screening tests or forensic DNA tests confront users with a basic problem, namely that of inferring the accuracy of a diagnostic test on the basis of statistical information about the test. The required statistics can be presented in a wide variety of ways. While these modes of presentation of statistical information are mathematically equivalent, psychologically they are not. In this chapter, we show that supplementing diagnostic technology with psychological knowledge about how people process frequency information provides us with a very simple—but powerful—method for improving diagnostic inferences.

New technologies often exceed the limits of our imagination. As a result, their advent can evoke unanticipated public responses, ranging from repulsion and hostility to fascination and even mystical attributions. Let us use the debut of the X-ray machine as an illustration. According to an historian of medicine, Joel Howell (1995, pp. 135–137), at first many people were merely fascinated by it. They lined up for one hour sittings to view their own bones. Coin-operated machines let people glimpse the insides of their hands and feet. Wealthy young women had X-ray pictures taken of themselves holding hands with their betrothed. Not everybody, however, responded so enthusiastically, and as we know in hindsight, so riskily. In particular, the power of the X-ray machine to invade privacy met with alarm. Coming close upon the heels of other technological inventions of the period, including the telephone and the photograph, the X-ray was sometimes viewed as evil in its ability to, in the words of one critic cited by Howell (1995, p. 140) 'render privacy a mere tradition of an unscientific past'.

What this mixture of fascination about and hostility toward technological innovation suggests is that new technologies often outstrip our ability to use them properly, let alone to comprehend them completely. If using the technology requires no specific expertise, however, then such lack of comprehension need not be of concern. In fact, many technological tools are designed to be easy to use. Think of the microwave oven or the television: few of us know how either actually works, but that does not mean we cannot use them properly—basically, all we need to know is how to switch them on and off. Sometimes, however, the proper use of a new technological tool or of its end product cannot be reduced to pushing a button but instead requires specific knowledge that is not necessarily part of the human mind's intuitive repertoire. Our argument is that when this is the case, technology needs psychology because the best technology is of little value if people do not comprehend what the results it produces mean.

Some technological tools yield results that need to be interpreted in the context of statistical information. We will show that understanding how laypeople and experts tend to process frequency and probability information can greatly improve people's comprehension of these results. The tools we are concerned with are (1) medical diagnostic tests (e.g. hemoccult test), which play an ever more important role in the diagnosis of specific diseases (e.g. colorectal cancer) and (2) forensic DNA analysis, which in the 1990s revolutionized the criminal investigation process. Note that in the present context the user of the tool is not the lab technician who conducts the actual analysis but, for example, the doctor, patient, judge, or juror. They receive information gleaned from the technology—for instance, a mammography test or a DNA match—and they have to determine what these results could possibly mean.

Medicine: how to improve the use of diagnostic tests

What does a positive medical test result—for instance, a positive mammography test—mean? Multiple studies suggest that physicians often do not properly infer the probability of a disease given a positive test result (Casscells *et al.* 1978). In a seminal article on statistical inferences based on results of mammography tests, David Eddy (1982) reported an informal study in which he provided physicians with information that can be summarized as follows (numbers are rounded): for a woman at age 40 who participates in routine screening, the probability of breast cancer is 0.01. If a woman has breast cancer, the probability is 0.8 that she will have a positive mammogram. If a woman does not have breast cancer, the probability is 0.1 that she will still have a positive mammogram. Now, imagine a randomly drawn woman from this age group with a positive mammogram. What is the probability that she actually has breast cancer?

This probability, also called the *positive predictive value* (PPV) of a test, can be calculated from Bayes' rule. This rule is named after Thomas Bayes (1702–1761), an English dissident minister, to whom the solution of the problem of how to make an inference from data to hypothesis is attributed (Stigler 1983). Equation 1 represents Bayes' rule, applied to the medical context:

$$\text{PPV} = \frac{p(\text{disease})\, p(\text{pos}|\text{disease})}{p(\text{disease})\, p(\text{pos}|\text{disease}) + p(\neg \text{disease})\, p(\text{pos}|\neg \text{disease})} \qquad (1)$$

In this equation, $p(\text{disease})$ is the *base rate* (or prevalence) of the disease (0.01 in Eddy's example); $p(\text{pos}|\text{disease})$ is the *hit rate* (or sensitivity) of the test, that is, the proportion of positive results among people suffering from the disease (0.8); and $p(\text{pos}|\neg\text{disease})$ is the *false-positive rate* of the test, that is, the proportion of positive results among people not suffering from the disease (0.1). Inserting the statistical information into Bayes' rule results in a positive predictive value for a mammography test of 0.075.

$$\text{PPV} = \frac{(0.01)(0.8)}{(0.01)(0.8) + (0.99)(0.1)} = 0.075$$

Yet most of the physicians in Eddy's study (95 out of 100) estimated the positive predictive value of the test to be between 0.7 and 0.8. That is, their estimates of the probability of breast cancer given a positive mammography test exceeded the correct value by a factor of 10. Eddy (1982) argued that the physicians drew the wrong inference based on mammography because they had confused the hit rate of the test with its positive predictive value. In his view, 'these errors threaten the quality of medical care' (p. 249). Given this and other demonstrations that physicians make errors when interpreting the outcomes of medical diagnostic tests (see also Windeler & Köbberling 1986), what can be done to improve their inferences? As we argue next, supplementing diagnostic technology with psychological knowledge about how people process probability and frequency information provides us with a very simple—but powerful—method for improving diagnostic inferences.

Natural frequencies help in making diagnostic inferences

Studies concluding that physicians (Berwick *et al.* 1981; Politser 1984) and laypeople (see Koehler 1996) make poor diagnostic inferences based on statistical information have typically presented information in the form of probabilities and percentages. From a mathematical viewpoint, it is irrelevant whether statistical information is presented in probabilities, percentages, absolute frequencies, or some other form because these different representations can be mapped onto one another in a one-to-one fashion. From a psychological viewpoint, however, the representation of information matters. Although different representations of statistical information are equivalent mathematically, psychologically they are not. This observation is central to our argument. In particular, we argue that a specific class of representations that we call *natural frequencies* (Hoffrage & Gigerenzer 1998) helps experts to make inferences the Bayesian way.

We now illustrate the difference between probabilities and natural frequencies using the diagnostic problem of inferring the presence of colorectal cancer (C) from a positive result of the hemoccult test (T), a standard diagnostic test. Let us assume that, in terms of *probabilities*, the base rate for colorectal cancer $p(\text{cancer})$ is 0.003; the test's hit rate, $p(\text{pos}|\text{cancer})$, is 0.5; and the false-positive rate, $p(\text{pos}|\neg\text{cancer})$, is 0.03. Armed with this information, people are then typically asked, 'What is the probability that a randomly drawn person who tested positive actually has colorectal cancer?'

In *natural frequencies*, in contrast, the same information would read 'Thirty out of every 10,000 people have colorectal cancer. Of these 30 people with colorectal cancer, 15 will have a positive hemoccult test. Of the remaining 9,970 people *without* colorectal cancer, 300 will still have a positive hemoccult test.' The question is, 'How many of the people who tested positive actually have colorectal cancer?'

What exactly are natural frequencies? They are absolute frequencies of events as directly experienced and they have *not been normalized with respect to the base rates* of the disease and its absence (Gigerenzer & Hoffrage 1995, 1999). For example, imagine an old, experienced physician in an illiterate society. She has no books or statistical surveys and therefore must rely solely on her direct experience. Her people, for instance, may have been afflicted by a previously unknown and severe disease. Fortunately, the physician has discovered a symptom that signals the disease, although not with certainty. In her lifetime, she has seen a large group of patients, few of whom had the disease. Of those who had the disease, some showed the symptoms; of those who were not afflicted, some also showed the symptoms. Thus, on the basis of her experience, the physician acquired representative information about the structure of her environment by sequentially encountering (randomly drawn) instances in the population. This is what we call natural sampling. The outcome of natural sampling is natural frequencies. Natural frequencies are not to be confused with probabilities, percentages, relative frequencies, or other representations where the underlying natural frequencies have been normalized with respect to these base rates.

Why and how should natural frequencies facilitate diagnostic inferences? There are two related explanations (for alternative views see Fiedler *et al.* 2000; Macchi 2000; for a discussion of those alternative views see Hoffrage *et al.* in press). The first is computational. Bayesian computations are simpler when the information is represented in natural frequencies than in probabilities, percentages, or relative frequencies (Christensen-Szalanski & Bushyhead 1981; Kleiter 1994). Consider the calculations when the information concerning colorectal cancer is represented in probabilities. A cognitive algorithm to compute the PPV of the hemoccult test based on probabilities amounts to Equation 1:

$$PPV = \frac{p(\text{cancer}) \, p(\text{pos}|\text{cancer})}{p(\text{cancer}) \, p(\text{pos}|\text{cancer}) + p(\neg\text{cancer}) \, p(\text{pos}|\neg\text{cancer})}$$

$$= \frac{(0.003)(0.5)}{(0.003)(0.5) + (0.997)(0.03)} = 0.048$$

Now compare these computations with those necessary when the same information is presented in natural frequencies. Because natural frequencies do not require figuring in base rates, the computations are much simpler—that is, fewer operations (multiplication, addition, or division) need to be performed, and the operations can be performed on natural numbers rather than fractions. Now the algorithm amounts to Equation 2:

$$PPV = \frac{\text{pos\&cancer}}{\text{pos\&cancer} + \text{pos\&}\neg\text{cancer}} = \frac{15}{15 + 300} = 0.048 \quad (2)$$

Equation 2 is Bayes' rule for natural frequencies, where pos&cancer is the number of cases with cancer and a positive test, and pos&¬cancer is the number of cases without cancer but with a positive test.

The second argument as to why natural frequencies facilitate diagnostic inferences complements the first. It suggests that people's cognitive algorithms are designed to make inferences from natural frequencies rather than from probabilities and percentages. This argument is based on the observation that for most of their existence, humans and animals have had to make inferences based on information encoded sequentially through their direct experience. Natural frequencies are the result of this process. Mathematical probability, in contrast, emerged only in the mid-seventeenth century (Daston 1988), and percentages seem to have become common representations only in the aftermath of the French revolution—when the metric system was adopted—mainly for calculating taxes and interest, and only very recently for expressing risk and uncertainty (Gigerenzer et al. 1989). Thus one might argue that minds have evolved to deal with natural frequencies rather than with probabilities. This argument is consistent with developmental studies indicating the primacy of reasoning about discrete numbers and counts over fractions, and with studies of adult humans and animals showing that they can monitor frequency information in their natural environments in fairly accurate and automatic ways (e.g. Gallistel & Gelman 1992; Jonides & Jones 1992; Real 1991; Sedlmeier et al. 1998).

Do natural frequencies improve physicians' statistical reasoning?

Can medical experts' use of diagnostic technologies be improved if they reason in terms of natural frequencies rather than probabilities? The following study with experienced physicians who treat real patients provides an answer. Hoffrage and Gigerenzer (1998) asked 51 physicians from Munich and Düsseldorf to participate. Three physicians did not give numerical responses to the diagnostic tasks they were asked to complete, either because they considered statistical information to be meaningless for medical diagnosis or because they said that they were unable to think with numbers. The remaining 48 physicians had practised medicine for an average of 14 years, had a mean age of 42 years, and worked in university hospitals, in private or public hospitals, or in private practice. The sample included internists, gynaecologists, dermatologists, and radiologists, among other specialists. The physicians' status ranged from directors of clinics to beginning physicians.

Each physician was given four diagnostic tasks. In two of them, the information was presented in probabilities, whereas in the other two it was presented in natural frequencies (task order was systematically varied). The four diagnostic tasks were to infer the presence of (1) colorectal cancer from a positive hemoccult test (see above), (2) breast cancer from a positive mammogram, (3) phenylketonuria from a positive Guthrie test, and (4) ankylosing spondylitis (Bekhterev's disease) from a positive HL-Antigen-B27 (HLA-B27) test (for the texts, see http://www-abc.mpib-berlin.mpg.de/users/hoffrage/papers/4tasks.html).

Each physician gave an estimate for each of the four diagnostic problems. Thus, the study yielded a total of 48 estimates for each problem and 24 estimates for each format of each problem. When a physician's estimate was within five percentage points (or the equivalent in frequencies) of the Bayesian answer, and additional information (from physicians' notes and/or an interview) indicated that the estimate was arrived at by Bayesian reasoning (or a shortcut thereof; see Gigerenzer & Hoffrage 1995) rather than by guessing or other means, then the response was classified as a Bayesian inference.

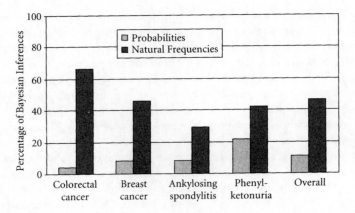

Figure 18.1 Physicians' percentage of Bayesian inferences in the probability and natural frequency versions of four diagnostic tasks (Hoffrage & Gigerenzer 1998).

Figure 18.1 shows that for each diagnostic problem, the physicians reasoned the Bayesian way more often when the information was communicated in natural frequencies than in probabilities. The size of this effect varied between problems, but even in the problem showing the smallest effect (phenylketonuria), the proportion of Bayesian answers was twice as large in the natural frequency than in the probability version. For the two cancer problems, natural frequencies increased Bayesian inferences by more than a factor of five. Across all problems, the physicians gave the Bayesian answer when provided with probabilities in only 10% of the cases; when they were provided with natural frequencies, this value increased to 46%.

As these results demonstrate, natural frequencies are a powerful tool for improving doctors' statistical inferences. Using natural frequencies also paid off in terms of time efficiency and an increased sense of self-efficacy. Given probabilities, the physicians in the study spent an average of 25% more time solving the diagnostic problems than they did when given natural frequencies. Comments made by the physicians revealed that they were more often nervous, tense, and uncertain about solving the tasks when they were working with probabilities than with natural frequencies. In addition, their spontaneous remarks revealed that they were less sceptical about the relevance of statistical information when it was expressed in natural frequencies. The physicians were aware of their better and faster performance with natural frequencies, as illustrated by comments such as the following: 'Now it's different. It's quite easy to imagine. There is a frequency—that's more visual' and 'a first grader could do this!'

Different expressions of statistical information affect how people process and understand information. Are the benefits of representing statistical information in terms of natural frequencies specific to medical inferences or do they generalize beyond the medical domain? We believe they do. A particularly important professional context—in which statistical inferences can literally determine whether a person lives or dies—is that of forensic DNA analysis. Since the early 1990s, this technology has dramatically altered the criminal investigation process in cases where biological evidence (e.g. semen, blood, saliva, hair) is found at the crime scene. Forensic evidence unearthed by this technology is currently challenging the practice of capital punishment in the United States. We next explore the nature of this technology-driven challenge and then ask how natural frequencies can improve statistical inferences drawn from DNA test results.

Forensic DNA analysis and the death penalty

In recent years, the US criminal justice system has applied its most severe penalty—the taking of the convict's life—ever more frequently. In 1999, ninety-eight prisoners were executed for capital crimes in the United States, more than in any other years since 1951 (*The Economist*, 10 June 2000). The increased rate of executions in the United States is particularly striking at a time when the US capital crime rate has been dropping and when, according to Amnesty International, over half of the countries in the world have abolished the death penalty in law or practice. Among the big democracies, only the United States, India, and Japan still put prisoners to death.

The increase in executions also contrasts with an important shift in public opinion documented in a Gallup survey conducted in February 2000. Though a majority of Americans still support the death penalty, the public support has been gradually decreasing since its high point in 1994 and at 66% has reached its lowest level since 1981.[1] Coinciding with the waning support for the death penalty is the growing belief—expressed by 91% of all respondents (i.e. even many of those who favour the death penalty)—that innocent people are at least occasionally wrongly sentenced to death. What fuels this belief in the fallibility of the criminal justice system, according to the Gallup researchers, is the advent of modern DNA technology (which was not used in US courts until 1989; see Wells *et al.* 2000)—specifically the fact that DNA testing has produced new evidence suggesting that innocent people were sentenced to death in American courts.

The public debate concerning the sentencing of innocent people to death has to a large extent been fueled by the *Innocence Project*. Founded in 1992 by two prominent New York criminal defence lawyers, Barry Scheck and Peter Neufeld, this organization at the Yeshiva University's Benjamin N. Cardozo School of Law provides *pro bono* legal assistance to people who challenge their convictions by using DNA testing. In the almost 10 years of its existence, the Innocence Project has represented or assisted in 36 of the 63 cases where convictions have been reversed or overturned in the United States over this period. In a recent book, Scheck, Neufeld, and *The New York Daily News* columnist Jim Dwyer (2000) recount the harrowing stories of men who were wrongfully convicted of crimes and, after many years and much resistance from prosecutors and judges, exonerated by DNA test results. In most of these cases, the charge was rape and the penalty a prison term, though some who were wrongfully convicted of murder were sentenced to death.

The Innocence Project and similar initiatives throughout the United States have stirred up a political debate about the death penalty (see Lifton & Mitchell 2000). In response to the mounting evidence of serious flaws in the capital punishment system, the governor of Illinois, for instance, declared a moratorium on executions in that state. And, for the first time in American legal history, a judge (at the Houston County Superior Court)

---

[1] For the detailed results of this survey, see http://www.gallup.com/poll/releases/pr000224.asp.

recently authorized DNA analysis of evidence from a death penalty case in which the convicted man has already been put to death. Prior to this order, advocates of the death penalty could argue that there exists no incontrovertible proof that an innocent person was ever wrongfully executed, as did Edwin Meese, Attorney General in the Reagan Administration: 'If a person is innocent of crime, then he is not a suspect' (quoted in Scheck et al. 2000, p. xi). The judge's order to analyse the DNA of the executed man will soon determine whether or not this argument is valid and can still be made in the future.

DNA tests—a gold standard for truth?

Wrongful convictions have been exposed in the past, but cases in which DNA analysis led to exoneration after conviction are particularly impressive, 'perhaps, because they all used a single, definitive technology to establish innocence' (Wells et al. 2000, p. 589). In Scheck et al.'s (2000) view:

DNA testing is to justice what the telescope is for the stars: not a lesson in biochemistry, not a display of the wonders of magnifying optical glass, but a way to see things as they really are. It is a revelation machine. (p. xv)

Given that Scheck and Neufeld have witnessed the exonerating power of DNA evidence first hand, their enthusiastic appraisal of DNA technology is more than understandable.[2] But it is nevertheless important to realize that, like medical testing, DNA testing is based on statistical information and does not necessarily produce incontrovertible proof. For this reason, insight into the aspects of DNA testing that call for statistical inference is crucial—not least because the belief that it is 'a gold standard for truth telling' (Scheck et al. 2000, p. 122) could itself become the source of wrongful convictions. In each of Scheck et al.'s exoneration cases, the lack of a match between a convict's DNA profile and traces found at the crime scene excluded him as the source of the trace. But what if a trace had matched? Would that have provided ironclad proof that the convict was the source? As with the outcome of a medical diagnostic test, interpreting the outcome of a DNA analysis requires statistical reasoning, and the statistical reasoning of judges, jurors, and possibly even DNA experts may be facilitated by natural frequencies.

How natural frequencies can help in interpreting DNA evidence

Although it is unlikely that a criminal suspect would coincidentally share a DNA profile with a piece of incriminating evidence, just how unlikely that coincidence is

---

[2] Scheck and Neufeld's unconditional support for DNA evidence may come as a surprise given the role they played in the O.J. Simpson defence team. There, they succeeded in thoroughly discrediting the DNA evidence presented by the prosecution. How do their past criticism of DNA testing and their present enthusiasm for it go together? In fact, Scheck and Neufeld's position on DNA technology actually has not changed, as a closer look at the Simpson trial transcripts (available on the Internet) shows. In his closing arguments, Scheck did not question the DNA technology per se—'DNA is a sophisticated technology. It is a wonderful technology'—but the evidence analysed in that case. Specifically, he argued that the DNA evidence against Simpson was mishandled and even fabricated.

depends on the frequency of a specific combination of genetic features in a specified reference class. This reference class may be a racial group, or an artificial probability space created by multiplying together the frequencies with which the individual genetic features of the profile appear in a population. The statistic usually reported at trial is the frequency with which the specific combination of genotypic features occurs in a specified reference class (e.g. an expert witness in the O.J. Simpson case stated that Simpon's DNA profile, which matched that of droplets of blood leading from the bodies, occurs in only 1 in 170 million people).[3]

This frequency may be interpreted as the chance that someone selected at random would have the profile in question. Unfortunately, this statistic seems to be widely misinterpreted by judges, jurors, and even DNA experts themselves. The low frequency of a DNA profile in some populations is sometimes misinterpreted, for example, as the likelihood that an accused person is innocent. While the estimated frequency of a DNA profile might be one in 5 billion, one DNA expert testifying in a US court misinterpreted this figure as 'a one in 5 billion chance that anybody else could have committed the crime' (see Koehler 1993, p. 32, for this and many other examples), and in Germany, the President of the *Deutschen Gesellschaft für Rechtsmedizin* claimed that a DNA match identifies a perpetrator with '100% certainty' (for this and other examples from the German legal system, see Krauss & Hertwig 2000). Other DNA experts have misinterpreted the profile frequency as the probability that the DNA evidence came from anyone other than the defendant, leading judges likewise to misuse it in their opinions as, for example, 'the probability of someone else leaving' the genetic trace.

Even if judges, jurors, and DNA experts could avoid such misinterpretations, the estimated frequency of a specific DNA profile would still be misleading for yet another important reason: it ignores the chance of a laboratory error. Despite occasional expert testimony declaring that laboratory errors are impossible, such laboratory errors do occur—with a frequency several orders of magnitude larger than the chance of a coincidental match (e.g. Koehler *et al.* 1995).[4] To see how laboratory errors affect the

---

[3] Much controversy about the use of DNA fingerprinting has centred on two questions: (1) what is the frequency of a particular type of DNA marker in the appropriate population, and (2) how should the frequencies of individual markers be combined to calculate the probability of a specific person's DNA profile (e.g. Lander 1989; Lander & Budowle 1994)? Each question hinges on the other. For instance, using the 'product rule' to combine the frequencies of the individual markers requires assuming that the individual alleles at different loci can be treated as statistically independent. This assumption has been hotly disputed (e.g. Lewontin & Hartl 1991). If a population (e.g. total US population) is made up of subpopulations with different gene frequencies (e.g. people of Italian, German, Vietnamese, African decent), then independence cannot be assumed. In response to this argument, a 'ceiling principle' has been proposed according to which the probability of a DNA profile is estimated by combining the largest known allele frequencies from a wide variety of populations (in other words, the figures are chosen from the range of probabilities that are most favourable to the accused). According to Kaye (1997), however, the product rule has recently experienced a comeback, at least in situations in which the class of plausible suspects is as broad as a racial group (see also NRC 1996).

[4] In analysing DNA evidence, technical and human errors can occur. On the technical side, enzyme failures, abnormal salt concentrations, and mischievous dirt spots can produce

calculations, let us look at the necessary Bayesian computations. We first provide the information in probabilities and then in natural frequencies.

In terms of probabilities, the information in this hypothetical case is as follows: first, the base rate of the DNA profile is 0.00001. Second, if someone has that DNA profile, a DNA analysis would reliably show it to match any samples that share it (i.e. $p(\text{match}|\text{profile}) = 1.0$). Third, if the chances of a false-positive laboratory error are as high as sometimes found, the probability of a false-positive match, $p(\text{match}|\neg\text{profile})$, could be 0.003. To calculate the probability of a person having a particular DNA profile given there is a match with the incriminating evidence, Bayes' rule is required:

$$p(\text{profile}|\text{match}) = \frac{p(\text{profile})p(\text{match}|\text{profile})}{p(\text{profile})\,p(\text{match}|\text{profile}) + p(\neg\text{profile})\,p(\text{match}|\neg\text{profile})}$$

Inserting the statistical information into Bayes' rule results in a probability that the person who matches actually has the profile of 0.003:

$$p(\text{profile}|\text{match}) = \frac{(0.00001)(1.0)}{(0.00001)(1.0) + (0.99999)(0.003)} = 0.003.$$

These computations are relatively complex, and, as shown earlier, they can be simplified when the information is presented in natural frequencies. In natural frequencies, the same information would be expressed as follows: first, in a population of 10 million, a particular DNA profile occurs with a frequency of 10 in a million. Thus, one might expect approximately 100 people in this population to have the DNA profile. Second, if someone has this profile, a DNA analysis would show it to match any samples that share it. Third, owing to false-positive laboratory errors, however, there could be up to 30,000 people in the population who do not have the same DNA profile but who would nonetheless be found to match in the DNA analysis.

To compute the probability of a person's having a particular DNA profile *given* a match, one requires merely a count of the people who actually have the profile out of all the people who match. The calculations amount to solving Equation 2 (here adapted to the DNA context):

$$p(\text{profile}|\text{match}) = \frac{\text{match\&\ profile}}{\text{match\&\ profile} + \text{match\&}\neg\text{profile}} = \frac{100}{100 + 30{,}000} = 0.003$$

The confusion over statistical evidence reviewed earlier suggests that judges, jurors, and sometimes even DNA experts do not spontaneously understand evidence presented

---

misleading DNA banding patterns. In terms of human errors, inadvertent switching, mixing, or cross-contamination of samples may lead to false positive errors. The likelihood of these and other errors are estimated on the basis of blind proficiency tests (described in more detail in Koehler 1993, pp. 24–25).

in terms of probabilities. Because Bayesian calculations are simpler when numbers are expressed as natural frequencies, these expressions may yield insight into the uncertainties of forensic scientific analyses with laboratory errors, but also with the choice of the reference class. Ultimately, choosing different ways to express the evidence could influence decisions about guilt and innocence.

Do natural frequencies improve statistical reasoning in the legal context?

Can legal experts profit from natural frequencies in making inferences, just as medical experts do? In a study conducted at the Free University in Berlin, we (Lindsey *et al.* in press; Hoffrage *et al.* 2000) asked 27 professionals who would soon qualify as judges ('jurists') and 127 advanced law students to evaluate two criminal court case files involving rape. In both cases, a DNA match was reported between a DNA sample from the defendant and one recovered from the victim. Aside from this evidence there was little reason to suspect that the defendant was the perpetrator. Expert testimony reported the frequency of the recovered DNA profile as 1 in 1,000,000 and then stated that it was practically certain that the analysis would show a match for a person who indeed had the DNA profile (i.e. the test's hit rate = 100%). The expert also reported the rates of technical and human errors that would lead to false-positive results.

The expert stated all the statistics as either probabilities or frequencies (see Appendix). Based upon these statistics all participants had to estimate two probabilities—that of having a particular DNA profile *given* a DNA match and that of being the source of the evidence *given* a DNA match. Immediately after their estimates, the participants rendered a verdict for the case: guilty or not guilty.[5] After reading one case file in one format, each participant was given a second case file with expert testimony in the other format. They then answered the same questions as before.

Similar to physicians' inferences, the estimates of the legal decision makers are strongly affected by how the statistical evidence was presented. Figure 18.2 shows the percentage of Bayesian inferences as a function of information format. We found a similar pattern across all estimates participants were asked to produce. Consider, for example, participants' estimates of the probability that the defendant was actually the source of the trace. When the statistics were expressed as probabilities, only 13% of the professionals and fewer than 1% of the law students made the correct inference. But when the

---

[5] The two probability judgments and the verdict correspond to the stages in the chain of inferences that arise when DNA evidence is presented in court. From a reported match, one may want to infer (1) the probability that the person for whom the match is reported actually has the DNA profile, (2) the probability that this person is the source of the trace recovered from the crime scene, and finally (3) the probability that the person is guilty. It is important to note that the first two probabilities are *not* sufficient to allow an inference of the probability of guilt. To see why, imagine one knows for certain that a particular person is the source of a DNA sample recovered from a crime scene. Thus, this probability (source given match) would equal 1. However, it is still possible that the person left the trace innocently either before or after the crime was committed or that someone else planted it there.

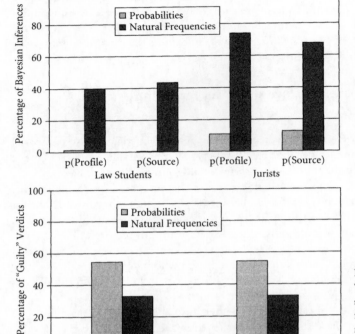

Figure 18.2 Law students' and jurists' percentage of Bayesian inferences in the probability and natural frequency versions (Lindsey et al. in press).

Figure 18.3 Percentage of 'guilty' verdicts of law students and jurists as a function of the representation of information (probability vs natural frequency).

identical statistics were stated as natural frequencies, 68% and 44% of these same participants made the correct inference.[6]

Participants' statistical reasoning also had a clear and important effect on judicial decision making. The mathematically identical statistical evidence led to a higher conviction rate in the probability format than in the natural frequency format. As Fig. 18.3 shows, in both participant samples the proportion of guilty verdicts was substantially higher in the probability format than in the frequency format (13 and 22 percentage points, respectively). Why does the probability format produce more guilty verdicts? There appears to be a simple answer. The estimates that participants calculated from the probability information far exceeded those computed from the frequency information. For instance, jurists, on average, estimated the probability of having the DNA profile in question given a DNA match to be 0.63. In contrast, the average estimated probability in the natural frequency format was 0.05. Thus, it is not surprising that the larger

---

[6] Based on the statistics reported in the Appendix, the probability that a person who is found to match in a DNA analysis (with the evidence from the crime scene) actually has the DNA profile in question is .09 (i.e. 10/110); the probability that a person is the source of the trace recovered from the crime scene given that he is found to match in the DNA analysis is 0.009 (i.e. 1/110).

estimates that were made based on the probability information—when taken as the source probability—led to a higher proportion of guilty verdicts.

Why did participants make higher estimates when given statistical evidence in probabilities? From participants' written explanations of how they derived their estimates we were able to identify two major non-Bayesian algorithms that participants used in the probability format. In the sample of jurists, for instance, about two-fifths (38%) of *all* responses (not just of those responses where a cognitive algorithm could be identified) were produced by *likelihood subtraction* and the *hit rate minus base rate* algorithm. Likelihood subtraction (see also Gigerenzer & Hoffrage 1995), which involves computing the difference $p(\text{match}|\text{profile}) - p(\text{match}|\neg\text{profile})$, makes no use of base rate information (here the base rate of the profile). In a context in which hit rate is very high, as in the case of forensic DNA analysis, and the false-positive and base rate are relatively low, both non-Bayesian algorithms will thus generate erroneously high probability estimates.

Where to go from here

Although uncertainty is deeply entrenched in many legal and medical decisions, the means to reckon with uncertainty have not necessarily been well understood or appreciated by professionals in either discipline. Take one of the first applications of probability theory to legal evidence in history as an example. The famous *Howland Will Case* figured the female heir to one of the greatest fortunes in the United States, Hetty Robinson, fighting to gain control of every last penny. The trial, which took place shortly after the end of the Civil War, turned into a protracted battle over a single piece of evidence: the signature of Hetty Robinson's aunt on a document that effectively left most of her property to her niece (for a description of the trial and its circumstances, see *The New Yorker*, 23 & 30 April 2001, pp. 62–70). The opposite side claimed that the aunt's signature had been forged. An impressive number of expert witnesses was enlisted to litigate the matter. Among them were Benjamin Peirce, Harvard professor, and his son, Charles Sanders Peirce (who later became a famous logician and philosopher).

To determine the validity of the signature in dispute, the Peirces identified thirty places in the aunt's verified signature where she had made a downstroke with her pen (thus forming a letter). When they superimposed the disputed signature on the verified signature, all thirty downstrokes started at exactly the same point on each letter. How likely is that to happen by chance? To estimate this likelihood they analysed previous signatures of Robinson's aunt and found that, on average, one out of five downstroke positions overlapped. With this base rate, they argued that the chance that Robinson's aunt could have unintentionally produced two signatures in which all thirty downstrokes overlapped, was 1 in 5^{30}—'so vast an improbability, is practically an impossibility .... It is utterly repugnant to sound reason to attribute this coincidence to any cause but design' (Benjamin Peirce in his deposition, as cited in the *The New Yorker*, 23 & 30 April 2001, p. 69). The time, however, was not ripe for such reasoning. Both the lawyers for Hetty Robinson and the public treated this argument as mathematical voodoo; it was ridiculed, and some even felt that it transgressed a boundary by applying the laws of probability to 'elements of will and desire unfit... for judgment by such laws' (*The New Yorker*, 23 & 30 April 2001, p. 69).

Are those reactions just reminders of a distant past in which both the legal profession and the public were not educated to think about uncertainties? Unfortunately, there is plenty of evidence that suggests that the justice system has not yet overcome the illusion of certainty. Take eyewitness identification as an example. Eyewitness identification can be highly persuasive to jurors, although it is the major source of wrongful convictions (Borchard 1932; Rattner 1988; Wells *et al.* 2000). Despite this knowledge, the criminal justice system, and in particular the prosecutors (see Wells *et al.* 2000), seem to be utterly reluctant to adopt new (and empirically validated) procedures in collecting eyewitness testimony, which were designed to increase the reliability of eyewitness identification (e.g. sequential lineups; see Wells *et al.* 2000).

Why do legal and medical decision-makers seem to have difficulties reckoning with uncertainties? The problem starts in their training. Medical schools teach statistics, but their focus is on methods of data analysis such as significance testing. But even if they are taught statistical procedures needed for risk assessment, students are typically instructed to mechanically insert probabilities into mathematical formulas such as Bayes' rule. In law, the case for statistical reasoning seems to be even worse. With few exceptions, law schools do not teach students how to reason on the basis of uncertain evidence—although virtually all evidence is uncertain. If, then, statistical illiteracy has its roots in training or in the lack thereof, what can be done? We suggest that the endeavour to bring about statistical literacy is more likely to succeed if training of statistical reasoning is based upon information representations that are suited to the human mind.

It is noteworthy that the beneficial effects of natural frequencies on statistical reasoning in the studies reported above occurred *without* any training or instruction. Naturally, this raises the hope that systematic training in the use of natural frequencies may improve people's ability to reason statistically even more dramatically. The key is to teach representations rather than rules; that is, to solve problems—such as the medical and legal ones described here—by translating probabilities into natural frequencies. In fact, Sedlmeier and Gigerenzer (2001) showed that such 'representation training' can make an enormous difference. In contrast to a traditional 'rule training', their two-hour representation training was both much more successful in improving people's performance in the short run and in keeping people from forgetting how to solve such problems in the long run. Thus these results suggest that the teaching of representations—in high schools, colleges, and universities—can be an important pedagogical tool to faster, more reliable, and more comprehensibly attainable statistical literacy.

Being able to reason statistically is important not only for professionals but also for their clientele, that is, us. During consultation with their patients, for instance, in the United States doctors are increasingly more likely to say 'I can't tell you what to do'. According to George J. Annas, the chairman of the health law department at Boston University's School of Public Health, 'many doctors are comfortable now saying, "It's not me, it's you, and you are the one who has to decide"' (quoted in *The New York Times*, 25 June 2000, Section 15, p. 1). Two factors seem to be driving this transfer of decision-making responsibility from the doctor to the patient. According to medical professors interviewed in a recent article in *The New York Times* (25 June 2000, Section 15, pp. 1, 10), one is the fear of lawsuits: 'Doctors, after all, can be sued over whatever decision they make. But if the patient makes the decision, who is to blame?'. The second factor is the burgeoning variety and complexity of technological tools in medicine: 'In a world where medical technology is getting all the more powered, and often accompanied by risks, nobody can decide for you'.

Radiation-based diagnostic procedures are a case in point. According to Horst Kuni (*Süddeutsche Zeitung*, 3 August 2000, p. B-2), professor of nuclear medicine at the University of Marbug, in Germany up to 50,000 people (!) per year fall ill with cancer because of radiation-based medical examinations of sometimes questionable utility (such as mammography, see Gotzsche & Olsen 2000). What this figure makes abundantly clear is that any exposure to radiation carries risk. This is a fact that neither the public nor scientific experts recognized when they celebrated the advent of the new technology in the early decades of the twentieth century (Howell 1995), but it is well known today. Thus, patients and health care providers must decide on a case-by-case basis whether the information gleaned from a medical diagnostic procedure using radiation justifies its use. Needless to say, it is therefore essential that patients have a proper understanding of the available statistical information (e.g. the positive predictive value of a test). We suggest such an understanding is more likely to be achieved if doctors communicate statistical information in terms of natural frequencies instead of probabilities.

Conclusion

Increasingly, modern technologies are shaping many aspects of our lives. Yet these rapid technological developments have not always delivered their intended benefits—often because of the limited understanding of how the results that the technologies produce ought to be communicated. It is here that technology needs psychology. For instance, to improve doctors' and patients' interpretation of a positive mammogram, we need to understand how the way this result is communicated relates to, and interacts with, the way the human mind works. We showed that psychological research on how people process frequencies allows us to improve the reasoning of those who need to make statistical inferences from the results of diagnostic technologies. Although representing those results one way or another may not make much of a difference for a mathematician, it can make one for a juror and a physician, and ultimately for a defendant and a patient.

Note

We are grateful to Tilmann Betsch, Valerie M. Chase, Steffi Kurzenhäuser, Peter Sedlmeier, and an anonymous reviewer for many helpful comments, and to the Deutsche Forschungsgemeinschaft for its financial support of the first author (Forschungstipendium He 2768/6-1) and the second author (Ho 1847/1).

Correspondence concerning this chapter should be addressed to Ralph Hertwig, Max Planck Institute for Human Development, Center for Adaptive Behavior and Cognition, Lentzeallee 94, 14195 Berlin, Germany. Electronic mail may be sent to hertwig@mpib-berlin.mpg.de.

References

Berwick, D. M., Fineberg, H. V. & Weinstein, M. C. (1981). When doctors meet numbers. *American Journal of Medicine, 71*:991–998.

Borchard, E. (1932). *Convicting the innocent: Errors of criminal justice.* New Haven: Yale University Press.

Casscells, W., Schoenberger, A. & Grayboys, T. (1978). Interpretation by physicians of clinical laboratory results. *New England Journal of Medicine, 299*:999–1001.

Christensen-Szalanski, J. J. J. & Bushyhead, J. B. (1981). Physicians' use of probabilistic information in a real clinical setting. *Journal of Experimental Psychology: Human Perception and Performance, 7*:928–935.

Daston, L. J. (1988). *Classical probability in the Enlightenment.* Princeton, NJ: Princeton University Press.

Eddy, D. M. (1982). Probabilistic reasoning in clinical medicine: Problems and opportunities. In D. Kahneman, P. Slovic and A. Tversky *Judgment under uncertainty: Heuristics and biases* (pp. 249–267). Cambridge, UK: Cambridge University Press.

Fiedler, K, Brinkmann, B., Betsch, T. & Wild, B. (2000). A sampling approach to biases in conditional probability judgments: Beyond base rate neglect and statistical format. *Journal of Experimental Psychology: General, 129*:399–418.

Gallistel, C. R. & Gelman, R. (1992). Preverbal and verbal counting and computation. *Cognition, 44*:43–74.

Gigerenzer, G. & Hoffrage, U. (1995). How to improve Bayesian reasoning without instruction: Frequency formats. *Psychological Review, 102*:684–704.

Gigerenzer, G. & Hoffrage, U. (1999). Overcoming difficulties in Bayesian reasoning: A reply to Lewis & Keren and Mellers & McGraw. *Psychological Review, 104*:425–430.

Gigerenzer, G., Swijtink, Z., Porter, T., Daston, L., Beatty, J. & Krüger, L. (1989). *The empire of chance: How probability changed science and everyday life.* Cambridge, UK: Cambridge University Press.

Gotzsche, P. C. & Olsen, O. (2000). Is screening for breast cancer with mammography justifiable? *Lancet, 355*:129–134.

Hoffrage, U. & Gigerenzer, G. (1998). Using natural frequencies to improve diagnostic inferences. *Academic Medicine, 73*:538–540.

Hoffrage, U., Gigerenzer, G., Krauss, S. & Martignon, L. (in press). Representation facilitates reasoning: what natural frequencies are and what they are not. *Cognition.*

Hoffrage, U., Lindsey, S., Hertwig, R. & Gigerenzer, G. (2000). Communicating statistical information. *Science, 290*:2261–2262.

Howell, J. D. (1995). *Technology in the hospital: Transforming patient care in the early twentieth century.* Baltimore: Johns Hopkins University Press.

Jonides, J. & Jones, C. M. (1992). Direct coding for frequency of occurrence. *Journal of Experimental Psychology: Learning, Memory, and Cognition, 18*:368–378.

Kaye, D. H. (1997). DNA identification in criminal cases: Lingering and emerging evidentiary issues. In *Proceedings of the Seventh International Symposium on Human Identification* (pp. 12–25). Madison: Promega Corp.

Kleiter, G. D. (1994). Natural sampling: Rationality without base rates. In G. H. Fischer & D. Laming (eds) *Contributions to mathematical psychology, psychometrics, and methodology* (pp. 375–388). New York: Springer.

Koehler, J. J. (1993). Error and exaggeration in the presentation of DNA evidence. *Jurimetrics Journal,* 34:21–39.

Koehler, J. J. (1996). The base rate fallacy reconsidered: Descriptive, normative and methodological challenges. *Behavioral and Brain Sciences,* 19:1–53.

Koehler, J. J., Chia, A. & Lindsey, S. (1995). The random match probability (RMP) in DNA evidence. *Jurimetrics Journal,* 35:201–219.

Krauss, S. & Hertwig, R. (2000). Muss DNA evidence schwer verständlich sein? Der Ausweg aus einem Kommunikationsproblem [Does DNA evidence need to be difficult to understand?]. *Monatszeitschrift für Kriminologie und Strafrechtsreform,* 83:155–162.

Lander, E. S. (1989). DNA fingerprinting on trial. *Nature,* 339:501–505.

Lander, E. S. & Budowle, B. (1994). DNA fingerprinting dispute laid to rest. *Nature,* 371:735–738.

Lewontin, R. C. & Hartl, D. L. (1991). Population genetics in forensic DNA typing. *Science,* 254:1745–1750.

Lifton, R. J. & Mitchell, G. (2000). *Who owns death? Capital punishment, the American conscience, and the end of executions.* New York: Morrow.

Lindsey, S., Hertwig, R. & Gigerenzer, G. (in press). Communicating statistical evidence. *Jurimetrics.*

Macchi, L. (2000). Partitive formulation of information in probabilistic problems: Beyond heuristics and frequency format explanations. *Organizational Behavior and Human Decision Processes,* 82:217–236.

NRC (National research council committee on DNA forensic science: An update) (1996). *The evaluation of forensic DNA evidence.* Washington: National Academy Press.

Politser, P. E. (1984). Explanations of statistical concepts: Can they penetrate the haze of Bayes? *Methods of Information in Medicine,* 23:99–108.

Rattner, A. (1988). Convicted but innocent: Wrongful conviction and the criminal justice system. *Law and Human Behavior,* 12:283–293.

Real, L. A. (1991). Animal choice behavior and the evolution of cognitive architecture. *Science,* 253:980–986.

Scheck, B., Neufeld, P. & Dwyer, J. (2000). *Actual innocence: Five days to execution, and other dispatches from the wrongly convicted.* New York: Doubleday.

Sedlmeier, P. & Gigerenzer, G. (2001). Teaching Bayesian reasoning in less than two hours. *Journal of Experimental Psychology: General,* 130:380–400.

Sedlmeier, P., Hertwig, R. & Gigerenzer, G. (1998). Are judgments of the positional frequencies of letters systematically biased due to availability? *Journal of Experimental Psychology: Learning, Memory, and Cognition,* 24:754–770.

Stigler, S. M. (1983). 'Who discovered Bayes' theorem?' *The American Statistician,* 37:296–325.

Wells, G. L., Malpass, R. S., Lindsay, R. C. L, Fisher, R. P., Turtle, J. W. & Fulero, S. M. (2000). From the lab to the police station: A successful application of eyewitness research. *American Psychologist,* 55:581–598.

Windeler, J. & Köbberling, J. (1986). Empirische Untersuchung zur Einschätzung diagnostischer Verfahren am Beispiel des Haemoccult-Tests. [An empirical study of the judgments about diagnostic procedures using the example of the hemoccult test.] *Klinische Wochenschrift,* 64:1106–1112.

Appendix

We present the text for the probability and natural frequency versions of one of the two cases involving forensic DNA analysis used by Lindsey *et al.* (in press). Each case description included the testimony of an expert who performed a DNA analysis. The expert testimony provided numerical information about the base rate of the DNA profile, and the analysis' hit rate and false-positive rate. The numerical information was presented in either probabilities or natural frequencies. Participants were asked to estimate two probabilities (or proportions): (1) the probability that a person who is found to match in a DNA analysis (with the evidence from the crime scene) actually has the DNA profile in question, and (2) the probability that a person is the source of the trace recovered from the crime scene given a match in the DNA analysis.

Probabilities

In a country the size of Germany there are as many as 10 million men who fit the description of the perpetrator. The probability of a randomly selected person having a DNA profile that matches the trace recovered from the crime scene is 0.0001%. If someone has this DNA profile it is practically certain that this kind of DNA analysis would show a match. The probability that someone who does not have this DNA profile would be shown to match in this type of DNA analysis is 0.001%. In the present case, the DNA profile of the sample from the defendant matches the DNA profile of the trace recovered from the crime scene.

Natural frequencies

In a country the size of Germany there are as many as 10 million men who fit the description of the perpetrator. Approximately 10 of these men would have a DNA profile that matches the trace recovered from the crime scene. If someone has this DNA profile it is practically certain that this kind of DNA analysis would show a match. Of the 9 999 990 people who do not have this DNA profile, approximately 100 would be shown to match in this type of DNA analysis. In the present case . . .

CHAPTER 19

FREQUENCY PROCESSING AND COGNITION: STOCK-TAKING AND OUTLOOK

TILMANN BETSCH AND PETER SEDLMEIER

Abstract

We summarize theories and research presented in the previous 17 Chapters of this volume. Compared to previous theoretical work, theorists today are less likely to neglect empirical results obtained outside their own field. They are starting to acknowledge that people can employ multiple judgemental strategies, and are placing a greater emphasis on factors moderating frequency processing and judgement based thereon. Capitalizing on a broad arsenal of research methods, a great deal of the research presented in this volume addresses the moderating conditions of accuracy and inaccuracy in frequency and probability judgement. Moreover, attempts to discover the factors that determine the selection of judgemental strategies are increasing. Insights into frequency processing and cognition are fruitfully applied to various fields such as marketing and the aiding of expert decision making. We close our review of the Chapters with an outline of some trajectories for future research and theorizing.

Introduction

The 17 Chapters of the present volume provide a representative sample of contemporary research and theorizing in the field of frequency processing and cognition. In this final Chapter we attempt to summarize what has been achieved in the field so far. The next three sections follow the line of the book. First, we examine progress in theoretical work. Second, we review the yield of research and briefly discuss methodological issues. Third, we discuss how insights gained from research on frequency processing and cognition may help to improve performance in judgement and decision making in applied settings. We close with some proposed guidelines for future research and theorizing.

Theory

The present volume gives a state-of-the-art overview of contemporary theoretical approaches to frequency processing and judgement. It is clearly beyond the scope of a summary Chapter to review the development of these theories, or to compare them with their predecessors. Nevertheless it is possible to diagnose trends in theory development if one adopts a broad time perspective. We consider theoretical work between the 1970s and 1980s as a level of comparison for current approaches published in the late 1990s. Such earlier theoretical work on frequency processing can be characterized by at least three features. First, theorists formerly had a more narrow focus in that they had considered only certain aspects of frequency processing and at the same time largely ignored theories and findings from other fields of psychology. Second, theoretical accounts had been primarily concentrating on a single judgement strategy. Third, the influence of context and task factors had often not been systematically spelled out in the models.

A look back: narrow research focus, single strategy models, neglect of context and task factors

In the 1970s two quite distinct approaches to frequency processing began to dominate the field. We will henceforth refer to these approaches as the *memory* and the *judgement-and decision-making* (JDM) approach, in order to highlight the different areas of psychology where these approaches originated. Rooted in the tradition of basic research in learning, memory and cognition, the first approach focused primarily on processes of encoding (learning) and memory representation (e.g. Hasher & Zacks 1979; Hintzman & Block 1971). Empirical evidence produced by the memory approach usually revealed a high degree of accuracy in human frequency judgement. In the field of judgement and decision making (JDM) a rather different notion emerged. JDM researchers concentrated primarily on the judgement process and spelled out different judgemental heuristics. Biases in frequency and probability judgement were attributed to these heuristics (e.g. Tversky & Kahneman 1973). The memory and the JDM approach co-existed over the last decades, and, as Dougherty and Franco-Watkins (Chapter 8) note, with relatively little cross-fertilization. Such a narrow focus orientation among researchers prevailed until the 1990s, as is evident from reference lists in publications. To some extent, this in-group focus was still echoed by the presentations and discussions of the inter-national symposium we organized in 1999, which motivated the writing of the present volume.

Former models of frequency processing and cognition assumed that a single strategy would be dominant in frequency judgement. Multiple-trace type models postulated that judges prompt their memory with a target (e.g. the category or event to be judged) and use echo strength (the similarity between the probe and the traces activated from memory) as a proxy to estimate frequency or probability (Hintzman 1988). Other memory models, which sided more with a trace–strength view, postulated that frequencies are recorded in tags or counters that are attached to category or event representations in memory (e.g. Alba *et al.* 1980). Accordingly, people were assumed to simply read out frequency from memory when forming quantitative estimates. Research on frequency processing in the field of JDM also favoured a single strategy for frequency estimation, namely the availability heuristic introduced by Tversky and Kahneman in the early 1970s. When applying the availability heuristic, individuals are assumed to assess the ease by which relevant instances come to mind (Tversky & Kahneman 1973). Ironically, favouring a single strategy violates one of the basic assumptions of the most influential metatheoretical framework of JDM research. As a pre-assumption, the heuristics and biases approach to judgement and decision making holds that people employ different judgemental strategies. In fact, this view is reflected in Tversky and Kahneman's original statement of the availability heuristic. They clearly stated that the availability heuristic is just one among other strategies people might use in frequency and probability judgement. However, this notion did not precipitate JDM research on frequency processing for a long time, because many researchers misinterpreted the original statement of the model and assumed the availability heuristic to be the predominant mechanism (Betsch & Pohl, Chapter 7). In some respects, this development was promoted by Tversky and Kahneman's reluctance to correct the widespread misinterpretation of their availability approach. As a consequence, JDM research hesitated to adopt a multiple strategy view and to seriously consider the conditions of strategy choice for a long time. In fact, it took the field until the mid-1990s to make a shift towards a multiple strategy perspective, which then eventually became visible in the literature (Brown 1995).

The narrow focus on one's own field and on a particular strategy of judgement were also responsible for the theoretical neglect of context and task factors. This is not to say that former models were generally blind to the learning and judgement context. The availability approach and the multiple-trace models jointly predicted that all factors which impact on memory processes would influence frequency judgement. However, a systematic theoretical consideration of those factors was rather the exception than the rule, although empirical evidence was available early, indicating that frequency encoding is affected by a number of variables (Sedlmeier et al., Chapter 1). Surprisingly, by the late 1980s there was little research on the influence of capacity constraints (e.g. time pressure) and motivation on frequency judgement. This is surprising because at the same time such factors received considerable attention, for instance in research on preferential decision making (Beach & Mitchell 1978) or in social psychology (Chaiken 1980; Fazio et al. 1982; Petty & Cacioppo 1986).

Stock-taking: broader focus, multiple strategy models, consideration of context and task factors

In the introductory chapter we lamented the lack of theoretical exchange between the different fields of frequency research in the past decades. The theoretical contributions compiled in this volume justify a more optimistic view of the future. Theorists are starting to recognize developments achieved in neighbouring fields (although the findings from animal research—see Gallistel, Chapter 10—are still not very widely known). This tendency manifests itself in attempts to form more integrative models and frameworks that are capable of covering a broader range of empirical evidence. However, it has to be noted that theories clearly differ with regard to their general scope. For example, Brown's multiple strategy framework (Chapter 3) is broad with respect to the phenomena it covers in frequency research but narrow in general scope compared to those models which explain frequency processing and judgement within the framework of general memory and cognition models (e.g. Dougherty & Franco-Watkins, Chapter 8; Sedlmeier, Chapter 9). Also, those models that refer to information theoretic accounts (Gallistel, Chapter 10), or draw on an evolutionary perspective (e.g. Gigerenzer, Chapter 4) extend far beyond the processing of frequency.

The majority of contemporary approaches dovetails in acknowledging that theories of frequency processing must face the fact that frequency judgements are sometimes accurate and sometimes prone to biases. Reviewing the evidence for their Automatic and Effortful encoding framework, Zacks and Hasher (Chapter 2) acknowledge that automatic encoding of frequencies does not protect judges from biases in frequency judgement. They suggest that the mixed evidence for accurate judgements on the one hand and biased estimates on the other can be modelled within their framework, if one realizes that different processes are at work during the encoding (learning) and the judgement phase. Although encoding of frequencies is an inevitable consequence of attending to events (in such this process is automatic), individuals do not necessarily rely only on frequency memory in their judgement. They propose that judgement processes are prone to be affected by multiple variables, which may cause distortions of automatically formed frequency knowledge.

In a similar vein, a number of other theories rooted in the memory paradigm face the fact that frequency judgements can be systematically biased. Dougherty and Franco-Watkins (Chapter 8) provide an easy-to-follow introduction to Minerva Decision Making (MINERVA DM) and explain both accuracy and biases in quantitative estimation. They show that findings formerly attributed to the work of the availability heuristic can be nicely modelled within their multiple-trace framework, which rests on a similarity based algorithm for frequency and probability estimation. Also starting from a multiple-trace perspective, Fiedler (Chapter 5) focuses on the sampling of information (from memory or the environment) before a judgement is made. Considering an impressive range of phenomena, Fiedler substantiates the notion that the structure of retrieval cues may have a pronounced influence on the sampling process and, as a consequence, on accuracy in frequency judgement. According to this notion, biases in judgements are likely to occur when the task evokes a highly differentiated structure of retrieval cues. Conversely, a less differentiated cue structure should yield more accurate estimates. Another theory that addresses both accuracy and biases in frequency judgement is suggested by Sedlmeier (Chapter 9). His Probability Associator (PASS) model belongs to the broad class of associative learning theories. It assumes that frequency learning involves the formation of associations between features. Associations between features increase as a function of the frequency by which events with overlapping features co-occur. At the time of judgement, memory is prompted by retrieval cues and associative strength is used as an informational base for the estimate. The simulations reveal that PASS can account for a number of effects in frequency, probability and confidence judgement, including prominence biases, sample size and regression effects.

From an even broader perspective, Gigerenzer (Chapter 4) focuses on conditions that allow people to make adequate judgements of relative frequencies and (conditional) probabilities. Gigerenzer raises an evolutionary argument by assuming that organisms must have built devices to (accurately) deal with frequentistic information in order to successfully adapt to their environment. Their capability to process frequencies is obscured, however, if people are exposed to single-event probabilities as they are asked to do in many laboratory settings. Gigerenzer claims that a probability format often entails unspecified reference classes, a high degree of polysemy and computational complexity which together are responsible for many biases documented in the JDM literature on quantitative estimation.

Another integrative theoretical framework is presented by Brown (Chapter 3). In line with recent memory models of frequency processing (Dougherty *et al.* 1999; Fiedler 1996; Sedlmeier 1999), Brown's multiple strategy perspective (MSP) covers the process from frequency encoding through representation to judgement performance. MSP makes unique assumptions about how processes at the early stage of encoding affect representation and how they constrain the breadth of choice for a judgement strategy. The chief assumption of his model is that people make use of a variety of strategies for quantitative estimation. Moreover, he provides a comprehensive classification of judgement strategies. The latter point straightforwardly leads to our next observation that a single-strategy view is overcome by modern approaches.

It took a long time before researchers in frequency processing started to elaborate on a multiple-strategy approach to frequency processing and judgement. Brown's work (e.g. Brown 1995) gave a strong impetus to the field. He was among those who initiated a trend in theorizing that is currently shaping an increasing number of theoretical approaches. This development is also mirrored in the present volume. Many contributors acknowledge that there are different strategies for frequency estimation. Aside from Brown's theoretical framework, advocates of memory models allow for the integration of a multiple-strategy view. Sedlmeier's PASS model (Chapter 9), for example, contains a cognitive algorithm (CA) module that is assumed to contain different algorithms or judgement strategies. In a similar vein, Fiedler (Chapter 5) and Zacks and Hasher (Chapter 2) suggest that individuals can use different techniques to transform retrieved information into judgements. The entire number of strategies identified in the book cannot be comprehensively reviewed in the present chapter. However, some strategies seem to be dominant and appear under different labels in various approaches. For instance, people can use the number of instances recalled as a base for frequency judgement, either by reproducing the count of instances or by extrapolating from this count. Both mental techniques are subsumed under the label of enumeration (Brown, Chapter 3), memory-based (Haberstroh & Betsch, Chapter 12) or recall–content based strategies (Schwarz & Wänke, Chapter 6). Another strategy, which also appears under different labels, involves the assessment of the experience of the retrieval process itself. A prominent example is the availability heuristic that can rest on the assessment of the ease of retrieval. An assessment of conceptual fluency (Reber & Zupanek, Chapter 11), similarity or echo strength (Dougherty & Franco-Watkins, Chapter 8; Fiedler, Chapter 5) may involve similar psychological mechanisms. Therefore, it seems justified to subsume such strategies under the broad label of memory-assessment strategies, as suggested by Brown. A further strategy would require the individual to read out explicitly or implicitly formed frequency tallies from memory (Brown, Chapter 3; Haberstroh & Betsch, Chapter 13).

One very important question refers to the mechanisms that guide strategy selection. Although there is no definite answer available yet, researchers are starting to address this issue both theoretically and empirically. Brown (Chapter 3) suggests adopting the principle of calculated strategy selection from preferential decision research (Beach & Mitchell 1978; March 1978; Payne et al. 1993), which rests on an effort–accuracy trade off. Schwarz and Wänke (Chapter 6) adopt the accessibility–applicability principle from social cognition. They suggest that judges choose the strategy that is applicable to the information and most accessible at that point in time.

By the 1990s, systematic consideration of context factors had scarcely filtered into theories on frequency processing. The theoretical contributions in the present volume indicate a trend shift. Context and task factors that impact on encoding, storage and retrieval of frequentistic information are discussed in each theoretical paper and play a major role in the reports on empirical findings. Most notably, however, theorists are beginning to implement those factors systematically in their models (e.g. Brown, Chapter 3; Haberstroh & Betsch, Chapter 13). Brown, for instance, spells out the way by which different conditions of encoding may affect memory formats of frequency knowledge and subsequent strategy choice. Favouring an effort–accuracy approach to strategy choice, task constraints and motivational factors at the time of judgement play a major role in his theoretical framework. The motivation to invest cognitive effort in the estimation task also plays a fundamental role in other models that side with a multiple strategy perspective. These models all converge in assuming that with increasing motivation, judges become more likely to chose an enumeration strategy or, in other terms, a strategy which capitalizes on recall content (Brown, Chapter 3; Haberstroh & Betsch, Chapter 13; Schwarz & Wänke, Chapter 6).

Research

The empirical work reviewed in this volume shows that contemporary research in frequency processing and judgement makes use of a broad variety of research methods. Although the major part of the data is still produced by laboratory experiments, researchers also conduct field studies (e.g. Schimmack, Chapter 12), surveys (Brown, Chapter 3; Schwarz & Wänke, Chapter 6), and computer simulations based on secondary data (Dougherty & Franco-Watkins, Chapter 8; Wettler, Chapter 17).

In laboratory research, experimenters frequently employ computer-controlled research tools. The advantage of such devices is that experimental control over the learning and the judgement phase can be maximized. Regularly, participants are presented with target stimuli on the computer screen and subsequently are requested to rate frequencies or probabilities of the events, features or categories that were presented before. A wide range of different stimuli is used in computer-controlled studies, for example, letters (Reber & Zupanek, Chapter 11), words and category labels (Brown, Chapter 3), consumer products and types of social behaviour (Fiedler, Chapter 5), reports of baseball results (Kleiter et al. Chapter 15), movements of aeroplanes on a radar (Baumann & Krems, Chapter 14) or fictitious creatures (Haberstroh & Betsch, Chapter 13). Moreover, in paper and pencil studies, people were asked, for instance, to rate the prevalence of emotional reactions in hypothetical scenarios (Schimmack,

Chapter 12). These examples show that the empirical evidence reported in this volume stems from a broad range of content domains.

In the following, we attempt to summarize recent important findings which have been presented throughout this volume. Since a number of chapters themselves provide reviews of research programmes, it is beyond the scope of this chapter to provide a comprehensive summary of the evidence. Rather, we attempt to identify converging themes in research interests and those advances that are pertinent to the theoretical issues addressed in the previous section. Specifically, we focus on the role that context factors play in accuracy and strategy choice. Moreover, we briefly consider methodological issues. An outline of trajectories for future research will follow in the final section of this chapter.

Moderating conditions of accuracy in frequency and probability judgement

In the introductory chapter, we referred to a debate on accuracy, which markedly reflected the antagonist views of JDM research on frequency processing on the one hand and memory research on the other. Over the last decades, advocates of JDM's heuristics and biases programme underlined the faulty nature of human judgement. Researchers were primarily concerned with biases in frequency and probability judgement. In contrast, advocates of the memory-paradigm in experimental psychology often focused on the assets of the cognitive system. Research conducted in this domain demonstrated that humans were remarkably good at encoding and storing frequency information. Until the mid-1990s researchers still attempted to favour one notion over the other (this is reflected, for example, in the debate on the 'reality of cognitive illusions', Gigerenzer [1996]; Kahneman & Tversky [1996]). It appears that the general consensus among most, if not all authors is that humans are highly sensitive to frequentistic information in the environment, and often produce estimates which quite accurately mirror actual ordinal differences and relative frequencies. However, relative or discriminative accuracy is often better than absolute accuracy (e.g. Alba, Chapter 16; Schimmack, Chapter 12). Accuracy is still a big issue today. In contrast to former research, contemporary empirical work is much more concerned with the *conditions* which promote accuracy or distortions in quantitative estimates (Baumann & Krems, Chapter 14; Gigerenzer, Chapter 4; Fiedler, Chapter 5; Haberstroh & Betsch, Chapter 13; Reber & Zupanek, Chapter 11; Schimmack, Chapter 12; Schwarz & Wänke, Chapter 6).

Moderating conditions of accuracy in frequency judgement

In the introductory chapter we pointed out the match between encoding and judgement categories as an important, if not necessary, precondition for accuracy in frequency judgement. There is some evidence that reliable memory records of frequencies can only be established for those aspects which are focused during encoding (Barsalou & Ross 1986; Betsch *et al.* 1999). Hence, if the categories to be judged differ from those that were dominantly activated during learning, estimates are likely to decrease in accuracy. This notion also dovetails with theoretical considerations. For example, multiple trace as well as associative memory models would assume that focused features are more likely to be copied into memory than non-focused ones. In turn, probing memory with previously focused cues will yield a more reliable memory sample than otherwise (e.g. Dougherty &

Franco-Watkins, Chapter 8). In accordance with this assumption, several authors report a decrease in accuracy when the retrieval cue structure deviates from the encoding situation. Such effects occur, for instance, when judgement categories are split into subordinates (Fiedler, Chapter 5) or when individuals have to estimate the frequency of those features which were not salient during encoding (see the discussion on encoding/test mismatch by Brown, Chapter 3).

Another important variable is the time and the effort people allocate to the estimation task. One might assume that more thorough thinking generally helps to increase accuracy in judgements and decisions (Hertwig & Ortmann 2001). In the domain of frequency judgement, some recent results seem to support this view. Aarts and Dijksterhuis (1999) demonstrated that an increase in motivation fostered by the prospect of monetary gains might inoculate judges against availability biases. However, empirical evidence presented in this volume leads us to doubt whether such findings could be generalized. Studies by Schwarz and Wänke (Chapter 6) and Haberstroh and Betsch (Chapter 13) jointly showed that high motivation or deliberation instructions lead participants to rely on recall content available at the time of judgement (see also Brown, Chapter 3). Therefore, under a deliberative mindset, all factors that affect recall should affect frequency judgement. This opens the door to systematic distortions and biases. Indeed, Haberstroh and Betsch found that accuracy was systematically lower under deliberation compared to a spontaneous guess condition. These results converge with findings obtained in other domains of judgement and decision making (Betsch & Haberstroh 2001).

One might argue, however, that in a deliberative mindset, judges have a greater chance of recognizing judgemental traps and taking measures against them. This, however, seems to be a rather optimistic view. Fiedler (Chapter 5) cites a couple of results suggesting that humans may at least sometimes lack the metacognitive insight into the mechanisms that are responsible for distortions in their quantitative estimates. On several occasions throughout the book, the authors highlight other factors yielding systematic distortions in frequency judgement. In line with a large body of prior research, some recent studies provide additional evidence for ease of retrieval effects (Reber & Zupanek, Chapter 11; Schimmack, Chapter 12; Schwarz & Wänke, Chapter 6). Accordingly, if instances of infrequent events can be easily processed, frequencies are likely to be overestimated. However, the susceptibility to availability biases may decrease with increasing sample size during learning (Sedlmeier, Chapter 9) or when people avoid deliberating before making a judgement (Haberstroh & Betsch, Chapter 13). Finally, the wording of the question in an estimation task, the design of the response scale and the specific question within a survey instrument provide a plethora of additional sources for biases. Schwarz and Wänke (Chapter 6) review the evidence and make a number of practical suggestions as to how survey researchers may cope with these problems.

Moderating conditions of accuracy in probability judgement

In principle, the above mentioned considerations about accuracy in frequency judgement also apply to probability judgements. However, there are some factors especially pertinent to accuracy in probabilistic reasoning. Major ones are natural sampling and information format. Gigerenzer advocates the notion that natural sampling (sequential and unconditional acquisition of frequencies) and a frequency format (in contrast to conveying information in terms of probabilities) are important facilitating conditions for accuracy in probabilistic reasoning (Gigerenzer, Chapter 4; Gigerenzer & Hoffrage 1995; see also Hertwig & Hoffrage, Chapter 18). There is also ample evidence that training programmes for many kinds of probability tasks are much more effective if they use frequency formats than probability formats (Sedlmeier 1999; Sedlmeier & Gigerenzer 2001). However, there is some counter-evidence indicating that, at least under certain conditions, biases are not reduced by natural sampling and frequency format. Reber and Zupanek (Chapter 11) varied information format (frequency vs. probability format) in two studies. Changing format conditions did not prevent participants from being biased by processing fluency (i.e. the ease by which information can be processed). In a number of studies, Fiedler (Chapter 5) had participants actively sample event frequencies by themselves. However, the subjectively rated probabilities that a woman who has received a positive mammography actually has breast cancer showed a pronounced deviation from normative standards when participants sampled information by the criterion (breast cancer) rather than by the predictor (mammography result). In a similar vein, Baumann and Krems (Chapter 14) allowed their subjects to sequentially sample frequencies of aeroplane movements in a simulated radar control station and examined whether such a procedure was capable of reducing the influence of order effects in belief updating. However, they found that natural sampling of frequencies does not prevent order effects in subsequent probability judgements. Therefore, it seems that despite the ample evidence that using natural frequencies helps in probability judgements, a closer look at conditions, kinds of tasks, and procedures involved in such judgements is warranted.

Assessment of accuracy

In order to determine the level of accuracy in judgement performance one needs a normative standard against which subjective probabilities can be evaluated. For conditional probability calculation, the normative standard is usually provided by the laws of probability (e.g. Bayes' theorem). One needs a different measure, however, to assess calibration of probability judgements. Kleiter and colleagues (Chapter 15) put forward a provocative argument. Based on recent development in probability theory, they criticize the way over- and underconfidence are usually assessed in calibration research. They outline a new formal method to assess calibration based on coherence. Reviewing some recent studies, they show that subjective probabilities can be well calibrated compared to suggested normative standards. Their contribution casts some doubt on the prevalence of overconfidence effects in probability judgement as presupposed by the JDM literature.

Moderating conditions of strategy choice

Strategy choice in frequency judgement seems to depend to a great degree on encoding and storage processes prior to judgement. Strategies capitalize on different sources of knowledge. In the absence of instance memory, for example, enumeration strategies cannot be applied. In such instances, memory marks boundaries to strategy selection by narrowing down the number of eligible strategies (Brown, Chapter 3). Thus the first step to understanding strategy selection is to understand memory processes. Important factors in this context are the presentation frequency and the similarity of instances. The higher the number of presentations and the more similar the instances are, the lower the likelihood that individuals can subsequently recall instances and base their frequency judgements on recall-based enumeration strategies or on a direct assessment of availability (Brown, Chapter 3; see also Tversky & Kahneman 1973, p. 221). The lack of instance memory fosters application of non-enumeration strategies such as memory assessment strategies ('mental magnitudes', trace strength, associative strength), schema or metacognitive-based inferences, or the reading out of tally numbers or rates from memory.

Even if enumeration is possible in principle, its application is further constrained by time and capacity limits at the point the judgement is made. Enumerative strategies rely on serial retrieval processes consuming time and mental effort. Thus time and resource restrictions, either due to the situation or due to low motivation, enhance the likelihood that individuals select non-enumeration strategies of frequency judgement (Brown, Chapter 3; Haberstroh & Betsch, Chapter 13; Schimmack, Chapter 12, Schwarz & Wänke, Chapter 6). One of these default strategies might require the individual to spontaneously rely on an intuitive feeling of quantity without using proxies such as ease of retrieval. These intuitions about frequency are often surprisingly accurate, indicating that frequencies are implicitly registered during encoding (Zacks & Hasher, Chapter 2).

There are several suggestions as to how such records are formed and represented in memory. Regardless whether one favours multiple trace, trace–strength or associative network models or proposes the existence of accumulative memory structures—all these views converge in assuming that impressions of quantity automatically arise when the memory system is probed with features, events or categories which have been encountered before. The capability of the memory system to readily provide information about quantity may indeed be as basic and inevitable as, for example, the penetrating directness with which affective reactions arise in response to familiar stimuli (Wundt 1907; Zajonc 1980; see also Zacks & Hasher, Chapter 2).

There are several other factors which moderate the selection of judgement strategies. Schwarz and Wänke (Chapter 6) highlight the adaptive nature of strategy choice. They report empirical evidence suggesting that people take a strategy's applicability into account. For example, if the informative nature of ease of retrieval is cast into doubt (e.g. due to the nature of the task or the lack of expertise) participants are likely to switch to enumerative strategies, which capitalize on recall content. The same authors also cite evidence for the notion that different mood states foster different kinds of judgement strategies. In a good mood, individuals tend to draw on their subjective experiences as a proxy to judgement, whereas in a bad mood, individuals are more likely to rely on recall content.

Taken together, research on frequency processing and cognition has meanwhile identified a couple of variables that impact on the selection of judgemental strategies. However, the studies reported in this volume are merely a first step towards a deeper understanding of the underlying mechanisms. Although we are on our way to being able to predict which type of strategies will predominate under certain conditions, we are still not able to predict particular strategy choices. Nevertheless, empirical work presented in this volume yields the first promising results and should encourage researchers to further pursue this line of research in the future.

Application

Frequencies play an important role in many areas of cognition and behaviour (Sedlmeier *et al.* Chapter 1; Zacks & Hasher, Chapter 2). Therefore, it is not surprising that insights into the mechanisms of frequency processing and cognition can be fruitfully applied in many domains of everyday life. Three contributions in this volume show how frequencies directly or indirectly can influence cognition and action.

Based on a broad review of the literature, Alba (Chapter 16) works out a plethora of implications that research on frequency processing has in consumer decision making. It was found, for instance, that frequent but modest price reductions lead to a higher perception of value and higher rates of purchase than infrequent but high price reductions. Another important effect is the sample size effect. Alba reports evidence indicating that, while keeping ratios of positive and negative aspects constant, people prefer an option with a higher absolute frequency of 'wins'. Such sample size effects in frequency processing have important implications, for instance, in the domain of advertisement and sales promotion. The remainder of his chapter describes other examples of how misperception of frequency information affects consumer choice. Moreover, he discusses the role that frequency based decision heuristics (e.g. majority of confirming dimensions rule) play in purchasing decisions.

Wettler (Chapter 17) also touches on the domain of marketing and consumer research in his chapter. Starting from contiguity theory of associative learning, he develops a technique to analyse the associative structure of the contents of advertisements and to estimate their success in subsequent campaigns. The basic argument is that intuitive impressions of the quality of products reflect the associations evoked by them. The likelihood that a certain product evokes positively evaluated concepts depends on the strength of the associative bonds between the features of the product and other concepts. In turn, the strength of a bond between two concepts reflects the frequency of prior co-occurrence. Although people differ with regard to their experiences, a great deal of variance systematically reflects shared cultural knowledge as conveyed by language. With the help of large machine-readable samples of texts, concept associations can be computed on the basis of word co-occurrences. This in turn allows the researcher to predict the evaluative meaning evoked by persuasive communications.

Hertwig and Hoffrage (Chapter 18) consider another field of application. Decision-making in proficient domains can be highly consequential for other people. Under some circumstances a person's life may depend on, for instance, the decision of a physician or a jury in a death sentence trial. These decisions often require a consideration of statistical or probabilistic information, such as the results of cancer screening or DNA tests. From a vast number of studies we know that people, even when being trained in statistics, are prone to systematically violating fundamental rules of probability theory. Hertwig and Hoffrage report studies substantiating their argument that natural frequencies help experts to avoid biases in diagnostic inferences. In contrast to probabilities or relative frequencies, natural frequencies are absolute frequencies, which have not been normalized with respect to base rates. Hertwig and Hoffrage converge with Gigerenzer (Chapter 4) in concluding that statistical reasoning outside of the laboratory can be improved substantially when information is presented as natural frequencies instead of probabilities.

Other chapters, although they are not included in the application section of the book, also refer to application issues. For example, Schimmack (Chapter 12) discusses the implications of his findings on emotion frequency for the domain of personality assessment. Schwarz and Wänke (Chapter 6) raise several suggestions about how survey researchers can circumvent traps in questionnaire construction to guard against distortions in quantitative estimation.

A look ahead: some trajectories to future research and theorising

It became apparent in many contributions to this volume that theorists are currently striving to achieve two things: (1) developing more integrative models covering a broader range of phenomena and (2) making their models more precise. We think that these attempts show the way for future research and propose to build computational models that take care of both integration and precision.

Integration

Integration can proceed in two ways, between frequency processing and other aspects of cognition, and within different aspects of frequency processing. One may attempt to embed frequency processing into larger frameworks that cover cognition in its entirety, that is, link frequency processing to perception, learning, memory, and reasoning. Another attempt at integration could consist of trying to develop theoretical (multiple strategy) frameworks that cover many different kinds of frequency-based judgements and cognitions, including the impact of contextual and task variables.

Several contributions in this volume demonstrate that a wide range of phenomena—though not all—can be accounted for with reference to basic principles of learning, memory and cognition. Dougherty and Franco-Watkins (Chapter 8) show how ease of retrieval effects and others which were formerly attributed to the work of different heuristics can be modelled within one integrative memory model (MINERVA DM). The Brunswikian Induction Algorithm for Social Cognition (BIAS) (Fiedler 1996) and PASS (Sedlmeier, Chapter 9) are also examples of memory models which are capable of covering a wide range of effects in frequency, probability and confidence judgement. It is not yet clear how far one can proceed with general models. Research from an evolutionary psychology perspective indicates that different specific processes may be involved for different domains and tasks (Cummins & Allen 1998; see also Gigerenzer, Chapter 4). If it turns out that specific influences play a role in frequency processing, it must be accounted for in such models (see also the paragraph on computational models, below).

Another way to achieve integration within the domain of frequency processing may involve the elaboration of a multiple strategy perspective. The advantage of this view is that virtually all possible estimation strategies can be handled within one meta-framework. However, this does not mean that general mechanisms underlying frequency judgement must no longer be sought out and identified. Without a precisely spelled-out metarule for strategy choice, the multiple strategy frameworks will fail to predict processes and outcome of quantitative estimation. There is one lesson to be learned from preferential decision research that also tackles the problem of strategy selection. From the point of hindsight (after a certain strategy has been chosen), strategy choice often reveals a high degree of adaptivity. Individuals seem to choose those strategies which are best applicable to the task, in such that they yield a satisficing compromise between tasks and system constraints and the individual's aspirations. Accordingly, since the late 1970s, decision theorists often propose a profitability or applicability based metachoice mechanism to underlie strategy selection. A global reference to profitability or applicability, however, has proved to be too vague to allow for clear-cut predictions. Today, preferential decision theory is still not able to deliver predictions of strategy selection on the level of distinct choice rules. Therefore it seems reasonable, as a first step, to isolate factors which reduce the set of eligible strategies and/or which enhance the likelihood that certain types of strategies will be employed. Some examples for that type of research strategy were presented throughout the book (e.g. Brown, Chapter 3; Haberstroh & Betsch, Chapter 13, Schimmack, Chapter 12; Schwarz & Wänke, Chapter 6).

Pursuing this approach further may gradually yield an increase of the predictive power of multiple strategy models. For that, it will be necessary to find effective theory-based methods that allow the researcher to recognize which strategy a given participant used. One good example for such a method is the use of reaction times to differentiate between enumerating and non-enumerating strategies (Brown, Chapter 3; Schimmack, Chapter 12). It may also be worthwhile to examine other effects of time for frequency judgements. As Gallistel (Chapter 10) shows, the number of events encountered during a given time interval and the length of the intervals between the events are essential to explain results in conditioning experiments with animals. Time may also turn out to play an important role in human frequency judgements.

Precision

Regardless of whether one initially sides more with the multiple strategy or the learning and memory perspective, future work should attempt to further increase the precision and predictive potential of our theories. Scientific progress requires as a precondition that theories are able to produce precise and restrictive predictions. If a theory does not prohibit anything but merely postulates the existence of phenomena, it has no predictive potential and, hence, cannot be falsified. For example, the original formulation of the availability approach (Tversky & Kahneman 1973) does not allow us to predict (a) under what conditions people will employ the availability heuristic, and (b) under what conditions people will use the retrieval-based or the schema-based version of the availability heuristic (cf. Betsch & Pohl, Chapter 7). The original version of the availability approach merely predicted that people *may* use the availability heuristic when forming frequency judgements—a statement which for logical reasons is immune against falsification (Popper 1961). In contrast, the theoretical papers in this volume clearly show that we are on our way to overcoming weak theories. Schwarz and Wänke (Chapter 6), for instance, demonstrate how the predictive potential of the availability approach can be improved. They extend the availability approach by additional assumptions about strategy choice and are thus able to more precisely predict processes and performance in quantitative estimation.

Computational models

In our view, the best means to increase theoretical precision is to formalize a theory as far as possible. Gallistel's (Chapter 10) information processing theory of conditioning may be seen as such a well-formalized theory. If a theory is thoroughly formalized, it is possible to describe it by a computational model that can in principle be implemented as a computer program. Models implemented as computer programs force the researcher to be precise—otherwise the program won't run. Computational models have a great advantage over static formalizations in that they can simulate not only the end products of cognitive and judgemental processes but also the *processes* themselves. Today, we do possess models that are elaborated on a high formal level so that one can run computer simulations to evaluate how these models fit (existing) judgement data (Dougherty & Franco-Watkins, Chapter 8; Sedlmeier, Chapter 9; Wettler, Chapter 17). However, one has to be careful about what a model fit means: successful simulations of empirical data do not suffice as a test of a theoretical model, because most simulation models incorporate several parameters which often cannot be precisely specified in the beginning. A fit between model prediction and data may be achieved just by playing with the parameter values long enough. Therefore, it is necessary to incorporate detailed theoretical assumptions, which also must be based on whatever is known from empirical research, into these models. Then, one should attempt to derive unique predictions that enable the researcher to set up competitive designs in order to compare the models' predictive power on an empirical basis. Discrepancies between empirical results and model predictions are then the basis to improve the model. Thus, computational modelling comprises the possibly of a never-ending interactive process between refining theories (expressed as computational models) and testing them in empirical research.

Besides having the potential to increase theoretical precision, computational models also allow for both of the types of integration discussed above. Exemplar and prototype models (see Chapters 1, 8, and 9) are integrative models of cognition: they contain learning rules, memory representations, and mostly implicit rules for reasoning and judgement. However, the research reported in this volume indicates that context and explicit deliberations can have a strong impact on judgements of frequency and probability and on frequency-based judgements. Research results from a multiple-strategy perspective can be the basis for specifying important factors and incorporating them into computational models. Research that explores how general the influences of these factors are across domains and tasks is needed. It may transpire that theoretical models may be adjusted for different domains. Of course, one cannot expect that everybody will make computer simulations, but we think that it would certainly help if research were conducted within a theoretical framework that could in principle be translated into such a computational model.

Conclusion

Without doubt, frequency information is one of the fundamental dimensions of experience, and thus deserves increasing attention in basic and applied research. In this volume we have attempted to compile research on frequency processing which is emerging from formerly rather unrelated areas of psychology. The contributions to this book sketch out recent developments achieved in experimental research, theory development and application. We hope that this volume will promote exchange between different areas of research on frequency processing and cognition. To those who are not yet active in the field, this book should be considered as an invitation to join this line of research on a fascinating topic.

Acknowledgement

We are grateful to Bronwyn Bosse and Frank Renkewitz for commenting on a draft version of this manuscript.

References

Aarts, H. & Dijksterhuis, A. (1999). How often did I do it? Experienced ease of retrieval and frequency estimates of past behavior. *Acta Psychologica, 103*:77–89.

Alba, J. W., Chromiak, W., Hasher, L. & Attig, M. S. (1980). Automatic encoding of category size information. *Journal of Experimental Psychology: Human Learning and Memory, 6*:370–378.

Barsalou, L. W. & Ross, B. H. (1986). The roles of automatic and strategic processing in sensitivity to superordinate and property frequency. *Journal of Experimental Psychology: Learning, Memory and Cognition, 12*:116–134.

Beach, L. R. & Mitchell, T. R. (1978). A contingency model for the selection of decision strategies. *Academy Management Review, 3*:439–449.

Betsch, T., Siebler, F., Marz, P., Hormuth, S. & Dickenberger, D. (1999). The moderating role of category salience and category focus in judgments of set size and frequency of occurrence. *Personality and Social Psychology Bulletin, 25*:463–481.

Brown, N. R. (1995). Estimation strategies and the judgment of event frequency. *Journal of Experimental Psychology: Learning, Memory and Cognition, 21*:1539–1553.

Chaiken, S. (1980). Heuristic versus systematic information processing and the use of source versus message cues in persuasion. *Journal of Personality and Social Psychology, 39*:752–766.

Cummins, D. D. & Allen, C. (eds) (1998). *The evolution of mind.* New York: Oxford University Press.

Dougherty, M. R.P., Gettys, C. F. & Ogden, E. (1999). Minerva-DM: A memory model for judgments of likelihood. *Psychological Review, 106*:180–209.

Fazio, R. H., Chen, J., McDonel, E. C. & Sherman, S. J. (1982). Attitude accessibility, attitude-behavior consistency, and the strength of the object-evaluation association. *Journal of Experimental Social Psychology, 18*:339–357.

Fiedler, K. (1996). Explaining and simulating judgment biases an an aggregation phenomenon in probabilistic, multiple-cue environments. *Psychological Review, 103*:193–214.

Gigerenzer, G. (1996). On narrow norms and vague heuristics: A reply to Kahneman and Tversky (1996). *Psychological Review, 103*:592–596.

Gigerenzer, G. & Hoffrage, U. (1995). How to improve Bayesian reasoning without instruction: Frequency formats. *Psychological Review, 102*:684–704.

Hasher, L. & Zacks, R. T. (1979). Automatic and effortful processes in memory. *Journal of Experimental Psychology: General, 108*:356–388.

Hasher, L. & Zacks, R. T. (1984). Automatic processing of fundamental information: The case of frequency of occurrence. *American Psychologist, 12*:1372–1388.

Hintzman, D. L. (1988). Judgments of frequency and recognition memory in a multiple trace model. *Psychological Review, 95*:528–551.

Hintzman, D. L. & Block, R. A. (1971). Repetition and memory: Evidence for a multiple-trace hypothesis. *Journal of Experimental Psychology, 88*:297–306.

Kahneman, D. & Tversky, A. (1996). On the reality of cognitive illusions. *Psychological Review, 103*:582–591.

March, J. G. (1978). Bounded rationality, ambiguity, and the engineering of choice. *The Bell Journal of Economics, 9*:587–603.

Payne, J. W., Bettman, J. R. & Johnson, E. J. (1993). *The adaptive decision maker.* Cambridge: Cambridge University Press.

Petty, R. E. & Cacioppo, J. T. (1986). The elaboration likelihood model of persuasion. In L. Berkowitz (ed.) *Advances in experimental social psychology* vol. 19 (pp. 123–205). San Diego, CA: Academic Press.

Popper, K. R. (1961). *The logic of scientific discovery.* New York: Science eds.

Sedlmeier, P. (1999). *Improving statistical reasoning: Theoretical models and practical implications.* Mahwah, NJ: Lawrence Erlbaum.

Sedlmeier, P. & Gigerenzer, G. (2001). Teaching Bayesian reasoning in less than two hours. *Journal of Experimental Psychology: General, 130*:380–400.

Tversky, A. & Kahneman, D. (1973). Availability: A heuristic for judging frequency and probability. *Cognitive Psychology, 5*:207–232.

Wundt, W. (1907). *Outlines of psychology.* Leipzig: Wilhelm Engelmann.

Zajonc, R. B. (1980). Feeling and thinking: Preferences need no inferences. *American Psychologist, 35*:151–175.

INDEX

ability 23
absolute accuracy 26, 190–2
absolute frequency, PASS model 148
absolute risk 64
accessibility of applicable inputs 104, 105
accuracy 26–7, 50, 190–8, 260–1, 309–11
activity value 273
activity vector 273
affect 9–10
   mere exposure effects 185–6
age 6, 22–3, 26
algorithms 67–8
animal behaviour 30
animal learning 2, 31
associationist approach 30–1
associative learning 137
   free word associations 272–3, 313
   frequency estimation 138–9
associative memory 9
associative strength 272
   assessment 112–13
attention 32
attentional capacity 25, 32
attitude 9
augmented embedding 243
automatic, use of term 33
Automatic and Effortful encoding framework 24–6, 306
automatic encoding 5, 25, 26–7, 33–4, 114, 185, 208, 306
automatic processes 25
automatic text understanding 2
availability 109–19
   informational value 115
   predictive power 110–13
   as a scientific theory 113
availability biases 131–4
availability heuristic 2, 5, 26, 27, 69, 71, 90–8, 109, 122, 176, 198, 199, 305, 315
   biases associated with 131–4
   conditions when it is employed 110–12, 115–16
   mechanism 112–13
   prevalence 113–14

'bait shyness' effect 31
baseball games 244–8
base rate 286

base rate fallacy 57, 81–4
base rate neglect 29–30, 45
Bayes' rule 9, 56, 63, 286
behavioural frequencies questions 39–40
behaviour repetition 10
belief–adjustment model 223–6
beliefs 58–9
   false beliefs 252
   prior beliefs 265
belief updating 221–37
   confidence and 227
   co-occurrences 222
   frequency learning 226–7
   order effects 222–3, 225, 235–6
BIAS model 149
biases 4–5, 69, 139–40, 210, 240, 241, 304
   correlated retrieval cues 84–5
   enumeration 49–50
   modelling 131–4, 145–7
binary events 240–2
blocked presentations 6
blocking phenomenon 159–60
brand quality 259
breast cancer screening 64–5, 81–4, 286–7

calibration 239–55
   baseball games 244–8
   coherence and 240
   definition 240
   hard–easy effect 250–3
   Lad argument 240–4
   Lens model 248–9
   psychophysics metaphor 239–40
calibration curve 240
CA-module 137
cancer screening 64–5, 81–4, 286–7, 288
categorization 3–4
category-split 70–4, 76
   spontaneous 78–9
causal relationships 138
ceiling principle 293
choice 9–10
classical conditioning 31
coffee, sales promotion 276–9
cognition 3–10, 27–8, 34
cognitive algorithms 2, 68, 72

Cognitive Aspects of Survey Methodology (CASM) 40
cognitive counter 111, 208
cognitive effort 215, 218
cognitive illusions 57, 58, 59
cognitive load 77
cognitive primitive 75
cognitive psychology 2
coherence 242–3
  and calibration 240
colorectal cancer screening 287, 288
competitive designs 114
computational models 316; see also specific models
computer-controlled studies 308
concurrent reports 99–100
conditional probabilities
  contingent frequencies and 167–8
  Minerva DM 125–8
conditional retrieval structure 81–4
conditioned emotional reaction 154
conditioning 31, 68, 168–9
  contingencies and 153–6, 160
  information theoretic perspective 156–62
  rapidity 164–7
confidence judgements 8
confirmatory tests 113–14
conjunction fallacy 57, 61, 69
conjunctive prompts 79–81
connectionist networks 4
connective retrieval 79–81
conservatism effect, modelling 129–30
constructive process 70
consumer decision making 30, 259–70, 313
context 7, 308
  frequency judgements of emotions 202
context–discrimination 132, 133–4
context memory 43
contextual information 98, 103, 104
context vectors 280
contiguity 271–2
contingencies and conditioning 153–6, 160
contingent frequencies 153, 168
  computing 162–4
  as conditional probabilities 167–8
continuous frequency 168
contrary items 250
convenience 50
co-occurrence 9
  belief updating 222
  word associations 272–5, 279–83
co-relations and word associations 279–83
counter-intuitive items 250
court proceedings 59, 63–4, 292, 293, 295–7

death, causes of 69
death penalty 291–2

decision making 9–10; see also consumer decision making
deep processing 7
deliberate judgement 211–12
  strategy selection 212–17
deliberative mindset 310
DeltaLQ 231
developmental approach 32
diagnostic inference 287–9, 313–14
diagnosticity 91, 92–3, 115
diagnostic tests 286–7, 299; see also cancer screening
diary data 190–1, 193
different-context lists 42
differential encoding 132–3
discrete frequency 168
discriminative accuracy across emotions 192–8
disjunctive prompt 79
distinctiveness 43
DNA analysis 291–7, 302
  laboratory error 293–4

echo intensity 125
  conditional 127
  encoding and 131–2
echo strength 70
ecology 62, 63
effort–accuracy trade-off 262
effortful encoding 25
egocentric bias 69
embedded constructions 272
emotions
  definition 189–90
  frequency judgements 189, 190–8, 200–1, 202
encoding 5, 22–4, 115–16, 185, 253–4, 306
  accuracy and price promotions 260–1
  echo intensity 131–2
  see also automatic encoding
Enlightenment mathematicians 56
enumeration and extrapolation 41
  PASS model 148–9
enumeration-based estimates 40–5, 70
enumeration bias 49–50
environment 2
estimation strategies 39, 40–8
event frequency 8–10, 37–53
event traces 42
everyday life 4, 5, 22
  salience effects 193
evolution 31, 62, 63
evolutionary psychology 2, 32
exemplar models 149
exemplars 4, 212, 213, 218
expectancy effect 178
expertise 95

extinction 160–2
  rapidity 164–7
eyewitness identification 298

false beliefs 252
false-positive rate 287
false predictions 252
famous names study 4–5, 131, 140
feedback 22, 94
FEN 137, 141, 143
fluency 176–8, 185, 186
  manipulating 178–86
free word associations 271–84
  associative learning of 272–3, 313
  co-occurrences and 272–5, 279–83
  co-relations 279–83
  sales promotion 276–9
  semantic organization 279–83
French Revolution 58
frequencies versus probabilities 8–9, 82, 167–8, 253, 287, 289; *see also* natural frequencies
frequency coding 185
frequency effect 61
frequency information, uses of 27–30
frequency judgements 2, 4–8
  assets and deficits 68–70
  factors affecting 6–7
frequency learning 226–7
frequency memory 74–85
frequency processing
  cognition and 3–10
  factors affecting 6–7
  historical aspects 1–2
frequency scales 98–102, 103
frequency tag 26
frequentist hypothesis 30

general impressions 49
general-knowledge questions 60
'general laws' 31, 32
generation effect 128

hard–easy effect 250–3
heartbeat 68
hedonic balance 201
hemoccult test 287, 288
heuristics 304; *see also* availability heuristic
Hierarchical Linear Modeling 5 (HLM5) 196–7
hit rate 286
hit rate minus base rate algorithm 297

Howland Will Case 297
How to lie with statistics (Huff) 56

illusory correlations 78–9
implicit learning 32–3
implicit tally 210–11
impression retrieval strategy 45
incoherence 240
individual differences approach 32
information 156
inhibitory conditioning 155
Innocence Project 291
innumeracy, future defeat of 55–66
instance memory 43
instance-only lists 44
instance-plus-property lists 43–4
instructions 6–7, 22
intentional learning 32–3
International Transparent Testimony Act 64
intrusion rates 41
irrational behaviour 57–8
item calibration 250

judgement and decision making 2, 304, 305
judgement domain 116
judgement-referral 267
judgement time 195, 197, 200

labelling 202
laboratory settings 4, 5, 22, 308
  salience effects 193–8
language acquisition 28–9
law courts 59, 63–4, 292, 293, 295–7
learning 32–3, 68; *see also* associative learning
least-effort approach 49
legal experts 295–7
Lens model 248–9
letter study 6, 140, 176, 177
likelihood subtraction 297
liking 2
linear constraints 241

magazine subscription incentives 263
magnitude effects 266–7
majority of confirming dimensions heuristic 264
mammographic screening 64–5, 81–4, 286–7
market share 259
matching paradigm 166–7

322 INDEX

MCD heuristic 264
medical profession 62, 64–5, 286–7, 289–91
memory 2, 74–85, 121–2, 267–8, 304–5
    attention and 32
    context and 43
memory assessment 48, 175
memory-based strategies 9, 206, 208–10, 213, 217
memory retrieval 46–8
memory traces 185
mere exposure 185–6
metacognition 74
metacognitive blindness 70
meta-memory 198–200
Minerva 2 model 7, 123–5, 149
    frequency judgements 128–9
Minerva DM (MDM) 122, 123–5, 134, 149, 175–6, 306
    availability biases 131–4
    conditional probability and frequency judgements 125–8
    conservatism effect 129–30
minority configurations 263–6
mood 97
motivation 95–7, 310
multiple strategies 104, 305–6
Multiple Strategy Perspective (MSP) 38, 39, 40, 48, 50–1, 307
multiple-trace memory models 123
multiple trace-type representation 26

natural frequencies 9, 62–4
    diagnostic inference and 287–9, 313–14
    DNA evidence interpretation 292–5
    statistical reasoning 289–91, 295–7
    training in 298
natural sampling 62, 288
natural schemes 45
neural networks 4
new technologies 285–6
nonenumeration strategies 45–8
nonnumerical strategies 46–8
numerical strategies 45–6
numerosity heuristic 74–8, 261–2

off-target enumeration 41, 43–5
on-line strategies 206, 207–8, 209–10, 212–13, 217
on-target enumeration 41–3
open response format 102
order effects, belief updating 222–3, 225, 235–6
organizational psychology 202
overconfidence 239, 240, 252
overestimation 48
    category split 76

cognitive algorithm 72
    regression and 74
overshadowing protocol 160

paradigmatic relatedness 280–2
parsing 29
PASS 8, 137, 139, 141–50, 306
    absolute frequency 148
    architecture and learning rules 141–3
    biased estimates 145–7
    enumeration 148–9
    probability estimates 147–8
    proportions 145
    regression effect 144–5
    rich environments 149
    sample size 145
    simulation of empirical results 143–7
    small frequencies 148–9
patients, decision making by 298–9
Pavlovian conditioning 153–5
perceived accuracy 50
perceived expertise 95
perceptual fluency 176, 185, 186
peripheral processing 263
personality assessment 189, 201–2
    accuracy 201
persuasion 261–8
pitch perception 67–8
politics 58, 59
polysemy 60, 61
positional letter frequencies 6, 140, 176, 177
positive predictive value 286
positivity bias 241
practice 22
precision 315–16
predictive events 138
presentation frequency 42
presentation time 5–6
price promotions 260–1
primacy effects, belief updating 223, 226, 232
prior beliefs 265
probabilities 4–8, 56–62, 66
    confusion caused by 57, 60–1
    extension 58–9
    frequencies versus 8–9, 82, 167–8, 253, 287, 289
    Minerva DM 125–8
    PASS 147–8
probability density function 208
probability function 241
probability learning paradigm 2
probability revision 9
probation officers 60
processing fluency 176–8
    manipulating 178–86
processing level 7
process-sensitive measures 40
product marketing 2

product rule 293
programmed reward rate 166
property frequencies 6
proprioceptive feedback 94
prototype models 149
prototype view 4
psychophysical function 240
psychophysics metaphor 239–40

qualitative retrieval strategy 46–8
quasi-probability 246
question interpretation 99–100

radiation risk 299
random rate process 157–8
rate-based strategies 45, 46
reasoning, statistical 29–30, 289–91, 295–7
recall 5–6
   diagnosticity 92–3
   ease 91–2
   experience and amount of recall 93–4
   fluency 176–7
   meta-memory and 198–200
recall-free judgement 200–1
recency effect, belief updating 223, 224, 225, 226, 227, 231–2, 234, 235
reconstructive process 70
reference class 59–60, 293
regression 71–4, 139
   in PASS 144–5
regression-like encoding 253–4
regression to the mean 179
reinforcement 9, 68
relative accuracy 27
relative frequency 2, 4–8, 26, 139, 262–8
relative risk 64
representations 3–4, 62
representation training 298
representativeness heuristic 111, 122
response distributions 46
response formats 191–2
retrieval
   ease 112
   qualitative 46–8
retrieval cues 68–9, 70–1, 73, 76, 78
   correlated, biases due to 84–5
retrieval split 73–4
retrieval structure 70–1, 79–84
   conditional 81–4
   connectives in 79–81
retrieved rate 45, 46
retrieved tally 45–6
retrospective reports 99–100

rich environments, PASS model 149
risk, absolute/relative 64
risk-averse 268
round numbers 48
rule training 298

sales promotion, word association role 276–9
salience effects 193–8, 201–2
SAM 210–12, 217, 218
same-context lists 42
sample size 8, 139, 262, 313
   in PASS 145
scenario-rating task 193–5, 198–9
schemas 111
scoring rules 243–4
self-generated items 6
semantic organization and word associations 279–83
semantic similarity 280–3
set inclusion 61
shallow processing 7
simple enumeration 41
Simpson, O.J., trial 63–4, 292, 293
Simpson's paradox 80
single-event probabilities 58–9, 60, 62
single-letter positional frequency 29
small frequencies, PASS model 148–9
spaced repetitions 7
spontaneous category split 78–9
spontaneous judgement 211–12
   strategy selection 212–17
standard sequence 243
statistical reasoning 29–30, 289–91, 295–7
stereotypical biases 69
Stigler's Law of Eponymy 57
Strategy Application Model (SAM) 210–12, 217, 218
strategy selection 49–50, 103–5, 212–17, 308, 311–13
strength-type representation 26
structural relationships 138
symbolic processes 272
syntagmatic relatedness 280–2

tally-based responses 45–6
taste aversion learning 31
temporal pairing 159
   and contingency 154
time
   availability heuristic 116
   of judgements 195, 197, 200
   recall 5–6
time frames 191–2
training 298
'treatments' 277

trick items 250
typicality 6

underconfidence 239, 252
underestimation 41, 42
unknown environment 213
unrealistic optimism 69

vague quantifiers 47, 102
validity 5
value-charged experiences 9

weather forecasts 57, 59
Weber's law 159
well-calibrated 239
window technique 274
word associations, *see* free word associations
word frequency 74
word segmentation 28–9
word superiority effect 29
workplace, emotions in 202
World Health Organisation (WHO) 65
written word decoding 29

X-ray machines 285

Printed and bound by CPI Group (UK) Ltd, Croydon, CR0 4YY